MANUAL OF
BUILT-UP
ROOF
SYSTEMS

C. W. Griffin

MANUAL OF BUILT-UP ROOF SYSTEMS

Second Edition

McGRAW-HILL BOOK COMPANY

New York   St. Louis   San Francisco   Auckland
Bogotá   Hamburg   Johannesburg   London   Madrid
Mexico   Montreal   New Delhi   Panama   Paris
São Paulo   Singapore   Sydney   Tokyo   Toronto

Library of Congress Cataloging in Publication Data

Griffin, C. W. (Charles William), date.
  Manual of built-up roof systems.

  Includes index.
  1. Roofs.   2. Roofing, Bituminous.   I. Title.
TH2450.G73 1982   690′.15   81-13732
ISBN 0-07-024783-8   AACR2

1 2 3 4 5 6 7 8 9 0   HDHD   8 9 8 7 6 5 4 3 2 1

ISBN 0-07-024783-8

The editors for this book were Joan Zseleczky, Alice Goehring,
and Margaret Lamb, the designer was Elliot Epstein, and the
production supervisor was Paul Malchow. It was set in Melior by
University Graphics, Inc.

Printed and bound by Halliday Lithograph.

Contents

Preface

This second edition, started many years ago, is both an expansion and updating of the first edition of the *Manual of Built-up Roof Systems*. There has been an increasing need to report on the latest roofing industry research developments, which have been accelerating over the past three or four years.

The twelve original chapters have expanded to twenty new chapters for two reasons:

- To give fuller, independent treatment to important subjects only briefly discussed in the original edition

- To acknowledge new research and developments that postdate the 1970 edition

Examples of chapters resulting from the need for expanded coverage and more intense focus are new Chapter 3, "Draining the Roof," Chapter 12, "Premature Membrane Failures," and Chapter 13, "Reroofing and Repair."

Chapter 3 is dedicated to the struggle against the nondraining, chronically ponded roof. Resistance to this fundamental rule of roof design stems from a combination of building owners' shortsighted economizing and designers' benighted lethargy. If we cannot eradicate this evil, we can at least reduce the excuses used to justify it.

Chapter 12, on premature roofing failures, contains an expanded discussion of common failure modes, notably membrane blistering and splitting, which received only superficial analysis in the first edition. Chapter 13 reflects the growing importance of renovation in the slowed United States economy and the frequent need to reroof buildings only a few years after they have been completed.

Chapter 10, "Protected Membrane Roofs," exemplifies the chapters added to report a new development, unmentioned in the first edition.

Over the past decade, the protected membrane roof has established itself as a valid roof concept, despite the widespread skepticism that greeted its entry into the United States roofing market.

Expanded revisions of original chapters retained in this revised edition include other new topics—notably the loose-laid, ballasted roof system, another well-established roof-system concept, discussed in Chapter 11, "Synthetic Single-Ply Membranes." Still another new topic is the self-drying concept, part of expanded Chapter 6, "Vapor Control."

Acknowledgments

There is no practicable way for me to acknowledge all the roofing industry experts who have helped, one way or another, with this revision. I owe special gratitude to Robert A. LaCosse, Technical Services Manager, National Roofing Contractors Association, for permission to reprint the NRCA's flashing details in Chapter 8. Use of these NRCA details could save the nation's building owners countless millions of dollars annually spent on repair and reroofing.

Notable among those who reviewed sections of the typescripts are W. C. Cullen, dean of United States roofing experts and former Deputy Director, National Bureau of Standards Office of Engineering Standards; NBS researchers R. G. Mathey and W. J. Rossiter; Carl G. Cash, partner in Simpson, Gumpertz & Heger; Justin A. Henshell, AIA; Richard L. Fricklas, Director, Roofing Industry Educational Institute; Haldor W. C. Aamot, consulting engineer; researchers H. O. Laaly and Rene M. Dupuis; Wayne Tobiasson, U.S. Army Cold Regions Laboratories; Edward T. Schreiber and Karl Potasnik, Construction Consultants, Inc.; R. J. Moore and Mike Dilts, ARMM Consultants; consultants Gerald C. Curtis, Sybe K. Bakker, and Charles McCurdy; and contractors Robert E. Linck, Paul L. Morris, John Ambrosio, Joe Lillo, Richard Baxter, William Kugler, and John D. Van Wagoner.

Obviously, none of these experts bears any responsibility for the contents of this book. "The buck stops here" applies to authors, if not to politicians.

C. W. Griffin

xi

About the Author

C. W. GRIFFIN, P.E., is a roof consultant. Following graduation from George Washington University, Mr. Griffin worked as a structural engineer and field inspector for 10 years before joining *Engineering News-Record*, where he became Senior Editor. Since 1967, he has been an independent consultant.

His books include *Energy Conservation in Buildings* (1974) and *The Systems Approach to School Construction* (1971). He is a member of the American Society for Testing and Materials, the American Society of Heating, Refrigerating, and Air-conditioning Engineers, and the American Arbitration Association.

ONE

Introduction

The volume of built-up roofing annually installed in the United States totals about 3 billion ft², enough to cover Washington, D.C., nearly twice. Of this vast volume, 4 to 5 percent fail prematurely, according to roofing contractors and manufacturers surveyed for a National Bureau of Standards (NBS) report; this same report notes that some industry experts estimate a still higher failure rate.[1] Regardless of the precise percentage, roofing failures constitute a major problem: a huge economic burden on building owners plagued with the high costs of tear-off-replacement and a hazardous legal threat to architects, contractors, and manufacturers involved in this grossly litigious sector of the building industry.

WHY ROOFS FAIL

Premature roofing failures are caused by both economic and technical factors. Economically, a building's roof system normally lags far behind the more architecturally glamorous building subsystems competing for the building owner's money. Pennywise, dollar-foolish decisions underlie many premature roofing failures: Whether through ignorance, laxity, or sheer perversity, many roof designers and building owners refuse to pay the slight additional cost for sloping the roof to avoid the ponding of rainwater (see Fig. 1-1).

Technically, the factors contributing to premature roof failures can be listed as follows:

- The extraordinary rigors of roof-performance requirements

- Proliferation of new materials

[1]H. W. Busching, R. G. Mathey, W. J. Rossiter, and W. C. Cullen, "Effects of Moisture in Builtup Roofing—a State-of-the-Art Literature Survey," NBS Tech. Note 965, July 1978, p. 2.

1

Fig. 1-1 This scene is not an irrigated wheat field; it is a level built-up roof system with ponded water. The roots of the flourishing vegetation can force their way through the membrane into the insulation, ultimately producing widespread leaks. (Sellers & Marquis Roofing Co.)

- Complexity of roof-system design
- Expanding roof dimensions
- Field-application problems
- The modern trend toward more flexible buildings

Roofs must withstand a much broader attack from natural forces than other building components. In some parts of the continental United States, roof surfaces experience annual temperature changes exceeding 200°F and daily changes exceeding 100°F. These temperature changes can occur rapidly, as when a summer shower suddenly cools the sun-baked membrane surface. Solar radiation heats the roof to extraordinary temperatures, up to 180°F for black surfaces; this heat greatly accelerates photochemical deterioration (see Fig. 1-2). Rain, snow, sleet, and hailstones pound the roof; acid mists and other airborne pollutants—even fungus—attack the roof.

Proliferation of new roof materials—new decks, insulations, vapor retarders, membrane and flashing materials, used in countless combinations—has complicated the field-manufacturing process and the designers' job of evaluating durability. Until a material has been field-tested in service, its durability remains highly unpredictable. Accelerated laboratory tests are better than nothing, but in-service performance is essential for proving a material's durability. Some materials enter the roofing market lacking even laboratory testing.

Many materials, satisfactory in themselves, prove disastrous when incorporated with other, incompatible materials in built-up roof systems. Insulation generally raises the probability of condensation entrapment within the roof system. The threat of condensation in turn creates the possible need for another roof component, a vapor retarder designed to intercept the flow of water vapor into the insulation, where it can cause membrane ridging, aggravated blistering, and destruction of the insulation itself. The vapor retarder may create the need for venting the insulation. And so it goes, with each solution creating its own subproblems.

Thus roof designers must never consider a component in isolation; they must always investigate its compatibility with other materials and its effects on the whole system. Far more important than the quality of the individual materials are their design and installation as compatible components of an integrated system.

Expanding roof-plan dimensions are a source of roofing troubles. Its greater size alone makes a large roof a more complex technical problem than a small roof. A roof more than 300 ft long generally should have one or more expansion joints to accommodate thermal contraction and expansion. On the other hand, a roof only 100 ft long should require no expansion joint. Membrane splitting from movement of unanchored insulation boards, the major cause of splitting, occurs more frequently

Fig. 1-2 Overweight mopping of surface asphalt produced this classic case of alligatoring (shrinkage cracking resulting from continued oxidation, erosion, and embrittlement). (Flintkote Co.)

on large roof surfaces because the boards have greater area to move and produce membrane stress concentrations. Peripheral venting is less effective in relieving vapor pressure within a large roof system because roof area quadruples as its perimeter doubles.

Large, level built-up roofs also run a greater risk of inadequate drainage—a major cause of roofing failure. Long structural framing spans deflect more than short spans, and these deflections increase the probability of ponding. Large plan dimensions increase the inevitable humps and depressions stemming from construction inaccuracies in column-base-plate elevations and fabricated column heights; dimensional variations in deck, insulation, and membrane thickness; and so on. Among other hazards, a ponded roof faces possible membrane delamination from freezing water that has penetrated the plies. Fungi growth, promoted by standing water, can deteriorate organic roofing materials. Irregular ponding can create a warping pattern of surface elongation and contraction, wrinkling the membrane. Water penetration into organic or asbestos felts drastically reduces their strength. Yet the lure of first-cost economy, achieved through the simpler fabrication of level framing, often seduces owners into accepting a "dead-level" roof.

Despite problems of bad roof design, poor fieldwork nonetheless accounts for most roof failures, according to the majority of roof experts. Some roofing contractors ignore the roof specifications, if they take the trouble to read them. Design errors rank next, with structural deficiencies and material failure further down the list.

Financial pressures cause much faulty fieldwork. Under the threat of liquidated damages extracted by an owner if the project is not completed on schedule, the general contractor often forces the roofing subcontractor to install the roof before the deck is ready, or in damp, rainy, or severely cold weather.

NEED FOR RESEARCH AND STANDARDS

The roofing industry has lagged in the promulgation of installation standards and test methods. Instead of focusing on the whole field-manufactured roof system, the true guide to roofing-system performance, the industry has concentrated on component-material quality. There are appropriate American Society for Testing and Materials (ASTM) and federal standards for testing important properties of surfacing aggregates, felts, bitumens, insulation, vapor retarders, and structural decks. But except for fire and wind-uplift resistance, there are no generally accepted tests for performance of the entire built-up roof system assembled from these components (see Fig. 1-3).

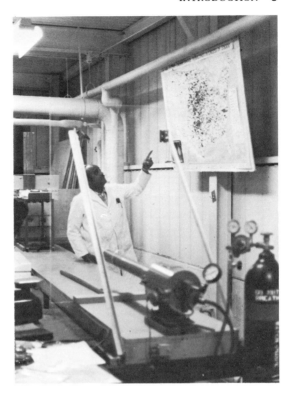

Fig. 1-3 The National Bureau of Standards' hailstone-impact testing apparatus comprises a compressed air gun (foreground) for launching "hailstones" (ice spheres), a timer for determining their velocity, and a target area (background) for positioning the tested roof-membrane specimen. Map near upper right-hand corner shows hailstorm frequency in continental United States. To satisfy the NBS-promulgated preliminary performance criteria, the tested specimen must withstand the impact of a 1½-in.-diameter hailstone traveling at 112 ft/s without allowing penetration of water (per "Hail-Resistance Test," *Build. Sci. Ser.* 23, NBS). Aggregate surfacing improves hailstone-impact resistance. A firm substrate is even more important.

Building code officials often fail to adapt their requirements to the increasing complexity of contemporary roof design. Few standard specifications recognize ponding as a problem requiring special deflection limits for the long spans common in contemporary roof framing. For example, the Open-Web Joist Specifications, High-Strength Series limits roof live-load deflection to ⅟₃₆₀ of the span for members supporting a plastered ceiling, $\frac{1}{240}$ of the span in other cases. The permitted 5-in. deflection on a 100-ft span joist could produce unstable, progressive deflection under rainwater load, resulting in roof ponding and even collapse (see Chapter 4, "Structural Deck").

AN EVOLVING CONSENSUS

Despite the many roofing industry controversies, there is an evolving consensus among experts as to good roofing practice. Intensified cooperative efforts among researchers, roofing contractors, and material manufacturers will slowly but surely build a solid, scientific basis for roofing technology. This manual sets forth this emerging consensus.

T W O

The Roof As a System

The built-up roof system is an assembly of interacting components designed, as part of the building envelope, to protect the building interior, its contents, and its human occupants from the weather. It is one of many other building subsystems, e.g., curtain walls; structural framing; heating, ventilating, and airconditioning (HVAC); ceiling lighting; and internal electronic communications, each similarly designed for a specific function.

DESIGN FACTORS

For each specific project, the best roof design is a synthesis of many factors. The most obvious and basic requirement is that the roof must satisfy the local building code. The roof probably will also have to satisfy an insurance company's requirements for wind and fire resistance. Beyond these mandatory requirements, the designer should pursue at least an informal life-cycle cost analysis. Ultimately, the roof specification should consider the following design factors:

- First cost and life-cycle (long-term) cost
- Value and vulnerability of building contents
- Required roof life
- Type of roof deck
- Climate
- Maintenance
- Availability of materials and component applicators
- Local practices

Designing the built-up roof system requires consultation with other members of the design team. The architect must confer with the mechanical engineer about heating and cooling loads to design the insulation and to keep roof penetrations to a minimum, thereby reducing the chances for flashing leaks. The architect must work with the structural engineer to ensure slope and framing stiffness that will avoid ponding and must advise the owner to institute a periodic maintenance and inspection program.

A major purpose of this maintenance and inspection program is to avoid clogged drains, which may pond rainwater to excessive depths and, in some well-publicized incidents, collapse the roof. Another purpose is to check the integrity of flashings and membrane, to ensure that splits, bare spots, blisters, and other repairable defects are corrected before they worsen and allow water to enter the building or degenerate to a point where expensive tearoff-replacement is necessary.

Perhaps the most important basic problem confronting the designer concerns roof slope. Through long, sad experience, roof experts have learned that ponded water is a major threat to the integrity of the roof system. In fact, deliberately ponded roofs, those kept perpetually flooded (except in winter), once enjoyed a minor vogue as a means of cooling the roof surface. Roof experts today reject this strategy; there are far less hazardous ways of saving cooling energy than deliberately ponded roofs. Moreover, even deliberately ponded roofs should be sloped for drainage in cold weather.

BASIC ROOF COMPONENTS

A modern built-up roof system has three basic components: structural deck, thermal insulation, and membrane. A fourth component, vapor retarder, is sometimes required for roofs over humid interiors in north-

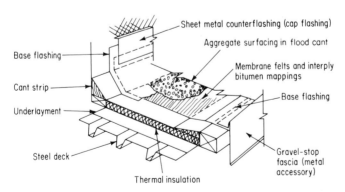

Fig. 2-1   Roof-system components.

Fig. 2-2 Solid anchorage of insulation to deck is essential to resist wind uplift and prevent membrane splitting in cold weather. (ES Products, Inc.)

ern climates. Flashing, although not a basic component of the built-up roof system, is an indispensable accessory. It seals joints wherever the membrane is either pierced or terminated—at gravel stops, walls, curbs, expansion joints, vents, and drains.

The built-up roof assembly, including flashing, functions as a system in which each component depends on the satisfactory performance of the other components. The integrity of the waterproof membrane depends on secure anchorage of all components, plus adequate shear strength between the deck and vapor retarder, between the vapor retarder and insulation, and between the insulation and membrane. The insulation's thermal resistance, which can be drastically reduced by liquid moisture, depends on the effectiveness of the vapor retarder and membrane. The integrity of the vapor retarder, insulation, and membrane depends on the stability of the structural deck (see Fig. 2-1).

Each component has its own unique primary function and also secondary functions that it must serve in conjunction with other component materials. The *structural deck* transmits gravity, earthquake, and wind forces to the roof framing. Its four major design factors are

- Deflection
- Component anchorage
- Dimensional stability
- Fire resistance

Decks can be classified as nailable or nonnailable (sometimes both) for purposes of anchoring the vapor retarder, insulation, or membrane to the deck (see Fig. 2-2). Some decks, e.g., timber or plywood, should only be nailed because of the threat of heated bitumen dripping through the joints. Poured concrete is limited to nonnailed anchorages, except in rare instances when wood nailers are cast into its surface.

Vapor retarders can be made of many various materials. A common vapor retarder, often known as a *vapor seal*, comprises three bituminous moppings with two plies of saturated felt or two bituminous moppings enclosing an asphalt-coated base sheet. Vapor-retarder materials also include various plastic sheets, aluminum foil, and laminated kraft paper sheets with bitumen sandwich filler, or bitumen-coated kraft paper.

Like insulation, a vapor retarder can cause problems, but unlike insulation, the vapor retarder may cause problems that outweigh its benefits. If the insulation contains moisture when installed, the vapor retarder helps prevent its escape because the retarder forms the bottom of a sandwich whose top is the membrane. Moreover, under the many field threats to its integrity—punctures, cutting of new roof openings, and so on—a vapor retarder almost always admits some water vapor. Thus the prudent designer specifying a vapor retarder will also design for the escape of water vapor: edge venting and, on large areas, stack venting to facilitate the escape of moisture from the roof sandwich and to reduce the threat of vapor-pressure buildup. Most roofing experts today reject the traditional advice, "If in doubt, use a vapor retarder" in favor of "If in doubt, omit the vapor retarder." However, the best advice appears to be, "If you need a vapor retarder, be sure it's a good one."

Thermal insulation cuts heating and cooling costs, increases interior comfort, and prevents condensation on interior building surfaces. Its secondary functions are almost as important. Through its horizontal shearing resistance, insulation helps relieve concentrated stresses transferred to the built-up membrane from movement in the structural deck. It also provides an acceptable substrate for the application of the membrane on a steel deck.

Insulation comes in many materials: rigid insulation prefabricated into boards, poured insulating concrete fills (sometimes topped with another, more efficient rigid board insulation), or dual-purpose structural deck and insulating plank. Fiberboard insulations are generally the most vulnerable to moisture, which eventually rots and weakens any fibrous organic vegetable board or organic plastic binder. But all insulation materials are vulnerable to some degree to moisture or freeze-thaw damage.

Insulation should have the following four structural properties:

- Good shearing strength, to distribute tensile stresses in the membrane and prevent splitting

- Compressive strength, to withstand traffic loads and (especially in the midwestern states) hailstone impact

- Adhesive and cohesive strength, to resist delamination under wind uplift

- Dimensional stability under thermal and moisture changes

Because of this host of demanding requirements, the design of thermal insulation and choice of materials is one of the roof designer's most complex tasks (see Fig. 2-3).

The built-up membrane, the weatherproofing component of the roof system, has three basic elements: *felts* and *bitumen,* alternated like a multideck sandwich, and a *surfacing,* normally of mineral aggregate. The membrane forms a semiflexible roof covering, with as few as two or as many as five plies of felt, custom-built to fit the contours of the deck (see Fig. 2-4).

The bitumen—coal tar pitch or asphalt—is the waterproofing agent. The felts stabilize and strengthen the bitumen, prevent excessive flow

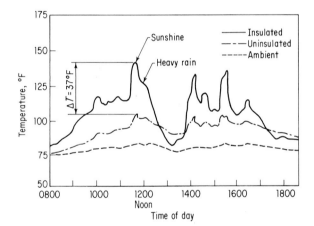

Fig. 2-3 Extreme roof temperatures produced by insulation sandwiched between deck and built-up roof membrane can make an insulated roof surface 40°F hotter in sunlight and 10°F colder at night than an uninsulated roof membrane. This extreme temperature cycling accelerates the membrane's deterioration, but in an era of constantly escalating energy prices, well-insulated roofs are essential. Moreover, the thinnest insulation produces nearly as extreme an effect on membrane temperature as the thickest, most efficient insulation (see Chapter 5, "Thermal Insulation"). (National Bureau of Standards Tech. Note 231.)

Fig. 2-4 The normal roofing practice of shingling felts (three shingled plies over a single-ply coated base sheet in the four-ply built-up membrane shown) prevents slippage between felts and interply moppings. Shingling also facilitates field application. (Flintkote Co.)

when the bitumen is warm and semifluid, and distribute contractive tensile stress when the bitumen is cold and glasslike.

Mineral aggregate surfacing, normally gravel, crushed rock, or blast-furnace slag, protects the bitumen flood from life-shortening solar radiation. Because of its damming action, aggregate permits the use of heavy, uniform pourings of bitumen (up to 75 lb/square), with consequent better waterproofing and longer membrane life. It also serves as a fire-resistive skin, preventing flame spread and protecting the bitumen from wind, rain, and foot traffic abrasion.

Surfacing aggregate is omitted on smooth-surfaced asphalt roofs with asbestos or glass-fiber felts. As its chief advantage, a smooth-surfaced membrane facilitates the detection and repair of leaks through membrane fissures obscured by aggregate. However, smooth-surfaced roofs are less durable and require more frequent maintenance than aggregate-surfaced roofs.

Designing the built-up membrane requires knowing what the membrane cannot do as well as what it can do. No membrane is strong enough to resist large movements in the deck, insulation, or other components in the built-up roofing assembly. Built-up membranes cannot resist puncture by sharp objects. Wherever more than occasional light foot traffic is expected on roofs, and especially where workers may drop steel tools, the designer should provide walkways.

Flashings are classified as *base flashings*, which form the upturned edges of the membrane where it is pierced or terminated, or *counterflashings* or *cap flashings*, which shield the exposed joints of the base flashing. Base flashings are normally made of bitumen-impregnated felts or fabrics, plastics, or other nonmetallic material. Counterflashings are often made of sheet metal: aluminum, copper, lead, stainless or galvanized steel. Generally recognized as the major source of roof leaks, flashing demands as much of the designer's attention as the basic builtup roof components.

DISASTROUS COMBINATIONS

The characteristic problems of roof-system designs are a combination of incompatible materials rather than isolated failures of single components. Two or more components may satisfy their individual material requirements to perfection and yet, in combination, fail disastrously. The art of roof-system design has lagged far behind the introduction of new materials. Because of this lag, some new materials have left a wake of litigation pressed by building owners plagued with failed roofs.

One of the worst system failures stemmed from the use of expanded polystyrene board insulation in conventional built-up roof systems, i.e., with the membrane on top of the insulation. These preformed board insulations offer excellent thermal resistance, with thermal conductivity k values ranging down to 0.20. But as noted in the previous section, insulation has vital secondary functions as well as primary functions. By its failure to perform these secondary functions, expanded polystyrene caused roof splits and even pulled flashings from their wall anchorages.

Polystyrene board is unsuitable as a substrate material for conventional built-up roof systems with felt-bitumen membranes mainly because it is virtually impossible to bond it to a deck with hot asphalt. It also has an extremely high thermal coefficient, roughly twice the thermal coefficient of a built-up membrane. Yet extruded poylstyrene has proved to be an excellent insulation for protected membrane roofs (PMRs), where it is placed on top of the membrane rather than under it.

Fire resistance also requires compatibility of materials. Through no intrinsic fault of the individual materials, certain combinations can be disastrous. When testing the fire resistance of a roof-ceiling assembly to qualify it for a given fire rating (in hours), one criterion is the assembly's resistance to heat flow. As a safeguard against the roof covering igniting, the average surface temperature rise during the furnace test must not exceed 250°F above the initial temperature. Because insulation retards heat flow, the designer might assume that adding more insulation to a rated roof-ceiling assembly must improve its fire performance.

The designer could be disastrously mistaken. Added insulation indeed depresses the roof-surface temperature, but if heat loss through the roof is excessively retarded, it can cause a structural collapse. A lower surface temperature means a *higher* ceiling plenum temperature (resulting from undissipated heat). This higher plenum temperature could buckle steel joists or ignite combustible structural members that otherwise would continue to carry their loads. The system designer, like a juggler, must watch more than one ball and must never assume that you cannot have too much of a good thing.

A third illustration of the system approach to roof design concerns the method of anchorage. The rising incidence of wind-uplift failures, in which all or part of a built-up roof system is ripped off the deck, makes anchorage of the roof components an especially important problem. But when balancing the advantages against the handicaps of the various methods of anchoring roof components, the designer can never focus solely on one troublesome factor, but must always consider several factors. When choosing among the various anchoring methods—nails or other mechanical anchors, hot-mopped bitumen, or cold-applied adhesives—the designer must consider fire resistance and membrane splitting as well as wind-uplift resistance.

CONSTRUCTING THE ROOF

Under the present organization on a normal roofing project, the work ideally proceeds as follows:

- The architect specifies the roofing components and installation procedures and submits progress and final inspection reports before acceptance.

- The roofing manufacturer furnishes products complying with the specifications and cooperates with the roofer to ensure that materials are dry when delivered to the site. Under the normal manufacturer's bonded roof program, the manufacturer inspects the work of the roofing subcontractor, who has previously demonstrated ability to install the manufacturer's materials.

- The general contractor schedules and coordinates the work of the roofing subcontractor and other subcontractors working on the deck (plumbers, electricians, HVAC workers, and so on), and makes sure the stored materials are kept dry. The general contractor also provides the roofing subcontractor with a satisfactory deck surface.

- The roofing subcontractor performs the actual fieldwork, coordinated by the general contractor, and installs the vapor retarder, insulation, built-up membrane, and, usually, the flashing.

Along with this apportionment of responsibility, a roofing job may be covered by a roofing manufacturer's *bond* or *guarantee*. A roofing bond, issued by the manufacturer and backed by a surety company, guarantees the owner that the surety company stands behind the manufacturer's liability to finance membrane repairs required to stop leaks caused by ordinary wear and roofing-application errors. Most bonds are written for limited liability, a small fraction of the cost of tearoff-replacement. (For more detailed discussion of roofing bonds, see Chapter 17, "Bonds and Guarantees.")

DIVIDED RESPONSIBILITY

Preceding a roofing failure on a typical job, where the manufacturer's bond may lull the architect and owner into a false sense of security because they have not studied the bond provisions, the following sequence of events may occur:

- The designer relies on the integrity of the prequalified roofer and the manufacturer and skimps on roofing-system specifications and details.

- The roof subcontractor exercises the option to select a cheaper roof specification by a manufacturer whose product still qualifies for the specified bond period.

- The general contractor, disregarding the roofer's qualifications and ignoring the application technique, selects the low roofing bid and relies on the manufacturer's inspection required under the bond.

- The manufacturer's inspector, often the sales representative who sold the materials to the roofer, is charged with inspecting the work of a customer on whose continued good will the sales representative depends for future material sales.

Assigning responsibility for roof leaks is a formidable challenge. Did the trouble stem from faulty design, careless fieldwork by the roofing subcontractor, defective materials supplied by the manufacturer, abuse by the owner, poor design by the architect, or an inadequate deck installed by the general contractor or another subcontractor? The technical complexity of modern built-up roofing systems, with proliferating material combinations creating new and sometimes unforeseen interactions between components, puts new stress on the creaking structure of responsibilities. Their diffused apportionment encourages a round or two of buck-passing.

UNIFIED RESPONSIBILITY

For intricate mechanical or electrical systems, a simpler, more rational apportionment of responsibilities has evolved. For an elevator subcontract, for example, the architect sets the general performance standards. The manufacturer advises the architect on hatchway and door clearances, access, power outlets, and other requirements. As the building rises, the elevator company's construction superintendent arranges for the installation of the rail brackets or inserts.

The elevator contract is essentially complete when the architect or owner accepts the installation, following the manufacturer's testing program. Under a contractor provision, the elevator manufacturer maintains the equipment, generally for a 3-month period. This more unified responsibility works better than the splintered responsibility of a typical roofing subcontract.

Several short steps toward unified responsibility in the roofing industry include the advent of "total" roof-system guarantees, which include insulation as well as the traditionally guaranteed membrane and flashing components. But the roofing system's intimate physical association with the rest of the building precludes the clean, isolated responsibilities assignable under an elevator subcontract.

PERFORMANCE CRITERIA

There has been some progress during the past decade in another aspect of systems design: the promulgation of performance criteria for roof-system components. The landmark was the 1974 publication of "Preliminary Performance Criteria for Bituminous Membrane Roofing," by National Bureau of Standards researchers R. G. Mathey and W. C. Cullen. (For a detailed discussion of this publication, see Chapter 7, "Elements of the Built-up Membrane.")

Based on extensive field experience with built-up membranes, this publication identifies 20 physical "attributes" required for satisfactory membrane service. As the most immediately obvious attribute, membrane tensile strength is required, to resist splitting stresses. Also required are *notch tensile strength* (to ensure that tensile strength is not significantly reduced by stress concentration) and *fatigue tensile strength* (to ensure the membrane's ability to resist cyclic stress caused by thermal, moisture, or building movement). *Impact resistance* is required to ensure the membrane's ability to resist such falling objects as hailstones, dropped tools, and tree limbs. Wind-uplift resistance and fire resistance are two more obvious attributes required by a built-up membrane.

The Mathey-Cullen study produced 10 quantified *preliminary* (emphasis added) performance criteria—for example, 200 lb/in. as the minimum tensile strength for a membrane tested in its weakest (cross-machine) direction, via ASTM D2523. Minimum tensile fatigue strength was set at 100,000 cycles of repeated force of 20 lb/in. at 73°F, 100 lb/in. at 0°F.

These membrane criteria set a pattern for the needed performance criteria for insulation. This more difficult job constitutes an even longer step toward the improbable dream of a truly scientific systems approach to roof design, in which the designer can rationally weigh the countless combinations of difficult roof-component materials for the

Fig. 2-5 "Dogbone" samples (bottom), cut from built-up membrane test samples in pattern of steel template (top), were tested for four "attributes"—tensile strength, tensile fatigue strength, flexural strength, and flexural fatigue strength—by NBS researchers R. G. Mathey and W. C. Cullen for their pioneering report, "Preliminary Performance Criteria for Bituminous Membrane Roofing," *Build. Sci. Ser.* 55, 1974.

most cost-effective system. Performance criteria for insulation are especially important now because of the continued escalation of energy costs pressing the need for thicker, more thermally efficient roof insulation.

Improved roof performance depends less on purely technological progress, manifested in the perennial search for new miracle materials, than on a deeper understanding of the roof as a complex system of interacting components. The correct combination of materials and good field application is more important than material quality as ingredients in a durable, weathertight roof.

THREE

Draining the Roof

If there is any single point on which roofing industry experts are unanimous, it is the necessity of draining a built-up roof. Several decades ago, ponded roofs enjoyed a minor vogue as a means of cooling the roof surface in hot summer weather. But that vogue has long since disappeared. There are now far safer ways of conserving cooling energy. Today's roofing industry consensus can be starkly stated: *If at all practicable, drain the roof. Take every practicable precaution to ensure that no accidental ponds are left after rainfall.*

Well-drained roofs' superior performances are confirmed by the industry's general experience. In a study of 86 randomly selected roofs, Montreal building consultant Donald J. Smith found that 67 buildings with roof slopes less than 2 percent (that is, ¼ in./ft) had a 58 percent leak rate, compared with an 11 percent leak rate for 19 roofs sloped at 2 percent or more. A nationwide survey of military authorities responsible for built-up roofs generally paralleled this finding. Although these general findings may lack scientific rigor, they represent an overwhelming empirical consensus. Owners and designers who ignore this experience proceed at their own peril.

Despite the long-standing industry advice to drain the roof, 18 percent of the roofs in an extensive National Roofing Contractors Association (NRCA) survey of member contractors had no slope whatsoever.[1] The nation's roof designers still display a durable obstinacy to assimilating this basic lesson of good roof design. To whatever it can contribute to the cause of eradicating this residual resistance to the fundamental rule of good design, this chapter is dedicated.

Ironically, the deliberately ponded, water-cooled roof argument boomerangs against those who cite it as a case against draining the roof. Since 1938, one major manufacturer has published a double-pour,

---

[1]"The Shape of Roof Construction," *Roofing Spec.*, November 1979, p. 40.

aggregate-surfaced, built-up roof specification for roofs that must resist ponded water (for water cooling the roof surface, for roofs with retarded drainage systems designed to relieve storm sewers, or for roofs subjected to periodic discharges from cooling towers or industrial processes). This specification contains the following three requirements:

- A 75 lb/square coal-tar-pitch flood coat, applied with 400 lb/square gravel or 300 lb/square slag, swept free of loose aggregate

- A second 85 lb/square coal-tar-pitch flood coat with 300 lb/square gravel or 200 lb/square slag, thoroughly broomed to remove loose aggregate

- Drainage of the roof prior to freezing weather (to avert the hazards of ice damage)

Few proponents of nonsloped ponded roofs apply the double-poured aggregate surface. And slope is required to fulfill the third requirement. (Otherwise, how can the water drain?) Thus collapses the defense of unsloped roofs via the deliberately ponded, water-cooled roof.

In 1963, with the unanimous concurrence of all member companies, the Asphalt Roofing Manufacturers Association (ARMA) Builtup Roofing Committee adopted a resolution recommending minimum ¼ in./ft slope. The ¼-in. figure contains a tolerance for the many inherent building industry imperfections. Flat roof surfaces inevitably contain humps and depressions resulting from a host of factors: column-foundation settlement; structural framing and deck deflection (elastic and long term creep); variations from plane surfaces in the deck and top flanges of structural members from inaccurate fabrication or inaccurate field finishing of slabs and topping; dimensional variations in insulation thickness; even variations in interply mopping thickness and flood coat, plus the irregularities of aggregate surfacing, forming dams and pockets impeding rainwater drainage.

In its 1978 *Manual for Builtup Roofs*, Johns-Manville recommends a minimum slope of ⁄₁₆ in. *after* construction. If designers specify ¼-in. slope and get ⁄₁₆-in. slope after construction, they can consider themselves fortunate.

Concurring with the ARMA recommendation of ¼-in. minimum slope are the NRCA, the Midwest Roofing Contractors Association (MRCA), and the Production Systems for Architects and Engineers (PSAE), sponsored by the American Institute of Architects (AIA). Such unanimity among manufacturer, contractor, and designer organizations is rare in the roofing industry.

Despite the importance of drainage and the overwhelming industry consensus, the rule is casually violated. The nation's high proportion of unsloped roofs is indicated by the ponded roofs observable from a jetliner landing at almost any urban airport in the United States. It's doubtful that half of the nation's flat roofs are properly drained. The prevalence of unsloped roofs has, in fact, been cited during roofing litigation as evidence that an architect who fails to slope a roof (even a smooth-surfaced roof) is not guilty of professional negligence because many of the architect's colleagues pursue the same benighted policy. With the Looking Glass logic characteristic of the legal world when it defends the indefensible, a lawyer defending an architect client attempts to establish incompetence as the prevailing community standard of competence by which to judge the client's performance.

Why, if this rule to drain the roof is so important, is it so frequently flouted? The most obvious explanation is first-cost economy. It is almost always cheaper (but only slightly) to build a "dead-level" roof than to slope the roof. A more respectable reason for a dead-level roof is to provide for future expansion requiring conversion of the roof into a floor. Perhaps an even more common explanation for undrained, unsloped roofs is designer ignorance and carelessness. Not only is it *cheaper* to design and construct a dead-level roof, it is *easier*. A nominally dead-level roof is of course never dead-level, but warped, humped, and depressed, with drains normally located near columns, which are usually high points (owing to the increasing deflection of roof framing members in proportion to their distance from columns). It takes care to ensure that openings, walls, expansion joints, and other flashed components are located at high points, or at least elevated above roof low points. Many designers and owners apparently rely on the slovenly logic that just as some cigarette smokers do not get lung cancer, so some ponded roofs do not fail prematurely.

As an additional point about ponding, note that ponding area and depth are important factors when assessing the hazard to the roof system in particular and to the building in general. A roof with 50 percent of its area chronically ponded is about 500 times worse than a roof with only a 1 percent chronically ponded area. Ponding furnishes a source of leakwater proportional to the volume of the contributory ponded area, and volume varies exponentially with ponding depth. If the contributory area of ponded water flowing through a membrane-leak opening is pyramidal, a doubling of ponding depth—say from ½ to 1 in. maximum—multiplies the leakwater volume by a factor of 8.*

---

*A pyramid's volume = ⅓ × base area × altitude. Thus for the same side slope, an inverted pyramidally shaped ponded area with ½-in. depth has only one-quarter of the base area of a pyramid with 1-in. depth.

WHY DRAIN THE ROOF?

Treatises on roof design usually attempt to convince readers of the need for positive drainage by citing long lists of reasons. Through their cumulative impact, these long lists are designed to mentally bludgeon the reader into acquiescence. This manual does the same. But before we get to this recitation, consider an anecdote that cuts to the heart of the issue.

Back in the early 1940s, during World War II, when such novelties were more impressive, one of my Navy colleagues always wore his "waterproof" wristwatch in the shower and the swimming pool. Asked to explain this cavalier attitude toward the hazards of water, he casually replied, "This watch is waterproof."

Some days later, when my friend appeared sans wristwatch, I inquired, "Where's the watch?"

"At the jeweler's," he replied. "It stopped when the works got wet."

The elementary lesson here springs from the question, "Why take unnecessary risks when the prospective losses far outweigh the gains?" Like the "waterproof" watch, a built-up membrane is commonly supposed to be waterproof. And so it is, under favorable conditions. If the design is perfect, the materials perfect, and the application perfect, or nearly so, then you can ignore slope in the roof without a worry in the world. You need worry only if any one of these critical factors—design, materials, or application technique—falls short of perfection. On the other hand, if you cannot count on a combination of perfect roof design with perfect materials and perfect application, you can trust in luck.

Now, to shift to a more technical viewpoint, the basic reason why you should drain the roof is the superiority of *water shedding* over *water resistance* as a means of keeping water out of the building. As anyone who has ever observed roofing application should know, practicable field techniques of applying a roof fall considerably short of laboratory precision. A thin spot in the top coat of bitumen; a fishmouth at a felt edge; a split caused by drying shrinkage of a wet felt, thermal contraction of unanchored insulation boards, or cracking of the substrate material; an interply void that later expands into a blister; or simply a puncture from a dropped tool—any of the foregoing can ultimately destroy the waterproof membrane. Because of its vulnerability to relative movement, a flashing joint submerged in ponded water poses an especially grave leak threat. However, if the roof sheds water, it can survive some imperfections without leaking. The advice to drain the roof is basically as simple as that.

For those unconvinced by this central argument, here is the previously promised list of ancillary reasons:

1. Structural roof collapses, sometimes with fatalities, are periodically caused by ponded water following heavy rains.

2. Ponded water is a source of moisture invasion into the membrane through any membrane imperfections: fishmouths, splits, cracks, bare felts, or even via an unsealed felt lap. Infiltrating liquid moisture can invade the insulation or leak into the building. Heated by the summer sun, entrapped moisture can accelerate growth of an interply void into a blister. As ice, moisture can delaminate the membrane as the freezing water expands, and subsequent freeze-thaw cycles can progressively enlarge these delaminated areas.

3. Perennial cycles of ponding and evaporation, corresponding with weather cycles of rainfall and sunshine, accelerate degradation of asphalt. Ultraviolet radiation degrades exposed asphalt chemically through photo-oxidation, which increases the number of high molecular-weight hydrocarbons (asphaltenes). It also increases the water-soluble products (OH groups on the HC chains, for which the H_2 molecule has a chemical affinity). Physically, the combination of heat and ultraviolet radiation is manifested in (a) migration of oily constituents to the surface (hence the slickness of wet asphalt pavements) and (b) hardening. Because it dissolves the photo-oxidized constituents and washes them away, water accelerates asphalt degradation by exposing fresh surfaces to photo-oxidative degradation.[1]

4. Moisture in insulation can rot organic fibers, weaken binders, and drastically reduce the insulation's thermal resistance.

5. Frozen ponded water moves with changing temperature. (Ice has an extremely high coefficient of thermal expansion-contraction.) Thermal movement of ice can erode the aggregate surfacing.

6. Standing water promotes the growth of vegetation and fungi, creating breeding places for insects and producing objectionable odors. Plant roots can penetrate the membrane and spread into insulation. When the plant dies, it leaves a large opening for entry (see Fig. 3-1).

7. Wide variations in roof-surface temperature between ponded and dry areas of the roof can range up to 60°F or more in summer, and these temperature differences can promote a warping pattern of surface elongation and con-

[1]P. G. Campbell, J. R. Wright, and P. B. Bowman, "The Effects of Temperature and Humidity on the Oxidation of Air-Blown Asphalt," *Mater. Res. Stand., ASTM*, vol. 2, no. 12, 1962, pp. 988ff.

Fig. 3-1 Ponded water can feed growth of rooftop plants, whose roots penetrate the membrane and spread into the wet insulation, fiberglass in the photo. When the plant dies, the opening penetrated by the plant roots is "unplugged," drastically increasing the leak potential.

traction, possibly wrinkling the membrane. Lateral migration of entrapped water from hot (high-pressure) to cold (low-pressure) areas can promote condensation that will enlarge the areas of wet insulation.

8. Prolonged repetition of wetting-drying cycles can reduce membrane tensile strength to less than 20 percent of its dry strength. (It can also rot organic felts and the organic fibers in asbestos felts.)

9. Randomly ponded water often occurs at flashed joints, which may provide easy access for standing water to enter the roof assembly.

10. Evidence of ponded water after a rainfall nullifies some manufacturers' roofing guarantees.

The collective weight of the foregoing items should convince the reader that any kind of dead-level built-up roof multiplies the odds on failure. But the risk also depends on another factor: the type of roof surfacing. If it is imprudent to specify an undrained aggregate-surfaced roof, it is stupid to specify a dead-level or very low-slope smooth-surfaced or mineral-surfaced cap-sheet surfaced roof. No statistical data are available to confirm these estimates, but in a locality with average United States rainfall, say, 30 in./year, it is highly doubtful that an undrained smooth-surfaced roof has better than an even chance of averting premature failure. According to a National Bureau of Standards (NBS)-conducted survey of military roofs at bases in the Northeast, middle Atlantic, Midwest, Alaska, and offshore Pacific islands (Hawaii and Guam), few built-up roofs, even aggregate-surfaced ones, that ponded water gave more than 10 to 15 years' service without major repairs and very expensive corrective maintenance. That finding applied to both asphalt and coal-tar-pitch membranes.

Regardless of the built-up roof's surfacing, even for the most favorable odds offered by an aggregate-surfaced roof, the no-drain policy becomes a ridiculously bad risk if you rationally weigh the initial cost savings gained from a no-drain policy against the probable cost of tearoff-replacement.

LIFE-CYCLE COSTING EXAMPLE

Let us examine the economic implications of gambling on a ponded roof. Assume that the probability of a tearoff-replacement rises from 5 to 15 percent (i.e., an increased failure probability of 10 percent). Assume further that this tearoff-replacement cost is incurred after the fourth year. Other assumptions: tearoff-replacement costs 50 percent more than original roof-construction cost of $1.50/ft², with roof-cost escalation rate 8 percent, interest rate 12 percent, and original cost of providing ¼-in. slope for drainage $0.10/ft². With failure probability rising by 10 percent, for 10 roofs there is a probable cost of $1.5 \times \$1.50 = \$2.25$ (current price) for one tearoff-replacement in exchange for a saving of $10 \times \$0.10 = \$1$ (cost of sloping 10 roofs).

Because this replacement roof has a salvage value (assumed straight-line depreciation) for 4 years of remaining service life at the end of the original roof's 20-year projected service life, this salvage value must be subtracted from the replacement roof's cost for accurate estimation of the future pretax return on investment in slope. This salvage value is the value 4 years from now of the replacement roof's remaining value when it is 16 years old.

Accordingly, for the left-hand term of our equation, we have the Future Worth (4 years from now, when the assumed tearoff-replacement becomes necessary) of the investment in sloping the 10 roofs. For the right-hand term, we have the projected cost of the tearoff-replacement, *minus* the replacement roof's Future Worth (4 years from now) when it is 16 years old. With straight-line depreciation, this value equals the projected tearoff-replacement cost $\times$ $\frac{4}{20}$ (to account for straight-line depreciation) divided by 1.12 raised to the 16th power, which discounts the projected 20-year Future Worth to a 4-year Future Worth.

Pretax return on investment is thus computed as follows:

$$\$1(1 + r)^4 = \$2.25 \times 1.08^4 - \frac{4}{20} \times \frac{\$2.25 \times 1.08^4}{1.12^{16}}$$

$$(1 + r)^4 = \$3.06 - 0.10$$
$$r = 2.96^{0.25}$$
$$= 0.31 \quad \text{or} \quad 31 \text{ percent}$$

A 31 percent return on investment is probably too conservative because the assumed $0.10/ft^2$ cost of sloping the deck is almost certainly too high. For a more realistic assumption of $0.05/ft^2$, the pretax rate of return is 56 percent.

There is, moreover, another dismal fact about tearoff-reroofing expenditures that tends to raise the rate of return on an investment in good drainage in the original design: A dead-level deck may be essentially uncorrectable. Providing proper slope for drainage may require raising peripheral or wall flashings to economically impracticable elevations, thus forcing a compromise on proper drainage. It is not merely more economical to do things right originally; it may be the only practicable time to do them right.

Still another factor may raise the pretax return of an original investment in a properly drained roof. If ponding promotes water invasion into the insulation, the moisture can substantially raise heating and cooling bills because water absorption can triple heat loss through the roof system.

If there are any countervailing positive factors economically favoring randomly ponded roofs, they remain a closely guarded secret.

HOW TO DRAIN THE ROOF

For the basic decision on how to drain the roof, the roof designer must decide between *interior* and *peripheral* drainage systems. In an interior drainage system, rainwater flows from elevated peripheral areas to interior roof drains. Leaders conduct the rainwater down through the building interior. Leaders in an interior drainage system are almost always located at columns. In a peripheral drainage system, rainwater flows in the opposite direction, from elevated interior areas to peripheral low points, to scuppers and leaders located outside the building.

Interior drainage has several notable advantages over peripheral drainage. Interior drain pipes, heated by the building interior, continue to conduct rainwater or melting snow through the cold winter weather. Peripheral drainage systems are totally subject to outside temperatures, freezing up and not functioning during cold winter weather. They also require more elaborate flashings, to protect gutter and scupper areas from erosion. Moreover, these peripheral areas are vulnerable to ice damming and metal distortion from freeze–thaw cycles. Peripheral drainage is obviously less troublesome in mild climates where winter temperatures remain above freezing.

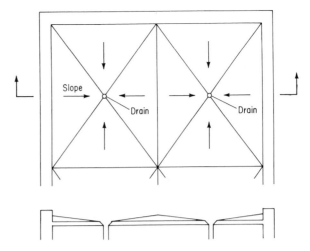

Fig. 3-2 Inverted pyramidal pattern with interior drains is the most dependable roof drainage.

Scuppers can, however, provide a safety feature on roofs with interior drainage systems and the threat of clogged drains. On roofs with parapets that could retain water to hazardous depths, peripheral scuppers can act as overflow valves. They should be located no more than 4 in. above general roof elevation. (A 4-in. average water depth results in an average uniform roof load of 21 psf.)

For roofs divisible into rectangular areas, the most positive drainage geometry comprises inverted pyramids, with four-way slope into each drain (see Fig. 3-2). Inverted pyramids can be created with field-built roof framing systems—wood and cast-in-place concrete. Tapered board insulation, wet insulating concrete fill, or dry-mixed, thermosetting asphaltic fill can also create these sloping pyramidal contours.

With prefabricated roof framing systems, e.g., structural steel or precast concrete, one-way slopes may prove more practicable and economical. Saddles and crickets, formed with tapered board insulation or poured fill, can complement one-way slope to provide positive drainage (see Figs. 3-3 and 3-4).

An important but often neglected aspect of roof design is flashing elevations. Locate flashings at a roof's high points, if practicable. In any event, keep flashings out of a roof's low points. (See Chapter 8, "Flashing," for further discussion of this topic.)

DRAINAGE DESIGN

To ensure good drainage, specify a minimum of two drains for a total roof area less than 10,000 ft². Add at least one additional drain for each

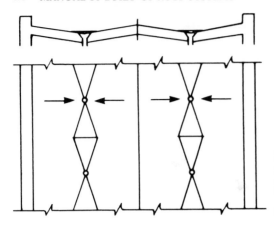

Fig. **3-3** Saddles, built by tapered insulation sections, can provide slope in valleys formed by decks sloped only in one cross section.

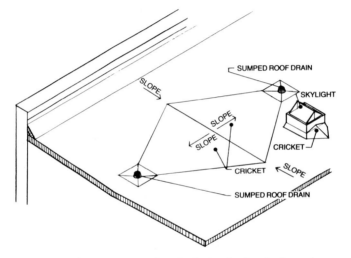

Fig. 3-4 Crickets are required on the high side of curbed openings to divert water flow around such obstacles. (National Roofing Contractors Association.)

additional 10,000 ft² of roof area, and limit the maximum spacing of drains in any direction to 75 ft. (That rule limits the contributing area to 5625 ft² in large, interior areas.)

Irregularly shaped roofs, with penthouses or other obstructions, require additional drains for good drainage. Do not expect water to turn corners as it flows toward drains. Pond-free drainage normally requires straight water flow from high points to a drain. The slight additional cost for several extra drains is cheap insurance against ponding. The drain's original installation costs much less than a later addition to relieve ponding. This slight additional drainage-assurance cost is trivial compared with that for tearoff-replacement to rectify a condition caused by ponding.

With uncambered roof framing, a drain's best location is near mid-span—ideally at the point of the roof framing's downward deflection. Avoid drain locations near columns or bearing walls, which are often high points regardless of the designer's intent. The extra cost for additional lateral leaders back to a column is well-justified. Leaders should start at a 90° angle with the drain head, to prevent deck deflection from dislodging drain rings installed over a rigid vertical drain leader. (Expansion joints can also accommodate this movement.)

Provide sumps at drains, recessing drains below immediately adjacent roof surfaces to prevent local ponding around the drain (see Fig. 3-5).

Required drain pipe size depends on three basic factors:

- Contributory area (maximum 10,000 ft^2)

- Rainfall rate (in./h)

- Roof slope (in./ft)

After you determine the contributing roof area per drain, consult local code authorities, the nearest rainfall recording station, or Table 3-1 for maximum rainfall rate.

A 1 in./h rainfall rate builds up at a 0.0104 gpm/ft^2 rate. [Because 1 gal = 231 in.^{3}, gpm/ft^2 = $^{144}/_{231}$ × 60 = 0.0104 gpm/(ft^2·in.·h) rainfall.] For Chicago, maximum rainfall rate = 3 in./h (from Table 3-1) and gpm/ft^2 = 3 × 0.0104 = 0.0312. For a 10,000-ft^2 contributing drain area, you need drain capacity = 0.0312 × 10,000 = 312 gpm. From Table 3-2, read drain pipe diameter of 5 in. (capacity = 360 gpm). Note also

Fig. 3-5 Drain sump pans, tack-welded in place, are furnished by steel-deck manufacturers. (Granco Steel Products Co.)

Table 3-1 Maximum Rainfall Rates (in./h)*

State & city	Rainfall	State & city	Rainfall	State & city	Rainfall	State & city	Rainfall	State & city	Rainfall
Alabama:		Macon	3.7	Escanaba	2.3	**North Carolina:**		Memphis	3.6
Anniston	3.6	Savannah	4.0	Grand Rapids	2.6	Asheville	3.2	Nashville	3.1
Birmingham	3.6	Thomasville	4.0	Marquette	2.2	Charlotte	3.4	**Texas:**	
Mobile	4.2	**Hawaii:**		Port Huron	2.4	Greensboro	3.4	Abilene	3.5
Montgomery	3.8	Hilo	3.0	Saginaw	2.4	Raleigh	3.7	Amarillo	3.4
Alaska:		Honolulu	3.2	Sault Ste Marie	2.0	Wilmington	4.0	Austin	4.0
Anchorage	1.0	**Idaho:**		**Minnesota:**		**North Dakota:**		Brownsville	4.5
Fairbanks	1.2	Boise	1.1	Duluth	2.6	Bismarck	2.8	Corpus Christi	4.5
Juneau	1.0	Lewiston	1.1	Minneapolis	3.0	Fargo	2.8	Dallas	4.0
Arizona:		Pocatello	1.3	St. Paul	3.0	**Ohio:**		Del Rio	4.5
Phoenix	2.5	**Illinois:**		**Mississippi:**		Cincinnati	2.7	El Paso	2.4
Tucson	3.0	Cairo	3.4	Jackson	3.8	Cleveland	3.0	Fort Worth	4.0
Arkansas:		Chicago	3.0	Meridian	3.8	Columbus	2.7	Galveston	4.6
Bentonville	3.8	Peoria	3.1	Vicksburg	3.7	Dayton	2.6	Houston	4.5
Fort Smith	3.9	Springfield	3.3	**Missouri:**		Sandusky	3.0	Port Arthur	4.5
Little Rock	3.7	**Indiana:**		Columbia	3.6	Toledo	3.0	San Antonio	4.0
California:		Evansville	3.0	Hannibal	3.4	Youngstown	2.6	Wichita Falls	3.8
Bakersfield	1.5	Fort Wayne	2.6	Kansas City	3.7	**Oklahoma:**		**Utah:**	
Eureka	1.6	Indianapolis	2.8	St. Joseph	3.6	Oklahoma City	3.9	Salt Lake City	1.3
Fresno	1.5	South Bend	2.8	St. Louis	3.4	Tulsa	3.9	**Vermont:**	
Los Angeles	2.0	Terre Haute	2.9	Springfield	3.7	**Oregon:**		Burlington	2.0
Sacramento	1.4	**Iowa:**		**Montana:**		Medford	1.4	**Virginia:**	
San Diego	1.5	Burlington	3.3	Billings	2.0	Pendleton	1.0	Lynchburg	3.4
San Francisco	1.6	Davenport	3.2	Havre	1.8	Portland	1.4	Norfolk	3.8

Location	Value	Location	Value	Location	Value	Location	Value	Location	Value
San Jose	1.6	Des Moines	3.3	Helena	1.5	*Pennsylvania:*		Richmond	3.6
Colorado:		Dubuque	3.1	Missoula	1.3	Allentown	3.0	Roanoke	3.3
Denver	2.5	Sioux City	3.3	*Nebraska:*		Erie	3.0	*Washington:*	
Durango	1.7	*Kansas:*		Lincoln	3.6	Harrisburg	2.9	Port Angeles	1.0
Grand Junction	1.6	Concordia	3.7	North Platte	3.3	Philadelphia	3.3	Seattle	1.0
Pueblo	2.6	Dodge City	3.6	Omaha	3.5	Pittsburgh	2.7	Spokane	1.0
Connecticut:		Topeka	3.7	*Nevada:*		Reading	3.0	Tacoma	1.1
Hartford	3.0	Wichita	3.7	Reno	1.2	Scranton	2.6	Walla Walla	1.0
New Haven	3.0	*Kentucky:*		*New Hampshire:*		*Puerto Rico:*		Yakima	1.0
Delaware:		Lexington	2.8	Concord	2.4	San Juan	4.0	*West Virginia:*	
Wilmington	3.5	Louisville	2.9	*New Jersey*		*Rhode Island:*		Charleston	2.8
District of Columbia:		*Louisiana*		Atlantic City	3.5	Block Island	3.0	Huntington	2.8
Washington	3.4	Lake Charles	4.4	Newark	3.0	Providence	3.0	Parkersburg	2.6
Florida:		New Orleans	4.5	Trenton	3.2	*South Carolina:*		*Wisconsin:*	
Apalachicola	4.3	Shreveport	4.0	*New Mexico:*		Charleston	4.1	Green Bay	2.6
Fort Myers	4.3	*Maine:*		Albuquerque	2.1	Columbia	3.6	La Crosse	2.9
Jacksonville	4.0	Portland	2.3	Roswell	2.6	Greenville	3.4	Madison	3.0
Key West	4.5	*Maryland:*		Santa Fe	2.2	*South Dakota:*		Milwaukee	2.7
Miami	4.6	Baltimore	3.4	*New York:*		Pierre	2.9	*Wyoming:*	
Pensacola	4.3	*Massachusetts:*		Albany	2.5	Rapid City	2.7	Casper	2.1
Tampa	4.2	Boston	2.6	Binghamton	2.5	Sioux Falls	3.3	Cheyenne	2.5
Georgia:		Nantucket	3.0	Buffalo	2.4	*Tennessee:*		Sheridan	2.1
Atlanta	3.5	*Michigan:*		New York	3.2	Chattanooga	3.3	Yellowstone	
Augusta	3.6	Alpena	2.2	Rochester	2.4	Knoxville	3.2	Park	1.5
Columbus	3.7	Detroit	3.0	Syracuse	2.3				

*Tabulated data from Josam Manufacturing Co.

31

Table 3-2 Pipe Sizing Data*

Pipe diameter, in.	Roof drain and vertical leaders	Horiz. storm drainage piping Slope, in./ft		
		⅛	¼	½
2	30			
2½	54			
3	92	34	48	69
4	192	78	110	157
5	360	139	197	278
6	563	223	315	446
8	1208	479	679	958
10		863	1217	1725
12		1388	1958	2775
15		2479	3500	4958

Flow Capacity for storm drainage systems, gpm

*Tabulated data from Josam Manufacturing Co.

from this same table that you need a 6-in.-diameter horizontal drainage pipe (capacity = 315 gpm) for ¼-in. slope.

DRAINING AN EXISTING ROOF

The toughest roof-drainage problem occurs with an existing roof that ponds water. This is an extremely common problem faced by many building owners with failed roofs and dead-level or inadequately sloped decks. Besides tapered insulation and sloped insulating fills (see Chapter 5, "Thermal Insulation"), another alternative for the existing deck is the *siphon drainage system*; it is, however, less satisfactory than sloping the roof.

The siphon drainage system is a set of pumps placed in the ponded areas of a roof, with plastic hoses leading to roof drains. The system recycles every hour, pumping only when ponds reach a ⅝-in. depth, when the units prime the drain lines. A siphoning action starts water flow down existing drains until each pond is reduced to a ⅛-in. depth. Evaporation then takes over, draining the small remnant of ponded water. (One obvious drawback to this electrically powered siphon-drainage system is that power failure during a rainstorm renders it temporarily useless.)

The *solar-powered drain* is still another technique for draining ponded roof areas and pumping the water to a roof drain or over the building's edge. Cycling radiant solar energy creates a siphon pumping action. Solar radiation absorbed on a black surface within the solar drain's transparent plastic dome increases the internal pressure within the solar drain and exhausts air through an outlet valve. Cooling reduces internal pressure, but the closed exhaust valve prevents pressure-equalizing outside air from entering. Vacuum pressure initiating each new cooling-heating cycle siphons water from the intake hose and pumps it through an exhaust hose.

ALERTS

Design

1. Provide minimum ¼ in./ft slope for aggregate-surfaced roofs, ½ in./ft for smooth-surfaced roofs. (For precisely controlled construction, ⅛ in./ft may be satisfactory for aggregate-surfaced roofs.)

2. Locate drains at low points (an obvious rule often overlooked).

3. Provide sumps at drains, or other means of recessing drain heads below the adjacent roof level.

4. Favor interior drainage systems over peripheral drainage, especially in cold climates.

5. Provide drain inlets with strainers, to prevent debris from clogging leaders.

6. Provide scupper-type receivers with flanges wide enough for anchorage to wood nailers.

7. When gutters form part of peripheral drainage systems, detail the gutters' outside vertical section at least 1 in. below roof elevation (to prevent water from standing on roof if downspouts are clogged).

8. Use tapered insulation boards to build slopes or crickets around rooftop equipment to keep water away from flashings, moving toward drains.

Maintenance

1. Inspect and clean drain screens regularly.

2. Keep roof clean. (Rubbish impedes drainage.)

FOUR

Structural Deck

The structural deck's major purpose—to resist gravity loads and lateral loading from wind and seismic forces—is beyond the scope of this manual. But in addition to its structural function as the base for the roof system, the structural deck must satisfy other design requirements:

- Deflection

- Component-anchorage technique

- Dimensional stability

- Fire resistance (see Chapter 15, "Fire Resistance")

- Surface character (continuous or jointed)

Rational deflection limits that prevent the formation of rainwater ponds can maintain positive drainage. Anchorage of the component layers above the deck is essential to prevent both delamination of the roof assembly by wind uplift or horizontal movement and membrane splitting. Deck dimensional stability depends on the coefficient of thermal expansion-contraction and, in organic fibrous materials, the degree of swelling accompanying moisture absorption. Excessive movement of the substrate can wrinkle or split the built-up membrane. Dimensional stability is especially important when thermal insulation is either omitted or placed below the deck. And the structural deck must carry its design loads through any fire-resistance tests.

Classified by the method used to anchor underlayment, insulation, or membrane, decks are either *nailable* or *nonnailable*. Wood (over ½ in. thick), structural wood-fiber, and gypsum decks are nailable; concrete and steel decks are nonnailable. Steel decks, however, take mechanical anchors, whereas concrete almost always has a layer of hot-mopped asphalt anchoring the next component (see Table 4-1).

35

Table 4-1 Anchorage to Structural Deck (Key: F = Mechanical Fasteners*, A = Adhesives†)

Component to be applied	Deck substrate					
	Wood (sawed, plywood)	Preformed wood fiber	Gypsum (poured, precast)	Metal	Concrete (poured, precast)	Asbestos cement
Vapor barrier:						
Felt type	F	F	F	A	A	A
Plastic sheet	F-A			F-A		
Metal foil	F-A					
Kraft paper	F-A	F-A	F-A	A	A	A
Insulation:						
Mineral aggregate board	F-A	F-A	F-A‡	F-A	A	A
Vegetable-fiber board	F-A	F-A	F-A‡	F-A	A	A
Glass-fiber board	A	A	F-A‡	F-A	A	A
Glass foam board	A	A	F-A‡	F-A	A	A
Plastic foam	A	A	F-A‡	F-A	A	A
Corkboard	A	A	F-A‡	F-A	A	A
Lightweight concrete	*Poured-in-place*					
Built-up membrane	F§	F§	F§		A	A

*Mechanical fasteners are nails or special fasteners. They are more reliable than adhesives, especially cold-applied ones.

†Adhesives are hot bitumen or fire-rated, cold-applied mixtures.

‡Apply coated sheet on gypsum substrate before installing insulation board.

§Base ply (or plies) only, unless slope requires back nailing.

DECK MATERIALS

The basic roof decks commonly used with built-up roofing systems are

- Steel—lightgage, cold-rolled sections
- Wood sheathing—sawed lumber or plywood
- Concrete—poured-in-place or precast
- Gypsum—precast or poured-in-place
- Preformed, mineralized wood fiber
- Composite decks of mesh-reinforced, lightweight insulating concrete on corrugated steel or formboards

Steel decks, by far the most popular, usually come in a narrow range of gages—22 gage (0.028 in.), 20 gage (0.034 in.), and 18 gage (0.045 in.)

thick (down to 28 gage for centering for lightweight insulating concrete fill). In cross section, steel deck is ribbed, with ribs generally spaced at 6 in. (on centers) and 1½ or 2 in. deep. The slope-sided ribs measure 1 in. across at the top for *narrow-rib* deck, 1¾ in. for *intermediate* deck, and 2½ in. for *wide-rib* deck. Prefabricated units are 18 in. to 3 ft wide, in 6-in. increments, up to 20 ft or more long (see Fig. 4-1).

Wood decks come in three basic types:

1. Board sheathing (up to 1-in. nominal thickness)

2. Plank (to 5-in. nominal thickness)

3. Plywood (minimum ½ in. thick, 4 × 8 ft panels)

Although wood decks are generally nailed or stapled, underlayments can be attached with hot bitumen or "cold-applied" adhesives.

Concrete decks come in two forms: cast-in-place and precast. A cast-in-place structural deck is continuous, except where interrupted by an expansion joint or another building component.

Precast concrete decks come in a variety of cross sections: single T,

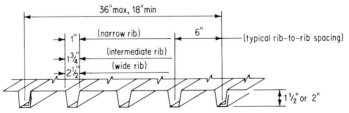

Fig. 4-1 Lightgage steel deck for roof systems comes in 1½-in. depth (occasionally 2-in.), in three rib widths. Actual rib width, measured from flange to flange, is about ½ in. greater than nominal width, which is measured between points inside rounded edges.

double T, inverted channels, solid or cored plank. Single or double Ts vary in width from about 2 to 8 ft, with spans up to 60 ft or so. Inverted channels come in shallow, 3-in. depths to 12 in., with spans ranging up to 20 ft or so. Planks, solid or cored, range in depth from 2 to 10 in., in 16- to 24-in. widths, and spans up to 25 ft or so. Because joints between precast deck units are usually uneven, they may require mortar troweling to smooth the surface for insulation.

Gypsum decks, like concrete, are either precast or cast-in-place. A cast-in-place gypsum deck is poured on gypsum formboards spanning flanges of closely spaced steel bulb tees (see Fig. 4-2). Like a poured concrete deck, a cast-in-place gypsum deck presents large, seamless expanses of roof surface (except where it cracks from thermal contraction or drying shrinkage).

Precast gypsum planks come in 2-in.-thick panels, with metal-bound tongue-and-groove edges. Because gypsum decks are nailable, insulation can be either nailed or bonded with hot bitumen.

Preformed, mineralized wood fiber, in 1½- to 3½-in.-thick planks, comprises long wood fibers bonded with a cement or resinous binder and formed under a combination of heat and pressure. Some preformed, mineralized wood-fiber planks have an integrally bonded urethane foam surface, for more efficient combination of this deck's dual structural-insulating function.

Fig. 4-2 Workers screed a fast-setting, wire-reinforced gypsum deck, poured on gypsum formboards spanning between closely spaced bulb tees. (U.S. Gypsum Co.)

DESIGN FACTORS

Design factors, considered in order, are

- Sloped framing for drainage
- Deflection
- Dimensional stability
- Moisture absorption
- Anchorage of components
- Deck surface and joints

Slope

The most dependable and often most economical way to provide slope on a new building's roof is to slope the structural framing and deck. This is relatively simple with cast-in-place concrete and wood-framed structures, which lend themselves to ready field sloping. Structural steel supporting steel decks is slightly more difficult to slope. It is very difficult to slope the dead-level plane of a ponded, failed roof. As demonstrated in Chapter 3, the economic returns from draining the roof and avoiding ponded-water problems should far outweigh the trivial cost of planning and building slope into the structure. This is usually better accomplished by sloping the framing than by using tapered insulation or tapered fills.

Compared with sloped framing and constant insulation thickness, tapered insulation has the following three disadvantages:

- Difficult design coordination for flashing heights and other roof details affected by varying insulation thickness
- Fairly complex field operations, involving special field drawings and coded insulation pieces for most tapered insulation systems
- Hazards of excessive insulation thickness—to 6, 7, or even 8 in. is sometimes required for adequate slope, with possible reduced membrane restraint and consequently increased splitting risk

Adding sloped insulating concrete fills sandwiched between the deck and membrane to provide slope carries three unique liabilities:

- Increased risk of entrapped moisture within the roof assembly, especially in ordinary steel decks that lack underside vent openings or over concrete

- Cracked substrate for the membrane, with a heightened splitting hazard from differential contraction or expansion rates

- Risk of a sliding membrane substrate where the fill tapers to a "feather edge" and breaks into fragments

Deflection

Deflection limits normally set in building codes, manufacturers' bulletins, and other guides or recommendations have little or no relevance to roof problems. The typical deflection limits—$\frac{1}{180}$, $\frac{1}{240}$, or even $\frac{1}{360}$ of the span for live, or even total load—may be excessive for long-span roofs, which are especially vulnerable to ponding, and too conservative for short-span decks. Structural failures, including sudden and total collapse, have occurred in some poorly designed level roofs. Most of these roofs satisfied the normal code design provisions.

Ponding caused by faulty design can be explained as follows: The deflection curve produced by an accumulating weight of rainwater may form a shallow basin in a roof. If the outflow does not prevent the capacity of the basin from increasing faster than the influx of rainwater, the roof is unstable, and a long, continued rainfall is structurally hazardous. Even if the structural deck withstands the ponding load, the standing water threatens the membrane, the insulation, and ultimately, if the roof leaks, the building contents.

As a rough safeguard against ponding, for level roofs designed with *less* than the minimum recommended slope of ¼ in./ft, the *sum* of the deflections of the supporting deck, purlins, girder, or truss under a 1-in. depth of water (5-psf load) should not exceed ½ in.

A proper code provision would specify a minimum stiffness, such as maximum deflection = ½ in. for live load = 5 psf (the weight of 1 in. of water). Thus the deflection would increase at roughly half the rate of the rainwater buildup. If the roof structure does not satisfy this requirement, the designer should present computations substantiating the safety of the slope used.[1]

When computing deflections, the designer must consider plastic flow in concrete and wood. Plastic flow (increasing deflection under constant prolonged loading) is inevitable in materials like reinforced or prestressed concrete. Laboratory tests indicate plastic strains ranging from 2½ to 7 times elastic strains in structural concrete. Because these larger extremes seldom occur in practice, the normal assumption is that

---

[1]For a more elaborate analysis of required roof slope and framing stiffness, see the *Timber Construction Manual,* John Wiley & Sons, Inc., New York, 1966, pp. 4-153ff and "Roof Deflection Caused by Rainwater Pools," *Civ. Eng.,* October 1962, p. 58.

plastic flow adds an increment of 2½ to 3 times elastic deflection. Thus the total deflection attributable to elastic strain + creep = 3 or 4 × instantaneous elastic deflection.[1]

Wood is similarly subject to long term plastic deformation. For glued laminated members (including plywood) and seasoned sawed members, residual plastic deformation is about 50 percent of elastic deformation. For unseasoned sawed members, plastic deformation rises to 100 percent, thus requiring a doubling of computed elastic deflection to approximate long term deflection.

In highly humidified interiors, preformed structural wood-fiber plank sometimes exhibits excessive, and erratically unpredictable, permanent inelastic deflection. These deflections are not a true creep or plastic flow, attributable solely to load-produced strain. A combination of thermal deformation and moisture absorption, which weakens certain cementitious binders used in wood-fiber materials, causes some preformed structural fiber planks to sag—up to ½ in. in a 4-ft span. Such excessive deflection threatens the entire roof assembly, increasing the chances of membrane splitting. Portland cement binder makes the preformed wood-fiber plank generally immune to excessive deflection.

Deflection and deck movement can also be aggravated by foundation settlement. In a roof designed for the minimum recommended ¼-in. slope, foundation settlement is normally insignificant. (A differential settlement of 1 in. in adjacent column footings 30 ft apart reduces a ¼-in. roof slope by only 13 percent.) Roof-deck vibration from seismic forces, traffic, or vibrating machinery may also require attention.

Dimensional Stability

The dimensional stability of a roof deck is determined largely by its coefficient of thermal expansion and dimensional change with changing moisture content. These factors vary greatly with different materials.

Wood has a relatively low thermal coefficient of expansion longitudinally, from 1.7×10^{-6} to $2.5 \times 10^{-6}/°F$ for sawed members of different species and about $3 \times 10^{-6}/°F$ for plywood. The average value is thus about one-sixth the value for aluminum, less than one-third the value for steel. Moisture is the greatest threat to a wood deck's dimensional stability. Under the extreme moisture variation normally antici-

---

[1]For more detailed analysis of the part played in the deflection of concrete structural members by shrinkage and temperature deformation as well as plastic flow, see *Proc. Am. Concr. Inst.*, vol. 57, 1960–1961, pp. 29–50; also see W. G. Plewes and G. K. Garden, "Deflections of Horizontal Structural Members," *Can. Build. Dig.*, no. 54, June, 1964.

pated in service, the expansion of plywood roughly equals the expansion of steel or reinforced concrete under a 150°F temperature rise.

Moisture Absorption

Sooner or later, as an expanding gas, expanding freezing liquid, or contracting solid, moisture retained in materials like wood, concrete, and gypsum damages the roof.

Opinions vary as to the time required for poured structural decks to dry. There should be a minimum of 24 h (depending on the weather) between the pouring of a gypsum deck and the application of roofing felts, according to some felt manufacturers and roofing contractors concerned about the hazards of entrapping water vapor released from inadequately dried deck surfaces. On the other hand, gypsum manufacturers permit almost immediate coverage.

When to apply insulation or roofing to a poured gypsum or concrete deck can be simply resolved, under normal conditions, by the U.S. Army Corps of Engineers' test for dry deck. Pour a small amount of hot bitumen on the deck. If, after the bitumen cools, you can readily remove it with your fingernails and hands, reject the deck as too wet for any applications. If the cooled bitumen sticks to the deck, too tight to be removed by your fingers, then accept the deck as dry enough for insulation or roofing application.

An alternative, or supplementary, test for properly vented poured concrete or gypsum decks is to apply hot bitumen as just described and watch for frothing or bubbling. If none occurs, the deck is probably dry enough for application.

Because of its high absorptivity, poured gypsum is not recommended for roof decks in buildings of high temperature and high relative humidity, e.g., laundries, bakeries, and textile mills. In highly humidified interiors, concrete and properly pressure-treated wood are the most generally satisfactory roof-deck materials. When properly maintained to control corrosion, steel decks are also satisfactory.

Wood decks can rot when exposed to high, unrelieved humidity. In the normal airconditioning range of 40 to 50% relative humidity (RH), wood's moisture content generally runs 8 or 9%, far below the 35 to 50% range required for active rot-fungus growth.[1] But in unvented ceiling plenums, where periodic condensation can keep relative humidity close to 100%, wood decks can rot.

---

[1] M. C. Baker, "Designing Wood Roofs to Prevent Decay," *Can. Build. Dig.*, no. 112, April 1975, pp. 112-1, 112-3.

Anchorage of Components

For anchoring built-up roofing or insulation to the structural deck, you have a choice of two basic techniques:

- Nailing or mechanical fastening

- Adhesives (hot-mopped bitumen or cold-applied)

Nailing or mechanical fastening is generally the most dependable method (see Fig. 4-3). Cold-applied adhesives have proved so generally undependable that for practical purposes they can now be considered a virtual violation of good roofing practice, rejected by several major roofing manufacturers and the National Roofing Contractors Association in its latest manual of roofing practice. (For further discussion, see Chapter 14, "Wind Uplift.") This leaves hot-mopped Type III asphalt as the sole recommended alternative to mechanical fastening. When the application procedure is properly carried out on a dry deck surface, hot-mopped asphalt generally provides the strongest adhesion between the deck and its contiguous component (insulation, vapor retarder, or built-up membrane).

However, mechanical fastening or nailing has a critical advantage over hot-mopped asphalt as a dependable anchorage technique because of its greater tolerance for field timing. According to Engineering Research Consultants, Madison, Wisconsin, "Hot asphalt applied at subfreezing temperatures to lightgage steel decks congeals so rapidly that mechanical fastening is probably the only effective technique for anchoring the insulation."[1]

Because of their extremely high heat conductivity, steel and concrete decks congeal hot asphalt much faster than a plywood deck, or, better yet, an insulation substrate.

To assure reliable anchorage, hot-mopped insulation adhesive literally requires split-second timing, proper tamping of the insulation, and good weather (moderate to high temperatures, with mild winds). Under the best of circumstances, you have only a few seconds, probably less than 10, to get the insulation board solidly embedded in the hot, fluid asphalt. A distracted worker failing to promptly place several insulation boards can cause a membrane split during the roof's first winter, when thermal contraction stresses and strains build up in the unrestrained membrane.

---

[1]R. M. Dupuis, J. W. Lee, and J. E. Robinson, "Field Measurements of Asphalt Temperatures During Cold Weather Construction of BUR Systems," *Proc. Symp. Roofing Technol.*, NBS-NRCA, September 1977, p. 272.

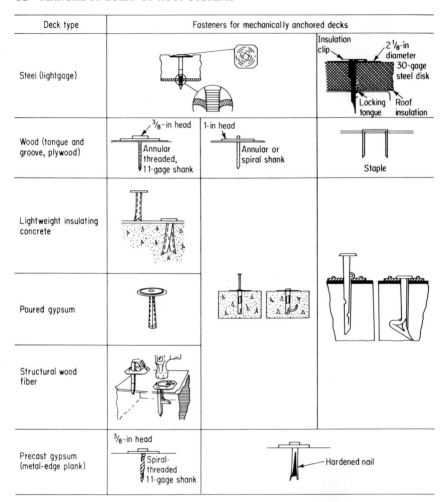

Deck type	Fasteners for mechanically anchored decks		
Steel (lightgage)			Insulation clip, Locking tongue, 2 1/8-in diameter, 30-gage steel disk, Roof insulation
Wood (tongue and groove, plywood)	3/8-in head — Annular threaded, 11-gage shank	1-in head — Annular or spiral shank	Staple
Lightweight insulating concrete			
Poured gypsum			
Structural wood fiber			
Precast gypsum (metal-edge plank)	3/8-in head — Spiral-threaded 11-gage shank	Hardened nail	

Fig. 4-3 Roofing nails and mechanical anchors come in assorted shapes and sizes, with varied anchorage devices—spiral or annular shanks, locking tongues, and hollow shanks—that expand when driven. Small-headed nails are driven through minimum 1-in.-diameter steel disk caps, minimum 30-gage thickness.

Other considerations can rule out mopping, especially solid mopping, as an anchoring method. On deck surfaces subject to shrinkage cracking, solid mopping increases the hazard of membrane splitting. By bonding two components throughout their contact area, solid mopping intensifies local stress concentration in a membrane directly over the deck crack. Solid mopping also seals in entrapped moisture, thereby promoting blisters and membrane wrinkling.

Strip or spot mopping alleviates both problems. By leaving areas unbonded to the membrane to distribute cracking strains, strip or spot mopping reduces membrane stress concentrations and possible splitting. Intermittent mopping provides lateral avenues of escape for water-vapor-pressure relief, which reduces the hazard of membrane blistering or wrinkling.

For poured gypsum and preformed, mineralized wood-fiber decks, which are subject to both shrinkage and high moisture absorption, nailing is mandatory. It provides better stress distribution in the membrane and better access for water-vapor-pressure relief to vents.

Deck Surface and Joints

As a working surface for the application of insulation or built-up roofing, poured-in-place decks have an advantage that at least partly offsets the moisture problems: They provide large deck areas without joints. Horizontal gaps between adjacent precast (or precut) units may allow bitumen to drip through, creating a potential fire hazard. Vertical misalignment between adjacent units can ruin the surface as a substrate for applying the insulation.

There are several techniques for closing the horizontal gaps between prefabricated units and leveling the ridges formed at joints. For wood sheathing, tongue-and-grooved instead of square-edged boards help in both respects: The tongue-and-grooved joint is much tighter than a square-edged butt joint, and the tongue-and-grooved boards do not warp as easily as square-edged boards. For plywood decks, when there is no supporting purlin or other structural member under the joint, H-shaped metal clips between adjacent units prevent unequal deflection and thus avoid the formation of vertical irregularities.

As precautions against bitumen drippage through the joints of wood decks, plywood can be stripped with felt, and board sheathing can be covered with 5-lb rosin-sized paper or 15-lb saturated felt. For glass-fiber-felt membranes on nailed wood decks, one manufacturer recommends a glass-fiber "combination" sheet: an asphalt-coated base sheet with kraft paper laminated to the underside.

Joints in preformed, mineralized wood-fiber or precast concrete decks should be pointed with cement mortar, which is itself subject to shrinkage cracking. Prestressed concrete units bent in a convex deflection curve by the prestressing may require a concrete topping to level a surface marred by extreme joint irregularities. In less serious instances, precast, prestressed units may require only local grouting along joint lines.

Low spots on concrete deck surfaces (more than ½ in. below level) should be filled with portland cement mortar. High spots should be ground down.

Before insulation, built-up roofing, or vapor retarder are applied, poured concrete decks should receive a surface primer of cutback bitumen. The primer is dual-purpose:

- To absorb dust inevitably remaining even after the surface is cleaned
- To provide a surface film to adhere the bituminous mopping

Uninsulated bulb tees used to support the gypsum and structural wood-fiber decks can create problems. Bulb tees without covering insulation act as thermal bridges, transferring heat through the highly conductive steel. (This thermal-bridge heat transfer can melt snow in parallel strips.) By creating varying membrane surface temperatures (lower in the exposed strips than in the insulated snow-covered areas), this melted-snow pattern can cause different thermal stresses in the membrane.

Thermal contraction or expansion of uninsulated steel bulb tees and their supporting joists can promote structural cracking of a poured gypsum deck, which is also subject to drying shrinkage (see Fig. 4-4). These thermal movements heighten the risk of membrane splitting in either a gypsum or structural preformed wood-fiber deck.

To avoid these hazards with uninsulated bulb tees, always specify a layer of insulation over the gypsum or structural wood-fiber decks. In this energy-conscious era, you will almost always need additional insulation for both its thermal resistance and membrane-protecting qualities.

STEEL DECKS

As the most popular deck material by far, used for about 65 percent of new industrial buildings and many other building types, lightgage, cold-rolled steel merits special attention. The steel deck's fluted (i.e., nonplanar) surface has a marked disadvantage compared with other deck materials as a roof-system substrate, and the steel deck's flexibility, its often distorted joints, and its general vulnerability to localized distortion make it a special problem for securely anchoring insulation.

A lightweight deck suitable for long-span joists, the steel deck is popular because it is economical. But the drive for economy sometimes goes too far. Some steel-deck sections have an actual thickness less than published values used to compute moment of inertia, according to Fac-

Fig. 4-4 Shrinkage cracking of grout over bulb tees supporting preformed, structural wood-fiber deck (top) requires special attention to the anchorage of components to avoid membrane splitting (see Chapter 12, "Premature Membrane Failures"). When built-up membrane is applied directly to deck, the base sheet should be nailed, to allow stress distribution in membrane when cracks occur between nails (bottom). (GAF Corporation.)

tory Mutual Research in its report to the Asphalt Roofing Manufacturers Association.[1] This petty economizing to save fractions of a penny per square foot is the root of steel-deck problems, as it is with many other roofing problems.

[1]Factory Mutual Research, *Steel Roof Deflection Study*, August 1975, p. 5.

Most problems with steel decks concern excessive deflection. Steel decks are subject not only to longitudinal but to transverse deflection (dishing or "rolling"; see Fig. 4-5). Live-load deflection in service *after* the insulation is in place can break the original bond if the deck and insulation do not deflect in perfect congruity.

Construction loads or handling stresses can cause even greater problems. Permanent deflection or distortion (beyond the steel's elastic limit) can prevent adhesive bonding of the insulation and deck at the time of application. Because construction loads normally exceed loads imposed during the building's service life, Factory Mutual researchers tested decks with four 250-lb stationary concentrated loads, each representing a 200-lb man carrying roofing felts or other material, located at midspan of one end span and the center span of a three-span continuous deck spanning up to 7 ft.

Steel-deck design requires special attention to three items:

- Side lap fastening

- End lap detail

- Top-flange V-groove stiffeners

Side lap fastening is vital to prevent differential deflection between adjacent steel-deck units. This differential deflection can result in:

- Permanent deck distortion (because the steel is locally stressed beyond the elastic limit)

- Breaking of adhesion with the insulation

- Fracture of insulation board and membrane

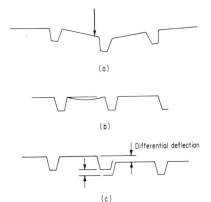

(a)

(b)

Differential deflection

(c)

Fig. 4-5   Deflection problems with steel deck include (*a*) "rolling" under concentrated loads, (*b*) "dishing" (especially common when deck flanges have stiffener V grooves), and (*c*) differential deflection from inadequate side lap fastening.

Correct end lapping of the steel sections is similarly necessary to avert differential deflection. If the ends are cantilevered, they can deflect differentially under concentrated loads, with the same consequences as inadequate side lap fastening.

Top-flange V-groove stiffeners pose three potential hazards:

- Interference with secure mechanical fastening (because they interrupt the plane flange surface usually necessary to engage the mechanical anchors' locking devices or serrated surfaces).

- Reduction of effective contact surface area required for adhesives (forming a trough for the adhesive to flow into, thus losing adhesive contact with the insulation).

- Possible warping of the flange's top surface. (The V groove stiffens the deck *longitudinally* while weakening it *transversely*.)

A minimum 20-gage (0.034-in.) deck section successfully resisted the rolling action of concentrated, moving loads under Factory Mutual tests (see Fig. 4-5). Use of minimum 20-gage steel deck is an advisable precaution against such local distortion.

Factory Mutual's Recommendations for Steel Decks

Here are seven of Factory Mutual's key recommendations:

1. Ban V-groove stiffeners in steel decks.

2. Plane top flanges must be flat, with no concavity or convexity greater than ⅟₁₆ in. across three contiguous flanges.

3. Deck thickness must at least equal the value upon which the manufacturer's moment of inertia calculation is based. Minimum acceptable thickness for published gage is

$$16 \text{ gage} = 0.057 \text{ in.}$$
$$18 \text{ gage} = 0.045 \text{ in.}$$
$$20 \text{ gage} = 0.034 \text{ in.}$$
$$22 \text{ gage} = 0.028 \text{ in.}$$

4. Provide structural steel framing around openings in steel decks.

5. Space side lap anchors (No. 8 self-drilling screws) at maximum 3-ft spacing (i.e., midspan location for spans up to 6 ft, two fasteners for spans over 6 ft, at one-third points).

6. End laps = 2-in. minimum, located over supports, with one end crimped to facilitate nesting. Maximum dimension between end lap surfaces = $\frac{1}{16}$ in.

7. Live-load deflection limit = $\frac{1}{240}$ span, computed by the following formulas:

$$\text{Single deck span: } D = 0.015 \frac{KPL^3}{EI}$$

$$\text{Two or more spans: } D = \frac{KPL^3}{48EI}$$

where K = 1 (unless substantiated as less than 1)
$\quad I$ = moment of inertia (in.^{4})
$\quad E$ = 29.5 $\times$ 10^6
$\quad L$ = span (in.)
$\quad P$ = 300 lb*

ALERTS

Design

1. Check deck and supporting structure for ponding deflection. Include creep or plastic flow factor when computing deflection of concrete or wood members. Also check for deflection under equipment wheel loads (see Fig. 4-6).

2. Design the roof to drain, with $\frac{1}{4}$-in. minimum slope, preferably by sloping deck and framing.

3. Locate drains near midspan, not near columns (unless beams are cambered to slope toward columns).

4. Check the method of attachment of next component above deck with insulation manufacturer, roofing manufacturer, and insurance agent (see Chaps. 14 and 15, "Wind Uplift" and "Fire Resistance," respectively.)

5. To reduce the threat of bitumen drippage when insulation, vapor retarder, or base sheet is hot mopped to deck, specify ASTM Type II or III asphalt.

*To represent a 300-lb *dynamic* load, simulating a 200-lb man carrying 100 lb of material, as opposed to a *static* load, P should be a minimum 600 lb because a dynamic load at least doubles the static-load deflection, in accordance with the following formula:

$$\Delta = \Delta_{st} + \sqrt{\Delta_{st}^2 + 2\,\Delta_{st}h}$$

where Δ = deflection under *dynamic* load
$\quad \Delta_{st}$ = deflection under *static* load
$\quad h$ = height (in.) from which dynamic load falls

Thus Factory Mutual's deflection formulas are not conservative enough to constitute a true representation of the dynamic construction loads.

Fig. 4-6 Decks should be checked for equipment wheel loads, which may deflect steel decks in particular and break the bond between deck and insulation. (Aeroil Products Co., Inc.)

Specify rosin-sized paper, felt, or tape over joints in wood sheathing, plywood, or other prefabricated units.

6. Check the fire resistance of entire roof-deck-ceiling assembly (see Chapter 15, "Fire Resistance").

7. Limit steel decks to minimum thickness of 22 gage.

8. Specify minimum ⅜-in.-diameter nail heads, driven through metal caps of minimum 1-in. diameter and 30-gage thickness.

9. Specify insulation over bulb tee steel-deck supports.

10. Specify a vapor barrier over a poured gypsum (or lightweight insulating concrete) deck, to prevent vapor from migrating into the insulation.

Field

1. Require a smooth, plane deck surface, with proper slope. For prefabricated deck units, tolerances are

> Vertical joints: ⅛-in. gap
> Horizontal joints: ¼-in. gap
> Flat surfaces of steel decks: ¹⁄₁₆ in.
> (between adjacent ribs)

When the vertical joint tolerance is exceeded, require leveling of the deck surface on a maximum 1 in./ft slope with grout or other approved fill materials.

When the built-up membrane is applied directly to the deck, require caulking or stripping of all joints. Where insulation is applied on monolithic poured-in-place decks, fill the low spots (more than ½ in. below true grade) and grind high spots. Cover openings over ¼ in. in wood decks with nailed sheet metal.

2. Permit no deformed side laps, broken or omitted side welds in metal decks.

3. Limit moisture content in deck materials to satisfactory levels—for example, 19 percent for wood and plywood.

4. Require waterproofed tarpaulin coverings over moisture-absorptive deck materials (wood, preformed wood fiber, precast concrete, precast gypsum) stockpiled at the site. Also require that these materials be blocked above the ground.

5. Prohibit placing of vapor retarder, base sheet, insulation, or roofing on a deck containing water, snow, or ice. Decks must be clean and dry (including flute openings of metal decks) before next roofing component is applied. A test, useful under normal circumstances, for dry gypsum or concrete deck surface to be hot mopped is to pour a small amount of hot bitumen on deck. If, after the bitumen cools, you can remove it with your fingernails and hands, reject this deck for application of insulation. If the bitumen sticks to the deck and cannot be removed by your fingers, then the deck is dry enough to receive the mopping. An alternative or supplementary test for poured-in-place concrete or gypsum decks is to apply hot bitumen, as described, and observe for frothing or bubbling. If none occurs, the deck is dry enough.

6. Place plywood walkways on a vulnerable deck surface, e.g., lightgage steel.

7. Forbid stacking materials on the deck in piles that exceed design live load. Also check for equipment wheel loads.

8. Set close tolerances for steel-column base plate elevations (because they determine roof framing and deck slope).

FIVE

Thermal Insulation

Thermal insulation in an airconditioned or merely heated building offers the greatest return on initial investment of any building material. In view of prospects of energy costs escalating 10 to 15 percent annually far into the foreseeable future, substantial thermal insulation is indispensable for occupied buildings.

Besides reducing energy costs, insulation also enhances building occupants' comfort. In cold weather it raises interior surface temperatures, thus reducing radiative losses that chill the occupants. In summer it reduces interior surface temperatures that retard occupants' heat rejection and make them feel uncomfortably warm.

Roof insulation has three other benefits:

- It can prevent condensation on interior surfaces.

- It generally furnishes a better substrate for the membrane than does the structural deck—notably steel.

- It stabilizes deck components by reducing their temperature variations and consequent expansion-contraction.

For a conventional roof system (membrane on top of insulation), insulation has three drawbacks:

- Although it reduces the probability of condensation on exposed interior surfaces, insulation increases the probability of condensation *within* the roof system. (Insulation creates a wider temperature range between the roof assembly's interior and exterior surfaces. During the heating season, insulation warms the interior surface and cools the exterior surface; it thus increases the chances of water vapor infiltrating from the warm side condensing within the roof's cross section.)

- Insulation raises a roof's surface temperatures in the summer, thereby accelerating the oxidizing chemical reactions that harden bitumen and make it more brittle and more subject to alligatoring, wrinkle cracking, blister growth, and general degradation.

- Insulation produces more rapid membrane thermal contraction, thus increasing the hazard of membrane splitting.

The protected membrane roof (PMR) concept, which places the insulation above the membrane, averts the last two problems but not the first.

Thus there are many problems, apart from thermal efficiency, associated with the selection of an insulation material, including water absorption, drying rate, dimensional stability, treatment of board joints, impact resistance, compressive and shearing strength. Despite these problems, thermal insulation is indispensable, and, as demonstrated in this chapter, many fears expressed about increased thermal efficiency are grossly exaggerated. The first ½-in. thickness of insulation in a roof system exerts a far greater effect on the surface temperature than does the next 4 in. Regardless of the problems created by insulation, few owners, and certainly not the nation at large, can afford to eliminate or skimp on thermal insulation.

THERMAL INSULATING MATERIALS

The insulation used in built-up roof systems falls into four categories:

- Rigid insulation, prefabricated into boards applied directly to the deck surface

- Dual-purpose structural deck and insulating planks

- Poured-in-place insulating concrete fills

- Sprayed-in-place plastic foam

Soft blanket and batt insulations obviously cannot serve as the substrate for a built-up roofing membrane. They must be located below the structural deck, normally on the ceiling, below a ventilated space. This arrangement is especially good in arid climates with daily temperature extremes. The ventilated space interposes an additional insulating medium that modulates the ceiling temperature, thus increasing occupants' comfort. It can also dissipate water vapor (see Chapter 6, "Vapor Control").

Below-deck insulation offers better acoustical control than conventional above-deck insulation. And low-density batt insulation costs only 20 percent or less than high-density, above-deck roof insulation of equal insulating value. However, below-deck insulation has several drawbacks that normally outweigh its advantages. Except in arid climates, below-deck insulation creates the risk of wetting from migrating water vapor condensing on a cold roof deck and dripping back into the ceiling-supported insulation. Below-deck insulation cannot substitute for above-deck insulation in a steel-deck roof assembly; a steel deck requires board insulation as a substrate for the membrane.

Combinations of above- and below-deck insulation are practicable, but the designer should check the dew-point location under design conditions.[1]

Winter venting of the ceiling plenum solves the moisture problem, but it sacrifices the thermal-insulating value of above-plenum roof components: deck, above-deck insulation, membrane, and outside air film. Whether these above-deck components' insulating value is totally lost remains an unanswered question, requiring extensive research. And although ventilation of a plenum space is a winter-heating energy loser, it is a summer-cooling energy winner.

In addition to the moisture problem, below-deck insulation has three other drawbacks:

- It increases building construction cost if additional building height is required for ventilating space.

- It does not protect utilities in a ventilated space from freezing.

- It does not restrict thermal expansion and contraction in the deck, whose movement may create problems.

Rigid Board Insulation

Far more common than ceiling insulation in modern built-up roofing systems is rigid board insulation, normally applied to the deck's top surface. Such insulation is classified chemically as organic or inorganic. Organic insulation includes the various vegetable-fiber boards and foamed plastics. Inorganic insulation includes glass fibers, perlite board, cellular glass, mineral fiberboard, and poured insulating concretes with lightweight aggregates.

---

[1]For detailed discussion, see R. L. Fricklas, "Insulation: Does Below-Deck Placement Increase Value?", *Specifying Engineer*, February 1977, pp. 66ff.

Physically, insulation is classified as cellular or fibrous. *Cellular* insulation includes foamglass and foamed plastics. The newest of these materials, the foamed plastics, are polystyrenes and, more recently, polyurethanes and isocyanurates. Air or other gas introduced into the material expands it as much as 40 times. Cells are formed in various patterns—open (interconnected) or closed (unconnected). Most rigid urethane foams are expanded with one of the halogen gases that, because of their extremely low thermal conductivity (about one-third that of air), give foamed urethane its high insulating quality. Gradual air diffusion into the cells replaces some of the original foaming gas and eventually lowers the thermal insulating quality by about 20 percent, but it nonetheless remains extremely high (see Table 5-1). Foamglass is expanded with hydrogen sulfide, another gas with low heat conductivity, about half that of air.

Fibrous insulation includes various fiberboards, which can be made of wood, cane, or vegetable fibers bonded with plastic binders. The materials are sometimes impregnated, or later coated, with asphalt to make them moisture-resistant. Fibrous glass insulation consists of non-absorbent fibers formed into boards with phenolic binders. It is surfaced with an asphalt-saturated, glass-fiber-reinforced organic material.

Perlite board, another fibrous insulation, contains both inorganic (expanded silicaceous volcanic glass) and organic (wood fibers) bonded with asphaltic binders.

Composite urethane boards are the newest insulation boards on the market. With urethane as the top layer, laminated to a bottom layer of perlite, fiberglass, or gypsum, these composite boards can qualify for Factory Mutual (FM) Class I steel-deck roof-construction fire rating. Some glass-fiber–plastic-foam insulation boards qualify for FM Class I without a second stratum of insulation. But composite boards offer another advantage: The dimensionally stable perlite or glass-fiber board reduces the composite board's coefficient of thermal expansion-contraction below urethane's high coefficient.

Dual-Purpose Structural Deck and Insulating Plank

Made of cement-coated wood fibers set and cured in molds, preformed structural wood-fiber decks also serve as insulation. Physically, they belong with the fibrous organic insulations (see Chapter 4, "Structural Deck").

Poured-in-Place Insulating Concrete Fill

The perlite or vermiculite aggregates normally used in lightweight insulating concrete contain varied-sized cells, which raise the concrete's thermal resistance to 14 to 20 times that of ordinary structural concrete.

Perlite is silicaceous volcanic glass; vermiculite is expanded mica. Both materials expand 15 to 20 times at high temperatures somewhere between 1400 to 2000°F. Normally, the concrete made from these aggregates has a portland cement binder, but perlite aggregate sometimes has an asphalt binder, a thermosetting fill discussed later (see Fig. 5-1).

Lightweight insulating concrete also comes with other aggregates (e.g., expanded polystyrene) and in cellular, foamed concrete, formed by adding a liquid-concentrate forming agent to the mix.

Still another type of insulating fill is thermosetting, produced by mixing a perlite aggregate with a hot, steep (Type III or IV) asphalt binder. Mixed and heated at the jobsite, thermosetting insulating fill is placed dry on a primed surface, screeded to the proper thickness (usually variable, to provide slope), and compacted with a hand tamper or roller to the specified density (about 20 pcf) (see Fig. 5-2).

Poured insulating concrete fills economically provide both insulation and a sloped roof surface for drainage. They also form a smooth substrate on top of steel decks, precast concrete, and other decks for application of the membrane or supplementary insulation boards, if required. Unlike some other insulating materials, they offer excellent compressive strength and fire resistance.

As one notable advantage over lighter materials, massive insulating materials like lightweight concrete fill (which may weigh 50 times as much as foamed plastic of an equivalent thermal resistance) stabilize

Fig. 5-1 Pumping lightweight insulating concrete mix to a steel deck removes buggy and other equipment loads that could cause roof-damaging deflections. (Pittsburgh Corning Corp.)

Table 5-1 Roof Insulation Material Properties*

Material	Density, pcf	Thermal conductivity† k, Btuh/(°F·ft²)	Thermal resistance† R, °F/(Btuh·ft²)	Vapor permeance, perm-in.	Coefficient of thermal expansion, (in.·in.)/°F × 10⁻⁶	Moisture expansion, % (50-97% RH)	Compr. strength, psi‡	Fire resistance
Board insulations:								
Perlite	10	0.36	2.8	25	10–13	0.2	35	Excellent
Glass fiber	12	0.24	4.2	40	5–8	<0.05	20	Excellent
Wood fiber	17	0.36	2.8	25	7–10	0.5	80	Poor
Foamglass	9	0.38	2.6	0	4.6	0	100	Excellent
Urethane foam	2	0.16	6.2	0.4–1.6	30	2–3	23	Poor
Isocyanurate foam	2.4	0.16	6.2			2–3	40	Fair
Polystyrene foam (extruded)	2.3	0.20	5.0	0.4	35	0.4	45	Poor
Polystyrene foam (beadboard)	1.0–1.3	0.26	3.8	3.0			12	Poor
Phenolic	3.0	0.21	4.8		17		25	Excellent

Material							
Composite boards:§							
Urethane-perlite						23	Excellent
Urethane-glass fiber						20	Excellent
Urethane-gypsum						23	Excellent
Sprayed-in-place:							
Urethane foam	2.5–3.0	0.16	6.2	1.2	30	45	Poor
Poured-in-place:							
Lightweight concrete, perlite aggregate	27	0.64	1.5	13	4.8	140+	Excellent
Lightweight concrete, vermiculite aggregate	25	0.80	1.3	22	3.0	125+	Excellent
Cellular concrete	30	0.70	1.4				Excellent
Compacted thermosetting fill	22	0.40	2.5			40	

*Tabulated data come from varied sources. Check with manufacturers and other sources for more complete, precise data.

†Thermal resistance and thermal conductivity are for oven-dry material at 75°F. Water-absorptive materials may suffer substantial losses in insulating value under typical service conditions.

‡Values for compressive strength may not be comparable because they vary in deformation at which ultimate strength is recorded.

§Because of varying thicknesses in composite boards, properties vary with varying proportions of each material.

Fig. 5-2 Thermosetting insulating fill, comprising perlite aggregate with an asphaltic binder, mixed and heated at the jobsite, can provide slope for dead-level decks (note pipe screeds, top photo). Roller (bottom) compacts fill to about 20-pcf density. (Silbrico Corp.)

the extreme fluctuations of an insulated roof's surface temperatures. A roof membrane directly on top of a light, highly efficient insulation reacts to the sudden heat of the sun like an empty pan on a burner. The metal in contact with a good insulator—in this case air—heats up faster than the metal of a pan filled with a poor insulator like water. A massive insulating material, like the water in the heated pan, has greater heat capacity than a lighter, more efficient insulator. Because they store and release heat slower than thinner, lighter insulations do, these <u>heav-</u>

ier materials tend to stabilize roof-surface temperatures. Thus membrane temperature reacts less quickly to the sudden heat of the sun emerging from behind a cloud or to the chill of a sudden rain shower. Massive structural decks—notably concrete or poured gypsum—stabilize roof-surface temperatures even more than insulating concrete (Fig. 5-3).

Sprayed-in-Place Polyurethane Foam

Sprayed-in-place polyurethane (SPUF), introduced in the late 1960s, claimed a steadily increasing share of the roof-insulation market, especially for reroofing, throughout the 1970s. SPUF has some advantages over prefabricated boards—chiefly the elimination of open joints that can form stress-concentration lines in the membrane. Its "aged" thermal resistance, R factor [6.25 Btuh/($°F \cdot ft^2$) per inch of thickness] makes it a highly effective insulation in the 1- to 3-in. thickness range to which

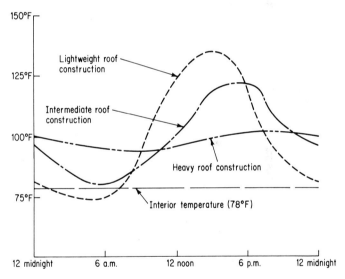

Fig. 5-3 These cooling-load temperature curves (for interior temperature of 78°F) show how increased roof-system mass reduces peak cooling load and smooths the cooling-load curves from solar heat gain. Heavy roof construction retards solar heat gain by absorbing the heat and slowly releasing it. Lightweight roof construction = steel deck + 1-in. (or 2-in.) insulation + suspended ceiling. Intermediate roof construction = 1-in. wood deck + 1-in. insulation + suspended ceiling. Heavy roof construction = 6-in. concrete deck + 1-in. (or 2-in.) insulation + suspended ceiling. (Plotting data for the curves from *ASHRAE Handbook and Product Directory, 1977 Fundamentals*, table 5, p. 25.7 for summer day with maximum 95°F air temperature.)

it is normally restricted. With its fluid-applied coatings (the only recommended membrane), it forms part of the modern single-ply synthetic systems discussed in Chapter 11.

SPUF is produced in the field by nozzle mixing two separate liquid streams: an A (isocyanate) component and a B (hydroxyl or polyol) component (see Fig. 5-4). Besides these basic chemical ingredients, the sprayed field mix also requires

- A *blowing agent* (to form foamed cells that expand the polyurethane foam resin's volume)

- A *surfactant* (to control cell size and cell-wall rigidity)

- A *catalyst* (to control the reaction rate between the two chemical components)

- *Fillers* (to cut material cost or alter a physical property)

The *blowing agent* for SPUF is normally a Freon, Refrigerant 11 (monofluorotrichloromethane, CCL_3F). During the highly exothermic polymerization process, the liquid Freon vaporizes and forms countless tiny polymeric cells. The *surfactant* is usually silicon or silicon-polyether copolymer.

The *catalyst* (tertiary amines and/or organotin compounds) controls the chemical reaction rate and also governs the polymer chain propagation, extension, and cross-linking. It also helps cure the foam, an important process to ensure attainment of full strength. *Fillers* are lightweight, bulky, inorganic materials that must be chemically compatible with the urethane polymer.

Fig. 5-4 Spraying polyurethane foam proceeds in multiple lifts, minimum ½-in. depths, on a typical reroofing project, over a spudded, aggregate-surfaced built-up membrane, which must be both solidly anchored and essentially dry.

Compared with conventional roof systems, SPUF offers these advantages:

- High thermal resistance per unit thickness

- Ultralightweight (less than 1 psf for 3-in.-thick foamed insulation plus membrane coating)

- Faster construction (generally two to three times the rate of conventional bituminous systems)

- Adaptability to curved and other irregular roof surfaces

- Possible elimination of an expensive, disruptive tearoff-replacement required for a conventional bituminous roof system with conventional board insulation

- Adaptability for slope on dead-level roofs

- Excellent adhesion, capable of producing 3000-psf uplift resistance when placed on a clean, dry, properly prepared substrate

- Simple flashing details

There are, as always, some offsetting disadvantages:

- Added cost of recoating the fluid-applied membrane periodically

- Greater difficulty in obtaining a level surface and uniform insulation thickness

- Extremely high dependence on the applicator's skill

- High dependence on good substrate preparation

- Reduced traffic and impact resistance

- High vulnerability to degradation in hot, humid climates

Half the items cited stem from the SPUF's high dependence on good fieldwork. In one sense SPUF is a regression: Instead of simplifying the fieldwork, the goal of most advances in roofing technology, SPUF complicates it. Application of the sprayed insulation and fluid-applied membrane is even more sensitive to weather conditions than built-up roof-system application. At best, a SPUF roof surface is somewhat irregular, and it has less impact resistance than a properly applied built-up roof system, a particular disadvantage in areas like the Midwest, with its many hailstorms.

Rough-surfaced foams are more easily damaged by foot traffic because the rough surfaces create a spongy, open-cell structure that readily holds water if it is wetted before the fluid-applied membrane is installed. Humidity is the big problem for SPUF roofs, so for that reason they are better suited to arid than humid climates, especially *hot,* humid climates.

Liquid moisture poses its greatest hazard on the substrate of a SPUF roof; on the roof surface it is a lesser hazard. Polyurethane foam's closed-cell structure makes it highly resistant to liquid water penetration; even if the foam's protective coating is broken, water seldom penetrates through the roof. However, standing water, in so-called "bird baths," softens the foam under these wet spots.

Liquid moisture on a substrate sprayed with polyurethane foam has several harmful effects. It impairs the chemical reaction between the isocyanate and polyol, producing a weaker, less thermally efficient material. Even more important, this hazardous practice can destroy the SPUF's adhesion to the substrate. Domed blisters at the deck-insulation interface can grow from an apparent combination of thermal expansion and moisture absorption (see Fig. 5-5). SPUF has a relatively high thermal coefficient, $22 - 44 \times 10^{-6}/°F$.*

Freeze-thaw damage is usually limited to little more than the area of original water penetration, *if* there is no significant water head (i.e., less than ½ in.). This limitation is one of several reasons for requiring positive drainage of SPUF roofs.

Water vapor diffusing into the SPUF cells impairs its strength, thermal resistance, and dimensional stability. The long term growth of polyurethane foam from water-vapor absorption apparently results from the plasticity of the cellular walls. If the material were perfectly elastic, contraction of the cellular walls when entrapped water vapor condensed with falling temperature would equal the expansion when an equal quantity of water evaporated with rising temperature. But the plastic material contracts less than it expands. As water vapor accumulates by diffusion into the cyclically expanding cells, there is a consequent long term tendency for dimensional growth (see later discussion under "Dimensional Stability").

Another factor possibly contributing to polyurethane foam's long term growth is slow continuation of the polymerization chemical reaction, accelerated by high temperatures. Application at high humidity combined with high temperature can cause the foam to expand and pull away from the substrate as a result of excessive heat from the exothermic chemical reaction. Irregular thickness, resulting from SPUF

---

*W. C. Cullen and W. J. Rossiter, "Guidelines for Selection of and Use of Foam Polyurethane Roofing Systems," NBS Tech. Note 778, May 1973, p. 11.

Fig. 5-5 Domed blister formed in sprayed polyurethane foam (SPUF) sprayed onto a wet substrate of a reroofing project (top). Several bird-pecked holes are extended through the foam cross section to water trapped between the reroofed substrate and the foam's bottom surface. The top coating was an asphaltic emulsion plus a fibrated aluminum asphalt cutback, an unsuitable membrane for a SPUF roof system. The original roof lacked slope, and no effort was made to slope the new SPUF roof (bottom).

application over steel deck, has produced bizarre effects, according to consultant Sybe K. Bakker, of Moisture Protection Systems Analysts, McLean, Virginia. Heat trapped in the thick sections apparently produced ½-in.-high ridges over the deck flutes, and the applicator had to grind down these ridges with a wire brush.

Compared with lower density foam, a relatively dense (3-pcf) foam has the following three advantages:

- Low water absorptivity (which varies inversely with density)

- Higher strength

- Lower gaseous diffusion rates

As a consequence of the lower gas diffusion rate in 3-pcf-density foam, the foam has improved dimensional stability and thermal resistance. Below 2 pcf, the foam "windows" (i.e., membranes connecting the ribs of adjacent foam cells) are too thin to retain the Freon and resist inward diffusion of water vapor and other atmospheric gases. But even under the best circumstances Freon gradually diffuses out, replaced by inwardly diffusing atmospheric gases.

This process naturally reduces thermal insulating value with aging because air is three times as conductive as Refrigerant 11. This aging proceeds most rapidly in the early stages and then abates.

SPUF's thermal resistance varies inversely with cell size. For a 1-in. thickness, R varies from about 8°F/(Btuh·ft²) for 100-μm (0.004-in.) cell size to 6°F/(Btuh·ft²) for 400-μm (0.016-in.) cell size. SPUF's R value is a maximum at 50°F; below that, the Freon begins to condense. And in common with other gases, its thermal conductivity increases with increasing temperature.

SPUF's vulnerability to moisture prompts two basic design rules:

1. Provide minimum ¼-in. slope for a SPUF roof system.

2. Avoid SPUF roof systems in hot, humid climates.

Specification of positive slope for drainage is even more important for a SPUF roof than for a conventional built-up roof system, according to NBS researchers Cullen and Rossiter.[1] They advocate an *absolute* minimum of ¼-in. specified slope because of the greater difficulty in controlling application and the resulting surface irregularities in the sprayed foam surface.

Membrane blistering, as well as domed blistering, can result from absorbed moisture under a highly impermeable membrane coating. Prolonged exposure to moisture may even adversely affect the foam's physical-chemical structure, especially when it is exposed to high temperature (95°F+). Despite a possibly dry appearance, the SPUF surface may contain absorbed water that can produce these membrane blisters. Because of its vapor permeability, SPUF normally requires a vapor retarder between the deck and the foam.

Spraying the polyurethane foam to a variable depth to provide slope is a hazardous practice. Within narrow limits—say from 1- to 3-in. maximum depth—this practice may be acceptable. But when foam depth approaches the 10-in. depths sometimes attempted, SPUF is seldom successful. Such thick layers of SPUF must be applied in stages, a method that multiplies the field-control problems beyond practicable solution. (Excessive heat buildup from the exothermic chemical reaction affects cell size, cell-wall structure, density, and other vital prop-

[1]Ibid., p. 20.

erties that must be controlled within narrow limits.) Spray application in multiple lifts (i.e., layers), sometimes with a day's wait between lifts, jeopardizes the SPUF's integrity. (Field problems include water entrapment from early-morning condensation and the sanding requirements to prepare the foam substrate for another lift.)

Application of variable-depth SPUF is thus a last resort. For new projects, where the deck can be sloped, specify constant-depth SPUF. For reroofing projects where the deck is level, limit maximum SPUF depth to 3 in., for a maximum 2-in. fall to its minimum 1-in. depth. An expensive alternative is to install tapered board insulation under the SPUF.

Although sometimes conceived as a total roof system, insulation-membrane all in one, SPUF requires a membrane coating like any other conventionally arranged insulation (as distinguished from the PMR concept). This membrane must protect the SPUF from three hazards: ultraviolet degradation; moisture invasion; and abrasion and impact forces of foot traffic, hailstones, dropped tools, and so on.

SPUF deteriorates rapidly when exposed to solar radiation; the yellowed exposed surface disintegrates into a rust-hued powder within a few days. (Chemically, ultraviolet degradation is believed to result from photochemical oxidation.)

SPUF forms a naturally advantageous continuous substrate for a fluid-applied membrane. This coating requires seven properties:

- Good adhesion
- Low-temperature susceptibility (i.e., viscous at high temperature but not brittle at low temperature)
- Abrasion resistance
- Weather resistance (to ultraviolet radiation, rain, and so forth)
- Maintainability (ease of repair when damaged, integration of repaired section with original material)
- Durability
- Strength, elasticity

By a broad consensus, silicone elastomers offer the best all-around performance for coating SPUF insulation. Silicone coatings performed best in both xenon arc test and in all-around weathering performance, according to a U.S. Bureau of Reclamation (USBR) survey of SPUF roof systems installed on bureau buildings.[1] No damage was observed in

---

[1]U.S. Bureau of Reclamation, *Laboratory and Field Investigations of New Materials for Roof Construction*, ser. PB259-635, April 1976, p. 28.

fabric-reinforced coatings after several hailstorms. Based on its experience, the USBR expects 20-year service from properly applied silicone membranes, which is far more than can be reasonably anticipated from other fluid-applied materials. Recoating with other materials may be required within 5 years.

Other satisfactory coatings include fluid-applied polyurethane, neoprane/Hypalon, and acrylic elastomers.

DESIGN FACTORS

The growing popularity of loose, ballasted, and protected membrane roof (PMR) systems complicates the task of setting performance requirements for roof insulation. Compressive strength, to resist traffic loads and hailstone impact, is a universal requirement for insulation in all above-deck locations. In its exposed location above the membrane in a PMR, however, insulation requires several properties—notably moisture resistance and freeze-thaw cycling resistance—that eliminate most insulating materials currently used in their conventional location under the membrane (see Chapter 10, "Protected Membrane Roofs," for detailed discussion of roof-insulation requirements in a PMR). In contrast with a PMR, a loose, ballasted system sets less rigorous requirements for insulation than does a conventional system with insulation sandwiched between the deck and membrane.

In the conventional sandwich-style roof assembly, insulation requires

- Compressive strength
- Cohesive strength to resist delamination under wind uplift
- Horizontal shear strength to maintain dimensional stability of the roof membrane under tensile contraction stress
- Resistance to damage from moisture absorption
- Dimensional stability under thermal changes and moisture absorption
- A surface absorptive enough to adhere the bituminous mopping but not so absorptive that it soaks up the bitumen

Strength Requirements

A compressive strength of 30 psi minimum is generally recommended in insulation to resist traffic loads, hailstones, dropped tools, and miscellaneous missiles' impact on the roof.

The *cohesive strength* within the insulation must at least equal the required wind-uplift resistance to prevent the insulation from breaking loose from the deck in high winds.

Horizontal shearing strength helps distribute membrane tensile stress, thereby preventing stress (or strain) concentrations that can split the membrane. Insulations in general use usually provide this horizontal shearing strength. However, thickened layers may create a problem with some materials.

Moisture Absorption

Water entrapped in insulation (during roof application, through leaks, or via condensation of infiltrating water vapor) can destroy the thermal-insulating value of some insulating materials. Water vapor can generally flow wherever air can flow—between fibers, through interconnected open cells, or where a closed-cell structure breaks down. Water can easily penetrate the larger enclosures of insulating concrete. Wherever water replaces air, the insulating value drops drastically because water's thermal conductivity exceeds that of air by roughly 20 times.

Fibrous organic insulations are especially vulnerable to moisture damage. Free water eventually damages any fibrous organic material or organic plastic binder. Fiberboard long exposed to moisture may warp or buckle and eventually decay. Expansion and contraction accompanying changing moisture content can seriously damage the built-up membrane.

Some preformed structural wood-fiber planks are subject to inelastic sag, attributable chiefly to moisture weakening of some (not all) cementitious binders. For that reason these planks are not recommended over highly humidified interiors.

Although less vulnerable to moisture, inorganic materials are not immune. Water penetrating into fiberglass insulation not only impairs the insulating value, it may also dissolve the binder. Foamed cellular glass neither absorbs water in its closed cells nor allows the passage of water vapor. But if moisture is trapped between the built-up roofing and the top of the insulation, water accumulates in open-surface cells. With roof-surface temperature alternating above and below freezing temperatures, ice formation can break down the walls between open-surface cells and the interior closed cells. Repeated freeze-thaw cycles progressively destroy the foamed glass, leaving a water-saturated gray-black dust. (Laboratory tests have indicated complete breakdown under as few as 20 such cycles.)

Lightweight insulating concrete poses an especially severe moisture-absorption problem because it may retain high moisture content for years. When laboratory tested, insulating concrete cores cut from some roofs in place for 10 or more years had more than 100 percent water content, measured on the so-called "dry" basis.[1]

A 100 percent dry-basis moisture content is the ideal moisture-content range for wet vermiculite concrete mix ready for placing. (Wet density ranges from 44 to 60 psf vs. 22 to 28 psf for dry density.) Unless it is properly vented, lightweight concrete fill can thus maintain a water content nearly equal to its original mixing water content throughout its service life.

A properly vented lightweight insulating concrete should attain a 10 to 25 percent minimum dry-basis moisture content after 2 to 5 years, with the low (10 percent) figure in arid places like Tucson, Arizona, and the 25 percent figure in the humid climates of the eastern United States. On its way to this 10 to 25 percent equilibrium moisture range, insulating concrete should lose from 25 to 40 percent of its moisture content from its drying period prior to membrane application, according to Tony Clapperton, manager of W. R. Grace's Zonolite Roof Applications. That slightly exceeds the 20 to 30 percent drying-period loss estimated by Professor C. E. Lund.[2]

Water entrapped within insulating concrete can result from a combination of the following three sources:

- Long term retention of original mixing water

- Condensate formed within the fill from vapor migration from the building interior

- Roof leaks

Although it can relieve vapor pressure within the insulating concrete, venting cannot significantly reduce moisture content. This is the conclusion of NBS researchers, citing both experimental evidence and theoretical calculations.[3] This NBS research makes it difficult to rec-

$$\text{Dry basis, \% moisture} = \frac{\text{wt. of moisture}}{\text{oven-dry wt.}} = \frac{\text{total wt.} - \text{oven-dry wt.}}{\text{oven-dry wt.}}$$

$$\text{Wet basis, \% moisture} = \frac{\text{wt. of moisture}}{\text{total wt.}} = \frac{\text{total wt.} - \text{oven-dry wt.}}{\text{total wt.}}$$

100% moisture content, dry basis = 50% on wet basis

[2]C. E. Lund, *The Performance of Perlite and Vermiculite Concrete Roof Decks*, paper delivered at 12th annual Midwest Roofing Contractors Association convention, Excelsior Springs, Mo., November 1961.

[3]Frank J. Powell and Henry E. Robinson, "The Effect of Moisture on the Heat Transfer Performance of Insulated Flat-Roof Constructions," *Build. Sci. Ser.* 37, NBS, October 1971, pp. 12, 13.

oncile claims that lightweight insulating concrete loses about 50 percent of its original moisture content within several years following membrane application. A prudent designer thus assumes greater moisture retention than the more optimistic estimates claimed for lightweight insulating concrete.

Excessive water entrapment in lightweight concrete fill has two harmful consequences:

- Shortened membrane life from increased probability of blistering (from vapor migrating upward into membrane voids) and splitting (from water weakening the membrane felts).

- Potentially drastic reduction in the roof system's thermal resistance and consequently increased heating and cooling bills. (With a dry-basis moisture content of 50 percent, a percentage sometimes exceeded in laboratory core samples cut from existing roof systems, vermiculite aggregate concrete can lose 75 percent of its dry-state thermal-insulating value.[1]

Moisture entrapped within insulation undergoes cyclic seasonal change. In winter, when the vapor pressure promotes upward vapor migration, water vapor starts condensing at an obstructive plane below the dew point, usually the underside of the membrane. So long as this condensed moisture is not excessive, it can remain in the upper levels of the insulation. The upward vapor-pressure differential tends to keep the moisture from dripping back through the insulation into the interior.

In summer conditions change drastically. With the roof-surface temperature soaring to 150°F or thereabouts, the trapped water evaporates. Vapor migration is then downward, impelled by the higher outdoor temperature and absolute humidity. Except in roof assemblies over refrigerated interiors, there is no condensation plane anywhere within the roof assembly; the temperatures are too high. Thus a roof assembly with a properly vented soffit tends to dry out in summer, from heat energy supplied by solar radiation.

Venting

Although backed by the conventional wisdom of the roofing industry, topside venting of insulation via stack vents may do more harm than good. Infrared moisture-detection surveys conducted by Wayne Tobiasson, research engineer with the U.S. Army Cold Regions Research and Engineering Laboratory (CRREL), have detected wet

[1]Ibid., table 2, p. 7.

areas around the overwhelming majority of breather vents. In many investigated cold-region roofs, Tobiasson found that almost all the wet insulation was attributable to exterior water entry at flashings and roof penetrations rather than to the condensation of water vapor flowing up from the building interior. These topside vents are apparently worthless as a means of releasing previously accumulated moisture or preventing moisture accumulation (see Figs. 5-6 and 5-7). And, on the negative side, they add vulnerable, flashed openings for exterior water penetration and objects that can be kicked over or otherwise damaged. [I have even seen a stack vent, mindlessly used as an anchor for a wind-bracing cable, overturned, with complete rupture of the surrounding membrane.]

If lightweight insulating concrete fill is poured onto a monolithic structural concrete deck, venting is mandatory even without a vapor retarder. A poured-in-place structural concrete deck is a poor substrate for lightweight insulating concrete, generally unacceptable to roofing manufacturers. Lightweight insulating concrete should be restricted to decks with underside venting: slotted steel decks, permeable formboards (e.g., fiberglass), or precast concrete sections with venting joints. If it is nonetheless impracticable to avoid using lightweight insulating concrete over a nonventing deck, nail a layer of permeable insulation (e.g., fiberglass) to the insulating concrete's top surface, or specify a venting base sheet (i.e., a heavy-coated felt with an underside grid of lateral venting grooves) between the fill's top surface and the membrane.

Underside venting of lightweight insulating concrete is superior to topside venting for the following reasons:

- Solar heat normally creates a steeper downward vapor-pressure gradient during the summer than the upward vapor-pressure gradient during the winter.

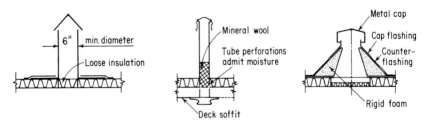

Fig. 5-6   A stack vent is inferior to the pyramidal vent because it leaves base flashing in the general roof plane, always a vulnerable location. The pyramidal vent also has a larger venting area and, because of its foam-insulated sides, suffers less risk of condensed moisture dripping back into the insulation.

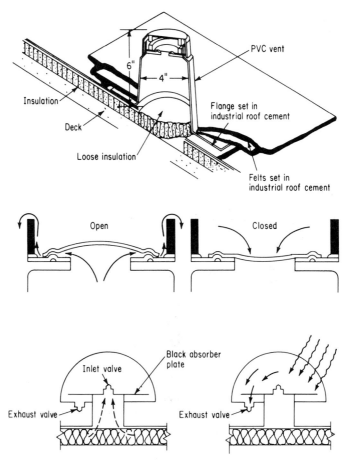

Fig. 5-7 Solar-powered roof vents feature one-way valves designed to promote cyclic pumping of air-vapor mixture from insulation to the ambient air with alternating sunshine and shade. (Johns-Manville Corp., top; B. F. Goodrich, bottom.)

- A properly vented underside normally offers a larger, better distributed surface area than the top surface, with its widely spaced vents.

- The topside is initially drier because it is open to the ambient air for several days before membrane application.

Five of eight major manufacturers associated with the Asphalt Roofing Manufacturers Association require both underside *and* topside venting of lightweight insulating concrete for bonded roofs. Two of the other manufacturers recommend both topside and underside venting, making for near unanimity. All eight demand topside venting.

Edge Venting Edge venting creates horizontal escape paths through grooved edge boards open to the edge of the insulation and placed around the roof periphery.

Underside Ventilation Underside ventilation for lightweight insulating concrete on steel decks consists of slotted holes in the flutes, providing a 2.5 percent open soffit area. Slotted holes are preferable to metal vent clips at longitudinal deck joints because they drastically reduce the distance, to one-third or less, that the vapor must travel through the insulating concrete to escape to the building interior. Dimpled side lap venting is totally ineffective.

Dimensional Stability

As the roofing sandwich filler between deck and built-up roof membrane, insulation must retain stable dimensions or it may damage the membrane. Contraction and expansion with changing moisture content can also distort some insulation materials beyond tolerable limits. Repeated expansion and contraction of insulation boards promote membrane splitting.

Insulation made of cellulosic fibers swells with moisture absorption and contracts on drying as much as 0.2 to 0.5 percent, with 50 to 90% RH changes. Under such a humidity change, a 4-ft-long cellulosic fiberboard may contract ¼ in., thus opening a joint.

Insulation materials with a high coefficient of thermal expansion can contribute to the splitting of roof membranes subjected to extreme temperature changes. Membrane splitting can occur when a roof membrane shrinks as temperature drops, or as wet felts dry out. Because the flexible membrane buckles under compressive stress, the membrane cannot expand with rising temperature, and contraction can proceed with a sort of ratchet action resulting in accumulating contraction, closing open joints between the insulation boards.

Under a 100°F temperature drop, a 4-ft-long polystyrene board contracts nearly ¾₆ in. Foamed urethane, with its thermal coefficient of roughly 30×10^{-6} in./(in.·°F), is nearly as bad.

Largely because of its dimensional instability, coupled with the difficulty of getting good adhesion with hot-mopped asphalt, extruded polystyrene insulation board is no longer used in conventional built-up roof systems (i.e., membrane on top of insulation) in the United States. However, this material is successfully marketed as insulation for protected membrane roofs.

This polystyrene foam insulation was involved in numerous membrane splitting failures (discussed in Chapter 12, "Premature Membrane Failures"). These failures emphasized the vital importance of anchoring insulation to the deck, an extremely difficult task with this insulation. (Its cell structure starts to collapse at 175 to 180°F, well below the temperature required for good adhesion with hot-mopped steep asphalt. Under hot asphalt at 400°F, it melts and collapses into craters.)

Urethane board insulation can pose similar dimensional stability problems. Urethane insulation—prefabricated boards and sprayed-in-place—exhibits long term growth that results from faster inward diffusion of small, atmospheric gaseous molecules—oxygen, nitrogen, water vapor, carbon dioxide, and so on—vs. the slower outward diffusion of the large Freon 11 molecules. Under the consequently increased internal pressure, the cellular walls yield plastically, with slow, permanent cell-enlarging deformations.

Warm, humid climates present the most favorable conditions for urethane growth and consequently the worst conditions for urethane use. Water vapor, experiencing condensation-evaporation cycles with changing temperature, is the most active gas promoting long term urethane growth. At constant pressure, evaporation of liquid water entails a 1500-fold volume increase. Trapped within foam cells, this water-vapor pressure increases exponentially with rising temperature as more liquid water evaporates. The foam's cellular structure expands under rising temperature and pressure, but it does not contract elastically with falling temperature and pressure.

Urethane swelling also varies inversely with density, providing yet another reason for setting minimum foam density at 3.0 pcf. As a performance criterion for urethane board expansion under the "humid-aging" (ASTM D2126) test, which subjects the board to 158°F temperature at 97% RH for 14 days, researcher Duane Davis, of GAF Corporation, has proposed the following limits: 1 percent increase in length and width, 7 percent increase in thickness. For a 4-ft-long board, a 1 percent increase in length is nearly ½ in. Even though the humid-aging test should be tougher than any roof system would normally endure in service, it nonetheless indicates the scope of the problem. It further indicates the greater importance of moisture over temperature as a factor in urethane dimensional stability. Even with its high coefficient of thermal expansion-contraction [30×10^{-6} in./(in.·°F)], a urethane board expands only 0.3 percent under a 100°F temperature rise, one-third of the 1 percent allowable expansion under the humid-aging test.

The warping-expansion sometimes observed between adjacent urethane insulation boards may result from moisture and thermal expansion coupled with top-surface contraction under high-temperature drying. The warping can also be explained simply as buckling under the compressive stress of adjacent moisture-expanded boards.

Tapered Insulation

With the increasing recognition of the importance of positive drainage, tapered roof insulation has become popular for reroofing projects, where it may provide the only practicable means of obtaining slope. There are two basic types of tapered insulation: (1) lightweight insulating concrete or asphaltic fills poured in place and screeded to sloped contour and (2) preformed insulation boards (cellular glass, perlite, plastic foam) with a tapered cross section designed to provide slopes of ⅟₁₆ in., ⅛ in., or ¼ in./ft.

A third method of tapering insulation is field grinding perlite board insulation placed in a terraced pattern of different thicknesses. Tapered-insulation-board systems require an extra step in the construction process: preparation of shop drawings indicating roof slope, drain location, drain height, and tapered block location. Extra care is necessary to place the coded blocks in the proper pattern, as shown on the shop drawings.

Anchoring Insulation to the Deck

In conventional, sandwich-style built-up roof systems, insulation must be firmly anchored to the deck to resist wind-uplift stresses and to prevent membrane splitting, especially in cold climates, distributing stresses that otherwise (i.e., with unanchored insulation) would become concentrated.

Of the three basic methods for adhering insulation—cold adhesives, hot-mopped bitumen, and mechanical fastening—mechanical fastening is by far the most generally dependable, especially on steel decks. Cold-applied adhesives are so unreliable that in 1978 three major manufacturers withdrew their approval of this method for bonded roofs. Hot-mopped steep asphalt, correctly applied with the insulation boards tamped firmly into the hot, fluid bitumen, provides excellent adhesion. But getting the insulation boards down into the hot asphalt in the few seconds available before it congeals into a nonbonding solid makes this a highly risky method, especially on steel decks, where the mopping requires extra care to limit its weight and keep it centered on the steel-deck flange. Hot-mopped asphalt adhesive is especially risky in cold, windy weather.

Overlooking the need for good anchorage at the deck-insulation interface, some designers of buildings planned for future expansion have specified insulation boards loosely laid on the deck, for easy removal when the roof is later converted to a floor. This first-cost economy may prove costly. Unanchored insulation boards vastly increase the risk of membrane splitting. Through internal stresses produced by thermal and moisture changes, a built-up membrane exerts a rachet action on a poorly anchored insulation. (The flexible membrane expands, compresses, and buckles in heat, and contracts and pulls in the cold, thus producing a cumulative rachet action toward the center of the roof area.) Compounded by aging and moisture changes in the felts, this rachet action sometimes pulls the insulation 2 or 3 in. from the roof edges, destroying the edge flashing.

The orientation of insulation boards with respect to the felt rolls affects the membrane strength. The conventional practice of unrolling felts parallel to the continuous longitudinal joints of insulation boards maximizes the chances of splitting. The continuous longitudinal joints in the insulation are lines of maximum stress concentration, and the felts must resist these stresses in their weakest (transverse) direction (see Fig. 5-11).

Orienting the felt rolls perpendicular to the continuous longitudinal joint, as in Fig. 5-11b and c, matches maximum membrane stress with maximum felt strength, thereby minimizing the chances of membrane splitting or wrinkling.

A practical objection to this orientation when the insulation boards are adhered is that application of a (hot-mopped or cold-applied) fluid adhesive across the insulation boards' generally 4-ft length is more difficult than over the normally narrower 2-ft width. Using two layers of insulation, with the bottom layer mechanically fastened and the top layer hot-mopped, largely removes the objection to the convenient parallel alignment because mechanical fasteners provide far more dependable anchorage and consequent membrane-splitting resistance than adhesives.

Solid Bearing for Insulation Boards

Solid bearing of insulation boards on steel-deck flanges is another anchorage rule sometimes violated in the field. Cantilevering insulation boards over deck flutes, especially wide-ribbed decks, can fracture brittle insulating materials, destroying the vital substrate support for the membrane and making it vulnerable to puncture. This practice can also promote membrane splitting by opening a joint between adjacent boards. The joint is subject to an extreme degree of separation and thermal warping (see Chapter 12, "Premature Membrane Failures"). An unsupported cantilevered section of insulation board is subjected to both additional longitudinal contraction and upward deflection from low top-surface temperature. This upward deflection can add membrane flexural stress to tensile stress.

To avoid this hazard, most insulation boards should have a minimum 1½-in. bearing on the steel-deck flange. Whenever the installation pattern reduces this dimension, the roofer should trim the board.

Refrigerated Interiors

The architect should not attempt to economize by insulating a refrigerated interior with insulation placed on top of the structural deck, for two main reasons:

- Placing the roof insulation between a cold, dry interior and (in summer) a hot, humid exterior increases the pressure of water-vapor migration down through the membrane. If the refrigerated space is maintained at 0°F, this pressure may always be downward. Constant downward vapor migration almost inevitably saturates the insulation and, at the least, aggravates membrane blistering, icing delamination, and the other problems associated with entrapped moisture.

- The thicker insulation required for a refrigerated space creates a less stable substrate for the membrane than does the normally thinner roof insulation. The inevitably wider joints and greater movement in a thickened substrate magnify the hazard of membrane splitting.

Providing a ventilated ceiling air space between the separately insulated refrigerated space and the roof deck relieves this threat to the roof. It greatly reduces the vapor-pressure differential between the roof and the interior, and it also reduces the chances of water penetrating into the refrigerated space, where it will freeze, reduce insulating efficiency, and raise operating costs.

Effects of Increased Insulation Thickness

Thickened roof insulation has raised fears, well-publicized in the technical press, about deleterious effects on the roof membrane. These fears, especially those concerning the effect of thickened insulation on accelerated degradation of the membrane, have been greatly exaggerated. As the following section shows, roof color is a much more important determinant of roof-surface temperature than insulation thickness, and roof-surface temperature is a prime determinant of the built-up membrane's degradation. The trivial effect of insulation on roof-surface temperature is established by both theoretical and empirical research.

Here, in summary, is the case against thickened insulation, which can expose the roof membrane to several specific life-shortening effects:

- Accelerated chemical degradation

- Increased risk of splitting

- Reduced impact resistance

- Increased risk of slippage

Note that most of these hazards result from the greater temperature range—hotter in summer, colder in winter—experienced by a heavily insulated membrane.

Chemical degradation of bitumen accelerates with the higher summer surface temperature resulting from thickened insulation because oxidation rate, chief agent of this chemical degradation, rises exponentially with temperature. But the temperature rise from increased insulation thickness is slight. An NBS study calculates a maximum temperature increase of 4°F, from 153°F for a black surface with 0.25 U factor (1-in. fiberboard insulation) to 157°F with a 0.066 U factor (5-in. fiberboard insulation).[1] Another theoretical study by Carl G. Cash and W. H. Gumpertz confirms the Rossiter and Mathey study's conclusions.[2] And a study based on field tests of membranes over varying thicknesses of fiberglass insulation (from ¾ to 7¼ in. in three layers) by D. E. Richards and Ed Mirra of Owens-Corning Fiberglas Corporation concludes that heavily insulated membranes experience only slightly higher membrane temperatures during hot weather.[3]

Roof color has a much greater effect on membrane temperature than insulation thickness. For roofs of equal insulation thickness, maximum calculated temperature difference between a black and gray surface is 15°F; between black and white it is 27°F, compared with the 4°F temperature difference between 1- and 5-in.-thick insulation. A black-surfaced membrane over 1-in. insulation gets 10°F hotter than a gray-surfaced membrane over 5-in. insulation (see Figs. 5-8 and 5-9).[4]

---

[1]W. J. Rossiter and R. G. Mathey, "Effects of Insulation on the Surface Temperature of Roof Membranes," NBSI Rept. 76-987, 1976, p. 11.

[2]Carl G. Cash and W. H. Gumpertz, "Economic and Performance Aspects of Increasing Insulation on the Temperature of Builtup Roofing Membranes," *J. Test. Eval.*, March 1977, pp. 124ff.

[3]D. E. Richards and E. J. Mirra, "Does More Roof Insulation Cause Premature Roofing Membrane Failure or Are Roofing Membranes Adequate?", *Proc. Symp. Roofing Technol.*, NBS-NRCA, September 1977, pp. 193ff.

[4]Rossiter and Mathey, op. cit.

Fig. 5-8 Even after it ages, aluminum-fibrated coating can substantially reduce peak roof-surface temperature, by about 20°F during hot, sunny weather. Reduced roof-surface temperature lengthens membrane service life and cuts cooling-energy consumption.

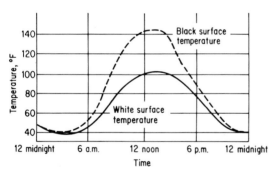

Fig. 5-9 Graph shows roof-surface temperatures recorded for 25-mm (3-in.)-thick sprayed-in-place urethane roof on sunny day. Heat flow lagged by about 1 h. (Graph from C. P. Hedlin, "Some Design Characteristics of Insulation in Flat Roofs Related to Temperature and Moisture," *Proc. 5th Conf. Roofing Technol.*, NBS-NRCA, April 1979, p. 21.)

Increased splitting risk results from several factors associated with thickened insulation. Most important is thickened insulation's reduced horizontal shearing resistance to membrane contraction, which can result from either a temperature drop or drying shrinkage of the felts. Horizontal shearing resistance can be assumed inversely proportional to insulation thickness; i.e., doubled insulation thickness reduces horizontal shearing resistance by half.

Remedies for this indeterminate increased splitting risk include mechanically anchoring insulation to the deck; specifying an additional felt ply to strengthen the membrane; more closely spacing expansion-contraction joints; and mechanically anchoring the base sheet to insulation, which can help reduce the slippage risk.

The splitting and impact hazards especially call for research into the structural properties of insulation materials. Insulation materials with low horizontal shear strength may require limitation in depth because the insulation must transmit membrane stresses to the deck.

Joint Taping

A perennial industry controversy concerns taping of insulation joints (see Fig. 5-10). According to Owens-Corning Fiberglas Corporation, which recommends this practice for fiberglass insulation, 6-in.-wide glass-fiber tapes bonded with steep asphalt at continuous joints between fiberglass insulation boards has several benefits:

- Prevention of asphalt drippage (and consequent bitumen loss) from membrane felt moppings through insulation-board joints

- Reduction in ridging over insulation joints

- Reduction in splitting hazard

The protection against membrane splitting results from the elimination—or at least attenuation—of stress concentration at the vulnerable

Fig. 5-10 Taped insulation joints—done manually on small jobs (left), mechanically on large jobs—can alleviate membrane stress concentrations and prevent bitumen dripping into insulation joints.

joint lines, where joint taping can (1) reduce differential movement between adjacent insulation boards and (2) ensure a more uniform membrane cross section (free of increased thermal contraction coefficient from membrane thickening at the insulation joint).

One disadvantage of joint taping is its requirement for aligning the insulation in both directions instead of staggering at least one line of joints. Staggered joints break lines of potential stress concentration in the membrane above.

Double-Layered Insulation

Double-layered insulation is a better method for reducing membrane stress concentration than joint taping, according to many roofing experts. According to C. G. Cash, of Simpson, Gumpertz & Heger, consulting engineers, of Cambridge, Massachusetts, double-layered insulation placed with vertically offset joints forms an almost continuous plane plate as opposed to the more likely warped surface of single-layered insulation. Double-layered insulation has the following advantages:

- Smoother top surface as membrane substrate because of better accommodation to deck irregularities.

- Elimination of thermal bridges and consequent heating- and cooling-energy leakage at insulation joints.

- Reduced ridging hazard (from reduced tendency of insulation-board warping because it is generally easier to get good adhesion between insulation boards than between insulation and deck).

- Reduced membrane splitting hazard because (1) there is less chance of insulation board warpage; (2) it is easier to get tight joints with thinner boards; and (3) there is reduced stress concentration from continuity in the insulation "plate," which according to one study reduces thermal contraction stress by 10 percent.[1]

- Blockage of vertical paths followed by upward-moving water vapor condensing on the cold underside of the membrane, where the felts can swell with the absorbed moisture, creating ridges.

- Better adaptability to mechanical anchorage.

[1]Owens-Corning Fiberglas Corporation, *The Whys of Double Layer*, March 1980.

Mechanical anchorage works better with double- rather than single-layered insulation for two reasons: (1) It reduces the required shank length by half the total insulation thickness because the fastener penetrates only the bottom layer of double-layered insulation. (Hot mopping to the bottom insulation layer can then provide good adhesion for the top insulation layer because insulation provides a better heat-retaining surface as well as an easier one for application than a deck, especially a steel deck.) (2) By limiting the fastener penetration to half the insulation depth, you also insulate the fastener, preventing its action as a possible thermal bridge that, at subfreezing outdoor temperatures, might cause condensation at its projection into the warm interior.

One caution concerning the mopped-asphalt adhesive film between the two insulation-board layers is the possibility of condensation forming at this asphalt plane, as a consequence of upwardly migrating water vapor being impeded at this plane. The adhesive film could function as a vapor retarder, and in cold weather, under many common conditions, its temperature may fall below dew-point temperature. A simple calculation, based on assumed interior and exterior relative humidity and temperature, can indicate whether such condensation poses a threat.

PRINCIPLES OF THERMAL INSULATION

Heat is transferred through (1) conduction, (2) convection, and (3) radiation. Conduction depends on direct contact between vibrating molecules transmitting kinetic (or internal heat) energy through a material medium. Convection requires an air or liquid current, or some moving medium, to transfer heat physically from one place to another. Radiation transmits heat through electromagnetic waves emitted by all bodies, at an intensity varying with the fourth power of the absolute temperature.

Radiation accounts for the extremes in roof-surface temperatures, above and below atmospheric temperature. Sun rays can raise the surface temperature of an insulated roof 75°F above air temperature, and on a clear night, without cloud cover to reflect radiated heat back to earth, roof-surface temperature can drop 10°F or more below the air temperature. Thus, in a climate with design temperature varying from a summer maximum of 95°F to a winter low of 0°F, the annual variation in roof temperature may be 160°F or more. Because it conducts less heat to or from the roof surface, insulation within the roof sandwich *increases* the extremes in roof-surface temperature; for a black roof surface the annual temperature differential could exceed 200°F. But insulation greatly reduces the daily or annual temperature difference experienced by the deck.

Heat flows through building materials primarily through conduction, less so by convection. Through an air space all three modes of heat transfer are at work. Thermal insulation resists all three, but primarily conduction.

Reflective insulation—normally with an aluminum foil or aluminum-pigmented heat-reflective coating—is common in walls with an enclosed air space. But the difficulty of accommodating such a space under a roof, plus the need to pierce the foil with numerous openings for pipes, conduit, ducts, and other mechanical items, makes reflective insulation generally impracticable for roofs. So does the difficulty of keeping reflective insulation from contacting other materials and thereby losing its effectiveness.

Good thermal insulators generally depend on the entrapment of air, a poor heat conductor, in millions of small cells or pockets; these arrest the transfer of heat by convection. Carbon steel, a good heat conductor, transfers heat nearly 2000 times as fast as an equal thickness of air. Cellular insulations, e.g., foamed glass or plastic, acquire their thermal-insulating value by establishing a temperature gradient through the cross section, with each tiny cell of entrapped air making its contribution to the total resistance to heat flow. Fibrous insulations exploit thin pockets of air between the fibers.

The efficiency of thermal insulation also depends on its capacity for impeding air flow, thus resisting heat flow by convection, and on the low thermal conductivity of its basic materials.

The quantity of heat transferred through a building component varies directly with the temperature differential, the exposed area, and the time during which the transfer takes place. It varies inversely with thickness. A good thermal-insulating material is a poor thermal conductor (and vice versa); it also blocks the passage of air, thus resisting convective heat flow. If it is sufficiently opaque, it resists the penetration of heat radiation.

Rising temperature reduces insulation's thermal resistance (i.e., increases its thermal conductivity). The R value on glass-fiber insulation, for example, drops from about $6.5°F/(Btuh·ft^2)$ at $-50°F$ to about 4 at 150°F, representing a loss of about 40 percent over the extreme temperature range that roofs experience.[1] This phenomenon is explained by the kinetic theory of gases: Thermal conductivity increases with faster movement of gaseous molecules transmitting heat energy from one surface to another. Temperature is an index of this internal molecular movement.

---

[1]C. J. Shirtcliffe, "Thermal Resistance of Building Insulation," *Can. Build. Dig.*, no. 149, April 1975, p. 149-4.

Moisture Reduction of Insulating Value

Published values for thermal conductivity or resistance are for *dry* materials. This policy permits accurate measurements of thermal conductivity because moisture within the tested sample would distort the conductivity measurements (chiefly by adding latent heat transfer resulting from condensation or evaporation to the purely conductive heat transfer that is desired). Dry thermal-conductivity values are thus an accurate basis for comparing different materials' thermal resistances under ideal conditions.

In the field, however, where most insulations inevitably contain significant quantities of liquid moisture, the insulation's dry thermal resistance (i.e., R factor) is reduced via two independent mechanisms:

- Sensible heat transfer, chiefly by conduction

- Latent heat transfer (i.e., changes of state from vapor to liquid or solid, or vice versa, which depend on the insulation's vapor permeability)[1]

Water filling an insulating material's interstices generally replaces air (thermal conductivity = 0.17 Btuh/(ft$^2\cdot$in.$\cdot$°F) at 40°F. Water's thermal conductivity is more than 20 times as much as air's at above-freezing temperatures (when k = 4 Btuh/(ft$^2\cdot$in.$\cdot$°F), nearly 100 times as much at subfreezing temperatures (the k value of ice = 15.6 Btuh/(ft$^2\cdot$in.$\cdot$°F).[*]

Vulnerability to moisture penetration is the chief determinant of an insulation's thermal resistance in the presence of moisture. Foamglass, the least vulnerable of roof-insulating materials in common use, has a water-resistant, closed-cell structure that protects it from significant moisture penetration. Even after being subjected to a simulated leak, foamglass retains nearly all (98 percent) its dry thermal resistance. But poured, lightweight insulating concrete fills lose about half their dry thermal-resistance values at 25 percent (dry basis) moisture content, up to 70 percent or more at 50 percent moisture content.[2]

Perlite board, fiberglass, and fiberboard insulation also suffer substantial losses from moisture absorption. Designers should consider these losses when computing total R factors through roof systems. A

[1]F. A. Joy, "Thermal Conductivity of Insulation Containing Moisture," ASTM STP217, 1957, p. 66.

[*]R. H. Perry, *Engineering Manual,* 2d ed., McGraw-Hill Book Company, New York, 1967, pp. 3-25, 3-27, 3-28.

[2]Powell and Robinson, op. cit., table 2, pp. 6, 7.

method for correlating moisture content with thermal resistance has been developed by Wayne Tobiasson and John Ricard, of CRREL, at Hanover, New Hampshire.[1] At 100 percent moisture content (by weight), most common insulations (e.g., perlite board, fiberboard, fiberglass, and felt-faced urethane) apparently lose from 40 to 60 percent of their insulating value.

Even when designers lack precise knowledge of the insulation's moisture content (the normal situation when designing a new building or even a reroofing project), they should estimate moisture-caused losses in thermal resistance, possibly based on equilibrium moisture content.

Trapped moisture in vapor-permeable insulations—fiberglass, fiberboard, perlite board, and lightweight insulating concrete—reduces the insulation's thermal resistance more during the summer-cooling season than during the winter-heating season. Latent heat transfer is the probable explanation for this empirically discovered phenomenon. Condensation of water vapor flowing upward toward a colder, drier exterior or downward toward a cooler, drier interior adds its high heat of evaporation (972 Btu/lb at standard atmospheric pressure) to the conductive heat loss through the roof. In winter, water-vapor flow remains upward, and condensate tends to remain in the upper, colder sections of insulation. However, in summer, vapor flow often reverses direction: downward from a solar-heated membrane during the day, upward at night, from an interior at 78°F toward an exterior at 65°F or so. Daily, or nearly daily, reversals in vapor-flow direction make the water vapor available for latent heat transfer in both directions, augmenting the already increased thermal conductivity resulting from the presence of liquid water.[2]

Heat-Flow Calculation

The simplified heat-transfer calculations normally used for building design require a knowledge of four indexes of heat transmission:

1. Thermal conductivity k = heat (Btu) transferred per hour (Btuh) through a 1-in.-thick, 1-ft^2 area of homogeneous mate-

---

[1]Wayne Tobiasson and John Ricard, "Moisture Gain and its Thermal Consequence for Common Roof Insulations," *Proc. Symp. Roofing Technol.*, NBS-NRCA, April 1979, p. 4-16.

[2]See Powell and Robinson, op. cit., pp. 15ff, and table 2, pp. 6, 7, for summer R factor losses.

rial per °F temperature difference from surface to surface. The unit for k is Btuh/(ft²·in.·°F). To qualify as thermal insulation, a material must have a k value of 0.5 or less.

2. Conductance $C = k$/thickness is the corresponding unit for a material of given thickness. (The unit for C is Btuh/(ft²·°F). For a 2-in.-thick plank of material whose $k = 0.20$, $C = 0.10$.

3. Thermal resistance $R = 1/C$ indicates a material's resistance to conductive heat flow. (For a material with $C = 0.20$, $R = 5.0$.) The unit for R is °F/(Btuh·ft²). That is, for a 5°F temperature difference surface-to-surface, 1 Btuh flows through a 1-ft² specimen.

4. Overall coefficient of transmission U is a unit like k and C, measured in Btu transmitted per hour (Btuh) through 1 ft² of construction per °F from air on one side to air on the other. However, it relates to the several component materials in a wall or roof. U is calculated from the following formula:

$$U = \frac{1}{\Sigma R} \qquad (5\text{-}1)$$

where R = sum of the thermal resistances of the components, plus the resistances of the inside and outside air films.

For example, to calculate the insulation's required thermal resistance R_i, the designer usually starts with a target U factor set by the mechanical engineer. For the other components, the designer merely tabulates the resistances R for all the materials, including inside and outside air films. (For the summer condition, when roof temperatures often rise to 60°F or more above outside air temperature, it is prudent to assume a roof-surface temperature of 150°F or so and omit the outside air-film resistance.) Data for conductances of various materials are available from the *ASHRAE Handbook and Product Directory, 1977 Fundamentals*. If not available in general tables, data for proprietary insulating materials should be furnished by manufacturers.

Consider the roof system shown in cross section, with target U factor of 0.085.

By Eq. (5-1),

$$U = \frac{1}{1.67 + R_i} = 0.085$$

$$R_i = \frac{1}{0.085} - 1.67 = 10.09$$

Two layers of 1%6-in.-thick fiberglass boards ($R = 10.52$) are satisfactory. (It might be prudent to reduce the insulation R factor by 10 to 15 percent to allow for moisture content.)

	R Value
a = Outside air film (7-1/2) mph wind	0.25
b = 3/8-in. BUR membrane	0.33
c = Insulation	R_i
d = 4-in. concrete slab	0.48
e = Inside air film (heat-flow up)	0.68
$\leq R$ =	$1.67 + R_i$

Heat Gain or Loss

Total rate of heat gain or loss Q (in Btuh) through a roof is computed by:

$$Q = U \times \text{roof area}$$
$$\times \text{temperature difference between inside and outside air} \quad (5\text{-}2)$$

To calculate the temperature at any parallel plane through the roof, use one of the following formulas:

$$T_x = \begin{cases} T_i - \dfrac{\Sigma R_x}{\Sigma R}(T_i - T_o) & \text{for winter conditions} \\[2mm] T_o - \dfrac{\Sigma R_x}{\Sigma R}(T_o - T_i) & \text{for summer conditions} \end{cases} \quad (5\text{-}3)$$

where T_x = temperature at plane X
T_i = inside temperature
T_o = outside *roof-surface* temperature
ΣR_x = sum of R values between warm side and plane X

ALERTS

Design

1. When practicable, align the continuous, longitudinal joints of insulation boards parallel to the *short* dimension of the roof and perpendicular to the rolled (longitudinal) direction of the felts (see Fig. 5-11).

2. Select two layers in preference to one. For multiple layers, specify staggered joints.

3. Investigate the need for taping insulation-board joints (to relieve membrane stresses).

4. Require bituminous impregnation of wood fiberboard and other organic fiberboard insulations.

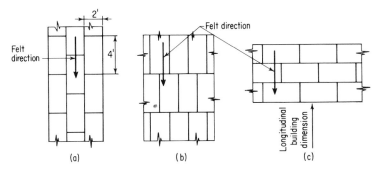

Fig. 5-11 The conventional pattern of unrolling felts parallel to the continuous longitudinal joints of insulation boards (*a*) heightens the chances of membrane splitting or ridging. The continuous, longitudinal joints form lines of maximum stress concentration, and the felts must resist these stresses in their weaker (transverse) direction. Ridging is also promoted by this pattern because wet felts expand more in the transverse than in the longitudinal direction. In pattern (*b*), a compromise, the felts resist the maximum stresses produced by the insulation joints in their stronger (longitudinal) direction. The longitudinal joints between insulation boards, however, become more critical and subject the felts to fairly high contraction stresses in their weak direction. In pattern (*c*), the best, the longitudinal joint again works against the strong felt axis, and the contraction stresses of the continuous joints in (*b*) are reduced, further relieving the stresses in the felts' weak direction. A practical argument against (*c*) is the difficulty of applying the insulation adhesive to a steel deck for this awkward board arrangement. However, with mechanical anchorage this practical difficulty is reduced. The superior insulation-board pattern is another argument, in addition to more reliable insulation anchorage, for specifying mechanical fasteners.

5. Install insulation for a refrigerated space below the roof deck.

6. Investigate the need for a vapor retarder to shield the insulation from water vapor migrating from a warm, humid interior (see Chap. 6, "Vapor Control").

7. To reduce the threat of bitumen drippage when insulation is hot-mopped to the deck, specify ASTM Type II or III asphalt. Specify rosin-sized paper, felt, or tape over joints in wood sheathing, plywood, or other prefabricated deck units to be nailed.

8. Check local building code, Factory Mutual, and Underwriters Laboratories listings for insulation's conformance with wind-uplift and fire-resistance requirements.

For lightweight insulating concrete:

1. Do not specify wet-fill materials (including perlite, vermiculite, or foamed concrete) over monolithic structural concrete, unvented metal deck, vapor retarders, or other substrates that do not permit underside venting.

2. Specify stack venting (one vent per maximum 900-ft² area) *and* perimeter edge venting (notched wood nailers or isolated flashing).

3. Specify slotted vent holes in steel deck used with insulating concrete.

4. Require minimum 2-in. thickness for lightweight insulating concrete (to allow embedment of mechanical fasteners).

5. Over lightweight insulating concrete fill, specify a coated base sheet for initial built-up membrane ply, *nailed* to the fill.

6. If impracticable to avoid a nonventing substrate under a wet-fill material, specify either a layer of porous insulation or a venting base sheet as bottom membrane ply to facilitate lateral venting.

7. Require a minimum 28-day compressive strength of 100 psi.

For sprayed polyurethane foam:

1. Minimum slope = ⅛ in./ft (preferably ¼ in.)

2. Require 1-in. minimum foam thickness, maximum = 3 in.

3. Density = 3.0 pcf ± 0.5 pcf (ASTM D1622)

4. Compressive strength = 40-psi minimum (ASTM D1621)

5. Tensile strength = 35-psi minimum (ASTM D1623)

6. Shear strength = 25-psi minimum (ASTM DC273)

7. Closed-cell content = 90 percent (ASTM D2856)

8. Minimum adhesion to substrate = 95 percent

Field

1. Require that stockpiled insulation stored at the jobsite be covered with waterproofed tarpaulins and raised above ground.

2. Require a smooth, plane deck surface [For specific tolerances, see Chap. 4, "Structural Deck" (Field Alert No. 1).]

3. Prohibit placing insulation on wet decks or snow- or ice-covered decks. Decks must be cleaned and dried (including flute openings of metal decks) before insulation is installed. Normally, test for dry deck surface as follows: Pour a small amount of hot bitumen on the deck. If, after the bitumen cools, you can remove it with your fingernails, reject this deck for application of insulation; the moisture content is too high. If the cooled bitumen sticks to the deck and cannot be removed by your fingers, the deck is dry enough to receive the mopping.

An alternative test for properly cured and dried cast-in-place decks (e.g., concrete, gypsum) is to apply hot bitumen as described and observe for frothing or bubbling. If none occurs, the deck is dry enough.

4. Limit moisture content to satisfactory levels, not to exceed equilibrium moisture content.

5. Limit daily application of insulation to an area that can be covered on the same day with built-up roofing membrane.

6. Place plywood walkways on insulation vulnerable to traffic damage.

7. Set adjacent units of prefabricated insulation with tightest possible joints. *Trim or discard units with broken corners or similar defects.*

8. Provide water cutoffs at the end of the day's work on all decks with insulation boards, cut along vertical face, and remove when work resumes (see Fig. 5-12).

9. Use mechanical anchors for attaching insulation boards to steel decks (see Fig. 5-13). For nonnailable decks, e.g., concrete, specify solid-mopped Type II or III asphalt.

10. Prohibit cantilevering of insulation boards over steel-deck ribs. Require solid bearing, 1½-in. minimum.

For lightweight insulating concrete:

1. Prohibit placing of wet insulating concrete fill if temperature is expected to drop below freezing within 48 h.

2. Let lightweight insulating concrete dry for at least the minimum time recommended by the manufacturer—preferably 5 days, plus an additional day for each rainy day occurring during the drying period.

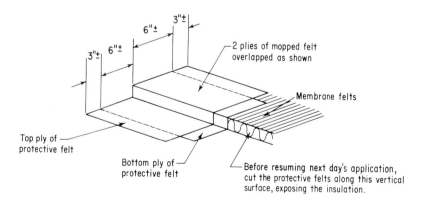

Fig. 5-12 Temporary insulation-board cutoff detail (to be used at end of day's work).

Fig. 5-13 Electrically powered driving tool engages fastener screw head and anchors it to steel-deck flange with pullout strength exceeding 300 lb for 22-gage steel. (Illinois Tool Works, Inc.)

For sprayed polyurethane foam:

1. Do not start application during damp or rainy weather, or under threat of rain. As further restrictions, start application only under the following conditions:

      Temperature = 50°F or above

      Relative humidity = 80 percent or less

      Wind velocity = 10 mph maximum (or shelter roof area from wind)

2. Foam surface must be free of voids, crevices, and pinholes. Maximum permissible roughness of foam surface is "verge of popcorn." Accept "orange peel." Reject "popcorn" and "tree bark."

3. Spray insulation with minimum ½-in.-thick lifts. If any foam surface stands exposed for more than 24 h, check it for evidence of oxidation ("rust" discoloration or powdering). If either occurs, skin the foam surface and apply new surface coating before applying either (a) another foam lift or (b) membrane coating.

4. Reject sprayed foam with internal separation at lift plane interface.

5. Allow a minimum 1 h, maximum 24 h, between completion of foam application and start of membrane base coating. If weather forces a delay beyond the 24-h maximum, inspect foam surface for rust or powdering. If either is discovered, skin foam surface and refoam before applying membrane base coat.

6. Take several 2-in.-diameter field-core samples during (or following) the spraying operations to check foam and membrane quality.

SIX

Vapor Control

Water vapor flowing upward into the insulation from the building interior causes severe hazards to the built-up roof assembly. Condensation of this vapor impairs the insulation's thermal resistance, and it may ultimately destroy the insulation itself. Liquid moisture may collect at the underside of the insulation until it flows into cracks or joints in the deck and leaks onto a suspended ceiling or directly into a room. Condensate trapped within voids at the membrane-insulation interface can reevaporate under solar heat, and the resulting internal pressure can aggravate blister growth. Trapped condensate can also freeze and expand, breaking the bond between insulation and base sheet or even breaking through the built-up roofing membrane.

Paradoxically, more efficient insulation heightens the problems of condensation. Insulation shifts the dew point (the surface temperature at which water vapor condenses) from under the roof system to within it (see Table 6-1 and Fig. 6-1). Thus, other factors being equal, the more efficient the insulation, the more need for a vapor retarder, subroof ventilation, or other means of preventing migrating water vapor from condensing within the built-up roof system.

FUNDAMENTALS OF VAPOR FLOW

In the generally temperate climate of the United States, water vapor normally flows *upward* through the roof, from a heated interior toward a colder, drier exterior (i.e., from high to low vapor pressure, along a vapor-pressure gradient). As the curves of Fig. 6-2 depict, this water-vapor pressure depends on two variables: temperature and relative humidity (RH).

The most important variable is temperature. At 0°F, even at 100% RH, vapor pressure is insignificant. But at 100°F, even at 20% RH, vapor pressure is more than 10 times as high as vapor pressure at 0°F, 100% RH. Mathematically, vapor pressure increases *exponentially*

Table 6-1 Dew-Point Temperature, °F*

RH, %	Dry-bulb temperature, °F														
	32	35	40	45	50	55	60	65	70	75	80	85	90	95	100
100	32	35	40	45	50	55	60	65	70	75	80	85	90	95	100
90	30	33	37	42	47	52	57	62	67	72	77	82	87	92	97
80	27	30	34	39	44	49	54	58	64	68	73	78	83	88	93
70	24	27	31	36	40	45	50	55	60	64	69	74	79	84	88
60	20	24	28	32	36	41	46	51	55	60	65	69	74	79	83
50	16	20	24	28	33	36	41	46	50	55	60	64	69	73	78
40	12	15	18	23	27	31	35	40	45	49	53	58	62	67	71
30	8	10	14	16	21	25	29	33	37	42	46	50	54	59	62
20	6	7	8	9	13	16	20	24	28	31	35	40	43	48	52
10	4	4	5	5	6	8	9	10	13	17	20	24	27	30	34

*For intermediate, untabulated combinations oɪ dry-bulb temperatures and RH, dew-point temperature can be interpolated on direct (i.e., linear) proportionality. For example, for 70°F, 35% RH interior temperature, dew point = (37 + 45)/2 = 41°F.

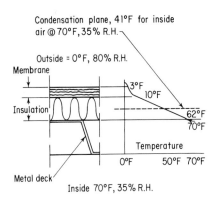

Condensation plane, 41°F for inside air @ 70°F, 35% R.H.

Outside = 0°F, 80% R.H.
Membrane
Insulation
3°F
10°F
62°F
70°F
Temperature
0°F   50°F 70°F
Metal deck
Inside 70°F, 35% R.H.

Fig. 6-1 The introduction of insulation between deck and membrane usually shifts the dew point from *below* the roof system to *within* the roof system. The dew point at interior conditions (70°F, 35% RH), is 41°F. Thus migrating water vapor will condense somewhere above the 41°F temperature plane, probably at the underside of the membrane. Without insulation, water vapor would condense on the steel-deck surface at about 35°F.

with increasing temperature, but only *linearly* with increasing relative humidity.

For roof-system designers, interested more in vapor flow than heating, ventilating, airconditioning (HVAC) system designers concerned with vapor quantities, relative humidity is most meaningfully defined as the ratio of actual vapor pressure to the vapor pressure of a saturated (that is, 100% RH) air-vapor mixture at constant temperature and overall atmospheric pressure. (For more detailed technical discussion, see the "Theory of Vapor Migration" section later in this chapter.)

These phenomena—exponential vapor-pressure increase with rising temperature, linear vapor pressure increase with rising RH—explain why cold climates are characterized by low exterior vapor pressure.

When the outside temperature drops to 0°F, regardless of whether outside RH is 100% or 0%, the vapor pressure is insignificant. If interior
conditions are 68°F, 50% RH, the vapor-flow problem is essentially the
same in cold, dry Montana as along the cold, humid Maine seacoast.
The colder the climate, the lower the outside vapor pressure and the
greater the pressure differential for a given interior temperature and
RH. Note in Fig. 6-2 that the vapor-pressure differential = 0 at 49°F,
100% RH, outside conditions. But at any outside temperature lower
than 49°F, even at 100% RH, the outside vapor pressure drops below
the interior vapor pressure, impelling vapor flow upward through the
roof.

The normal direction of vapor flow reverses in the warm, humid climates of the southeastern or south central United States and in Hawaii,
where vapor flow is normally downward through a roof toward an airconditioned interior. Referring again to Fig. 6-2, note that for the com-

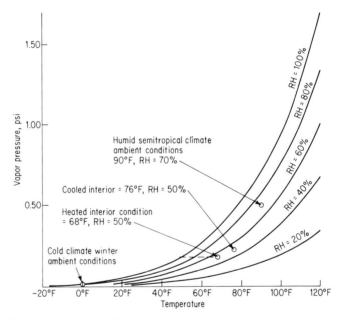

Fig. 6-2 This graph, which plots vapor pressure for a given temperature and relative humidity (RH), is a handy technique for determining the direction of water-vapor flow through a roof system. When
the ordinate for outside ambient conditions is *below* the ordinate representing interior conditions, the vapor-flow direction is *upward*.
(The vapor-pressure differential for a heated interior at 68°F, 50%
RH with 0°F, 50% RH ambient, is about 0.16 psi.) Conversely, if the
ordinate for outside ambient conditions is *above* the ordinate representing interior conditions, the vapor flows *downward*. (The downward vapor-pressure differential for a cooled interior at 76°F, 50%
RH, with 90°F, 70% RH outside, is about 0.27 psi.)

mon Corpus Christi, Texas, ambient conditions of 90°F, 70% RH, the outside vapor pressure greatly exceeds the interior vapor pressure corresponding to a typical cooled airconditioned interior (76°F, 50% RH). Note further that for normal occupancies—residences, office, or commercial buildings—the vapor pressure impelling downward vapor flow in humid, tropical, or semitropical climates may be unidirectional, or nearly so, year-round. In northern locations with climatic extremes, like our north central states, vapor flow changes direction, to predominantly downward in summer.

Although a vapor retarder may be required in a cold climate, there is no such requirement in warm, humid climates. In these locations, the roof membrane itself must perform double duty, resisting downward vapor flow as well as liquid moisture. (Low membrane permeability is thus beneficial in a warm, humid climate, but it may be detrimental in a cold climate.) A vapor retarder in its conventional location between deck and insulation is detrimental in a warm, humid climate. It traps water vapor within the insulation, where it may condense and impair insulation efficiency instead of letting the vapor escape into the interior, where it can be vented or condensed out by airconditioning equipment.

Water vapor penetrates a built-up roof system via air leakage and diffusion. Depending on local conditions, air leakage and diffusion vary in relative importance, but they normally reinforce one another; i.e., both tend to force the vapor in the same direction. In the normal case— heated, humid interior and cold, dry exterior—the atmospheric pressure under the roof exceeds the atmospheric pressure on the roof, and the air escaping through deck joints and other small openings conveys water vapor into the roof system from below. Because warm air can hold more water vapor than cold outside air, the water-vapor pressure of a heated, humidified interior exceeds the outside vapor pressure and thus promotes outward, upward vapor migration.

Yet the two mechanisms are physically independent. For example, it is possible to have an inward air leakage associated with outward vapor diffusion. On a hot, windless day in an arid climate, a cooled, highly humidified plant might produce this phenomenon. Because of the reverse "chimney effect" created by the cooled interior, the outside atmospheric pressure could exceed the interior pressure, and the water-vapor-pressure imbalance from the humidified interior could produce upward water-vapor diffusion. (For a more detailed discussion, plus vapor-migration calculation, see the section on "Theory of Vapor Migration" at end of chapter.)

TECHNIQUES FOR CONTROLLING MOISTURE

As the basic strategy for limiting moisture within the roof assembly, roof designers can choose one of three methods:

- General ventilation, designed to carry interior moisture out on air drafts circulating through a space left under the roof

- Vapor retarders, designed as low-permeability membranes blocking the passage of water vapor into the built-up roof system

- The "self-drying" concept, which relies on heat energy supplied by the spring and summer sun to evaporate winter-accumulated moisture and drive it downward through a vapor-permeable deck into the building interior

General Ventilation

Properly designed subroof ventilation is the most certain technique for preventing water-vapor infiltration into a built-up roof. But for low-sloped roofs, it poses the most difficult set of design conditions. Because natural convection decreases with diminishing roof height, ventilation is far less effective for low-sloped roofs than for steeply sloped roofs. Thus under a level roof, the mechanisms left for dissipating moisture are diffusion and wind-induced ventilation. Exterior wind-produced pressure differentials tend to either force moist air out of the loft space or displace it with colder, drier air, but with minimal benefits, particularly for a large level roof area.

There are, moreover, practical obstacles to general ventilation in commercial, industrial, and residential apartment construction. In these buildings, ducts and pipes located in loft space pierce the ceiling, thus making an effective air seal difficult. Because warm, humid air will probably infiltrate the cold loft space and condense on cold surfaces, ducts and pipes must be insulated. Especially in multistory buildings, where the chimney effect increases pressures in upper parts of the building and promotes upward air leakage, the ceiling air seal becomes essential. A further practical objection to general ventilation is the increased building height required to provide the ventilation space. General ventilation is thus the highest first-cost method of controlling vapor migration. However, as a compensating advantage, it is the most foolproof method.

Vapor Retarder

According to conventional vapor-retarder theory, the vapor retarder forms an essentially impermeable surface on the warm, humid side of the roof sandwich, blocking the entry of water vapor. Vapor retarders are used only in roofing systems where the insulation is sandwiched between the structural deck and the built-up membrane. A properly designed and installed vapor retarder can prevent condensation from forming within the built-up roofing system (see Fig. 6-3).

Water-Vapor Transfer   To qualify as a vapor retarder, a material's vapor permeance rating should not exceed 0.1 perm. A material rated at 1 perm admits 1 grain of water vapor per hour through 1 ft² of material under a pressure differential of 1 in.Hg (0.491 psi). Vapor resistance is the reciprocal of permeance.

For perm ratings of selected vapor-retarder materials, see Table 6.2. Due to variations in some composite materials and other difficulties in measuring vapor-transmission rates, perm ratings are generally imprecise. Thus sophisticated designers do not rely on overly refined calculations but design vapor retarders conservatively.

Another index of a vapor retarder's permeability, the *perm-inch*, or permeance of unit thickness, sometimes enables the designer to make a direct comparison between materials of equal thickness. (For films less than 0.020 in. thick, the perm-inch index is impracticable.)

For many uses the 1-perm rating formerly set as the maximum permeance permissible for a vapor retarder is not good enough. Under a high differential vapor pressure, or even under low differential pressure over long time intervals (as in a refrigerated warehouse), a vapor retarder rated at 0.1 perm admits too much water vapor. For certain kinds of industrial buildings subjected to continual high temperatures and high relative humidity, vapor retarders of less than 0.1-perm rating are recommended. In severely cold northern locations, where the

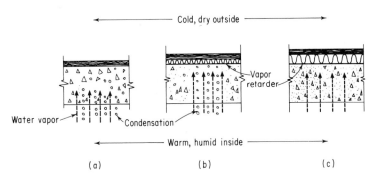

Fig. 6-3 How a vapor retarder prevents condensation. In (a), an uninsulated concrete roof system, migrating water vapor condenses when it reaches the dew-point temperature somewhere in the roof cross section. In (b), an inadequately insulated system in which the vapor retarder is at or below the dew-point temperature, migrating water vapor again condenses and drips. In (c), an adequately insulated system, the vapor-retarder temperature is above the dew point, preventing condensation.

Table 6-2   Perm Ratings of Selected Vapor-Retarder Materials*

Material	Permeance	
	Dry cup	Wet cup
Aluminum foil, 1 mil	0†	
Aluminum foil, 0.35 mil	0.05†	
Polyethylene, 4 mil	0.08†	
Polyethylene, 6 mil	0.06†	
Polyester, 1 mil	0.07†	
Saturated and coated roll roofing	0.05‡	0.24‡
Reinforced kraft and asphalt-laminated paper	0.3‡	1.8‡
Asphalt-saturated and coated vapor-barrier paper	0.2–0.3†	0.6‡
15-lb tarred felt	4.0‡	18.2‡
15-lb asphalt felt	1.0‡	5.6‡
Asphalt (12.5 lb/sq)	0.5‡	
Asphalt (22 lb/sq)	0.1‡	
Built-up membrane (hot-mopped)	0‡	

*Values from *ASHRAE Handbook and Product Directory, 1977 Fundamentals*, table 1, pp. 20.4, 20.5.

†Per ASTM E96-66 ("Water Vapor Transmission of Materials in Sheet Form").

‡Per ASTM C355-64 ("Water Vapor Transmission of Thick Material").

vapor-pressure differential may persist in the same upward direction for weeks, a virtually impermeable vapor barrier is needed to prevent a destructive moisture buildup within the built-up roofing system.

Vapor-Retarder Materials   An essentially 0-perm rating is attainable by the most common vapor retarder: two or three moppings of bitumen and two plies of saturated felt. A modern version of this vapor retarder uses one coated base sheet in combination with one or two bituminous moppings. Other materials that may qualify as vapor retarders include certain plastic sheets (vinyl, polyethylene film, polyvinyl chloride sheets), black vulcanized rubber, glass, aluminum foil, and laminated kraft paper sheets with bitumen sandwich filler or bitumen-coated kraft paper. Steel decks with caulked joints may also qualify as vapor retarders (see Table 6-2).

As another layer in the multideck roofing sandwich, the vapor retarder must be solidly anchored to the deck below and to the insulation above. It must resist wind-uplift stresses and horizontal shearing stresses produced by thermal stresses in the membrane and transferred through the insulation or produced by thermal or moisture-induced swelling or shrinkage in the insulation itself.

Vapor-retarder application varies with the materials. Sheets may be mechanically attached to the deck with nails or other deck-puncturing devices, bonded with cold adhesives, or mopped with steep asphalt. The bonding agent used to anchor the vapor retarder to the deck is frequently used to anchor the insulation to the vapor retarder.

The vapor-retarder material must satisfy secondary requirements, such as practicable installation. Although they have low permeance ratings, plastic sheets, for example, create field problems that impair their overall performance as components in the roof system. On windy days, these light, flexible sheets are difficult to install. Billowing and fluttering heighten the risk of tearing and make it difficult to flatten the sheet on the deck. Once in place, these sheets are vulnerable to traffic damage (see Fig. 6-4).

In addition, some plastic sheets have other disadvantages. Their shrinkage (after application) may tear the material where it spans steel-deck flutes. Hot bitumen leaking through the insulation joints when the roof membrane is being applied may soften or even melt the sheets.

Fig. 6-4 This mechanical applicator spreads ribbons of cold adhesive on the steel deck, lays the plastic vapor-retarder sheet, and then spreads another set of ribbons on top to adhere the insulation boards. The adhesive ribbons should parallel the deck flutes. The worker installing insulation boards allows a time lag between the placing of the adhesive and the placing of the insulation, to permit the volatile solvent to escape and ensure better adhesion. (Lexsuco, Inc.)

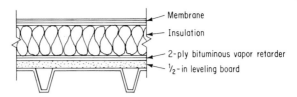

— Membrane
— Insulation
— 2-ply bituminous vapor retarder
— $\frac{1}{2}$-in leveling board

Fig. 6-5 This steel-deck roof assembly is a far more dependable roof system than one with a plastic sheet vapor retarder applied directly to the steel deck with cold adhesive. When a vapor retarder is really needed in a steel-deck roof assembly, this detail should be considered.

Bonding insulation to thin sheets can create problems. The cold adhesive should be applied parallel to the ribs of a steel deck on the flange surfaces. Thus the rigid insulation, which spans from high point to high point of adjacent flanges, may make inadequate contact with the adhesive in a deck depression, resulting in poor attachment to the deck. This flaw can reduce the system's wind-uplift resistance. It can even reduce the membrane's splitting resistance, which depends on the transfer of horizontal shearing stresses through the membrane-insulation interface and the insulation-deck (or vapor-retarder) interface.

To avoid these problems with plastic sheet vapor retarders over steel decks, you can substitute a conventional two-ply vapor retarder placed on top of a thin layer (½- or ¾-in.) board insulation mechanically fastened to the deck (see Fig. 6-5). A coated base sheet forms the bottom ply of this two-ply vapor retarder. Thicker insulation boards are mopped to the vapor retarder, providing most of the roof system's thermal resistance and maintaining the vapor-retarder temperature above the dew point. When you really need a vapor retarder, the additional expense for this system, with superior permeability rating and wind-uplift resistance and reduced vulnerability to field damage, can pay off its investment many times over.

Flashing the Vapor Retarder Field punctures and unsealed joints are not the only weak spots in a vapor retarder. To effectively block vapor flow into the roof system, a vapor retarder must be flashed and enveloped at roof openings and penetrations. For a hot-mopped vapor retarder, set the vapor retarder's felt edges in plastic cement and flash the edges to roof-penetrating components with a 15-lb felt collar sandwiched between two plastic-cement sealer trowelings. Extend the vapor retarder at all vertical surfaces.

Vapor-Retarder Controversy In the past, vapor retarders were recommended in (1) areas where the January temperature averaged 40°F or less or (2) wherever building occupancy or use created high relative humidity (see Tables 6-3 and 6-4).

This approach is oversimplified; each project requires its own analysis. Based on much sad experience, the efficacy of vapor retarders has been attacked by researchers and engineers administering vast building construction programs. A new consensus developing among many roofing experts reverses the conventional policy for vapor retarders. According to conventional vapor retarder theory, the advice is, "If in doubt, use a vapor retarder." But according to the more sophisticated modern view, the advice is, "If in doubt, omit the vapor retarder."

Table 6-3 Winter Relative Humidities

Dry occupancies
Relative humidity expected to range under 20%, closely related to prevailing outdoor relative humidity and indoor temperature

Aircraft hangars and assembly plants (except paint shops)
Automobile display rooms, assembly shops (except paint shops)
Factories, millwork, furniture (except plywoods and finishing units)
Foundries
Garages, service and storage
Shops, machine, metalworking (except pickling and finishing)
Stores, dry goods, electrical supplies and hardware
Warehouses, dry goods, furniture, hardware, machinery and metals

Medium moisture occupancies:
Relative humidity expected to range from 20 to 45%, varying with outdoor relative humidity, with moisture content increased by indoor activities, equipment, and operations

Auditoriums, gymnasiums, theaters
Bakeries, confectioners, lunchrooms
Churches, schools, hospitals
Dwellings, including houses, apartments, and hotels (highest relative humidity in kitchens, baths, laundries)
Factories, general manufacturing, except wet processes
Offices and banks
Stores, general and department

High moisture occupancies:
Relative humidity over 45%, determined primarily by processes, not climate

Chemical, pharmaceutical plants
Breweries, bakeries, food processing, and food storage
Kitchens, laundries
Paint and finishing shops
Plating, pickling, finishing of metals
Public bath and shower rooms, club locker rooms, swimming pools
Textile mills, paper mills, synthetic fiber processing plants
Photographic printing, cigar manufacturing

Behind this skeptical view lies much field experience, followed by successful changes in practice and more rigorous analysis of moisture problems. Plagued with numerous blistered roofs in a vast building program, one multiple building owner revised its former policy of generally using vapor retarders. This company now generally omits vapor retarders in buildings located south of 41°N latitude and where interior relative humidity is less than 45%.

This successful policy reversal springs from a reevaluation of what is practicable in the field, not from a challenge to the basic theory of vapor migration. The key question is whether vapor retarders can be built soundly enough to form the virtually impermeable membrane required to do the assigned job. Field threats to the vapor retarder's integrity—punctures and cutting of new roof openings—make the construction of highly impermeable vapor barriers extremely difficult.

An inefficient vapor retarder is not merely inefficient; it becomes a definite liability. The ideal vapor retarder is a jointless material with one-way permeability—letting water and water vapor leave but not enter the roof sandwich. Lacking such a miraculous material, the roof designer must accept the vapor retarder's liabilities along with its assets. Together with the built-up membrane, the vapor retarder seals entrapped moisture in the insulation instead of letting it escape.

In summary, here is the case for using a vapor retarder:

- A vapor retarder can ensure the continued thermal resistance of insulation sandwiched between the vapor barrier and built-up membrane.

- A vapor retarder is a good safeguard against vapor migration if a building's use changes from a "dry" to a "wet" use.

- A vapor retarder is advisable (required by at least one major manufacturer) over wet decks—poured gypsum and lightweight insulating concrete—to prevent vapor flow upward into the insulation. For this purpose the vapor retarder should be a venting base sheet, with mineral granule underside surfacing, mechanically fastened to the deck.

Here is the basic case against using a vapor retarder:

- The vapor retarder, together with the built-up roofing membrane, inevitably seals within the roof sandwich entrapped moisture that can eventually destroy the insulation, help split or wrinkle the built-up membrane, or, in gaseous form, blister it.

Table 6-4   Recommended Relative Humidities and Temperatures for Various Industries*

Industry	Temperature, °F	RH, %	Industry	Temperature, °F	RH, %
Baking:			Laboratories	As required	
Cake mixing	70–75	65	Leather:		
Crackers and biscuits	60–65	50	Chrome tanned		
Fermenting	75–80	70–75	(drying)	120	45
Flour storage	65–80	50–65	Vegetable tanned		
Loaf cooling	70	60–70	(drying)	70	75
Makeup room	75–80	65–70	Storage	50–60	40–60
Mixer-bread dough	75–80	40–50	Paint application:		
Proof box	90–95	80–90	Air-drying lacquers	70–90	60
Yeast storage	32–45	60–75	Air-drying oil paint	60–90	60
Cereal Packaging	75–80	45–50	Paper products:		
Confectionary:			Binding	70	50–65
Chocolate covering	62–65	50–65	Folding	75	50–65
Centers for coating	80–85	40–50	Printing	75–80	45–55
Hard candy	70–80	30–50	Storage	75–80	40–60
Storage	60–68	50–65	Textiles:		
Cork processing	80	45	Cotton carding	75–80	50–55

106

Application			Application		
Data processing	75	45–55	Cotton spinning	60–80	50–70
Electrical:			Cotton weaving	68–75	85
Manufacturing cotton-covered wire	60–80	60–70	Rayon spinning	80	50–60
Storage, general	60–80	35–50	Rayon throwing	70	85
Food:			Rayon	70	60
Apple storage	30–32	75–85	Silk processing	75–80	65–70
Banana ripening	68	90–95	Woolens—carding	80–85	65–70
Banana storage	60	85–89	Woolens—spinning	80–85	50–60
Citrus fruit storage	60	85	Woolens—weaving	80–85	60
Egg storage	35–55	75–80	Tobacco:		
Grain storage	60	30–45	Cigar and Cigarette	70–75	55–65
Meat, beef aging	40	80	Other processing and storage	75	70–75
Mushroom storage	32–35	80–85	Casing room	90	88–95
Mushroom, growing stages	Various	60–85	Woodworking:		
Potato storage	40–60	85–90	Finished products	65–70	35–40
Produce	Various	70–95	Gluing	70–75	40–50
Sugar	80	30	Manufacture	65–75	35–40
Tomato storage	34	85	Painting lacquer (static control)	70–90	60 min.
Tomato ripening	70	85			

*From A. L. Kaschub, "Industrial Humidification—Psychrometric Considerations and Effects of Humidification," Plant Eng. Mar. 22, 1979.

- In event of a roof leak through the membrane, the vapor retarder traps the water below the insulation and releases it through punctures that may be some lateral distance from the roof leak, thus making it more difficult to locate the leak. A large area of insulation may be saturated before the punctured membrane can be repaired.

- A vapor retarder is a disadvantage in summer, when vapor migration is generally downward through the roof. (Hot, humid air can infiltrate the roofing sandwich through the vents, or through diffusion through the membrane. It may condense on the vapor retarder itself.)

- A vapor retarder may be the weakest horizontal shear plane in the roofing sandwich. Failure at the vapor-retarder-insulation interface can split the membrane. At the least, the vapor retarder introduces an additional component where shear resistance may be critical to the membrane's integrity.

To use a vapor retarder as a temporary roof is to invite failure. Under the best conditions, installing an effective vapor retarder is a difficult job, but to subject it to the punishment of days or weeks of roof traffic and field operations makes a difficult job virtually impossible. (For discussion of the controversial topic of venting a roof system with a vapor retarder, see Chapter 5, "Thermal Insulation.")

Self-Drying Roof System

Omission of a vapor retarder is the key to a self-drying roof, which exploits the heat of the summer sun to evaporate accumulated moisture and drive it down through a vapor-permeable deck into the building interior. As demonstrated by an elaborate, 6-year research program conducted by National Bureau of Standards (NBS) researchers Frank J. Powell and the late Henry E. Robinson, a scientifically designed, self-drying roof system can dissipate construction moisture, winter-gained moisture, and even large amounts of liquid moisture (10 percent by volume) simulating a leak. *To qualify as a self-drying roof, the system must dissipate this moisture during a laboratory-simulated summer season fast enough to (1) reestablish its equilibrium moisture (i.e., show no trend toward long term moisture gain) and (2) suffer only minor loss in thermal-insulating value.*

The NBS test program proceeded as follows: 27 roof-assembly test specimens, roughly 2-ft square with calculated dry $U = 0.12$, were subjected to temperatures and humidities simulating seasonal conditions. Summer-simulated temperatures were 75°F nighttime and 138°F peak daytime; winter-simulated temperatures were 38°F nighttime and 75°F daytime [allowing for a roughly 30°F peak temperature gain for winter solar heat (see Fig. 6-6)]. Spring and fall simulation was 56°F nighttime, 106°F daytime. "Indoor" conditions—90°F and 30% RH—correspond to 75°F, 49% RH, a typical airconditioned environment.

Each simulated season lasted up to 16 weeks. With the specimens originally containing roughly the same quantity of moisture, the first summer established each specimen's effective (as opposed to theoretically dry) thermal resistance, as well as its self-drying capacity. "Spring" and "fall" indicated whether winterlike wetting or summerlike drying would occur in these intermediate seasons. The first winter exposure indicated the rate at which the specimen gained moisture. And the second summer served a twofold purpose: (1) to show how fast the specimen would expel winter-gained moisture, and (2) to establish a drying rate after the introduction of liquid water (10 percent of specimen volume) simulating a leak (see Fig. 6-7).

From the results of these self-drying roof experiments, the NBS researchers derived a set of general rules, plus some mathematical criteria. First, the general rules:

1. Do not use a vapor retarder.

2. Use at least one material with water-absorptive capacity.

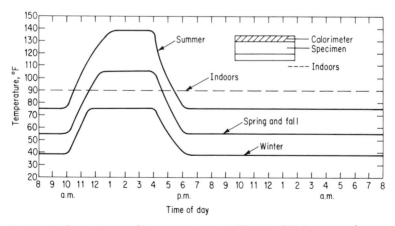

Fig. 6-6 With interior conditions constant at 90°F, 30% RH (corresponding to water-vapor pressure exerted at 75°F, 49% RH), the NBS test program varied "outdoor" temperature as shown by the curves for the four seasons. Daily temperature rises simulate membrane heat gain from solar radiation. (Graph from "The Effect. of Moisture on the Heat Transfer Performance of Insulated Flat-Roof Constructions," *Build. Sci. Ser. 37*, NSB, 1971, p. 48.)

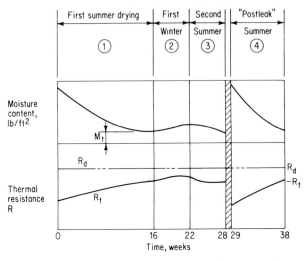

Fig. 6-7 The diagrams show laboratory performance of a test specimen satisfying NBS-suggested criteria for a self-drying roof system. (1) The first summer's 16-week drying removes substantial quantities of initially contained moisture. By definition, M_t = average evaporable moisture content (lb/ft²) after this 16-week summer drying. M_t must equal at most twice the hygroscopic moisture at 75°F, 90% RH. (2) During the 6-week "winter" exposure, average moisture gain must not exceed 0.05 lb/ft². (3) After the second summer (6 weeks), the system must have vented all moisture gained during the winter (i.e., moisture at end of second summer must not exceed M_t). (4) Leak-simulating water (10 percent by volume) at start of 28th week (following second summer) must be dissipated in a third summer exposure (9 weeks), with water again limited to M_t at end of the summer drying period. As a further condition, average thermal resistance R_t, calculated for 37 of the 38 test weeks (exclusive of the week following addition of the leak-simulating water), must equal or exceed 0.6 R_d (specimen's dry thermal resistance). Moreover, the final R_t measurement should at least equal the original R_t.

3. If more than one insulating material is used, locate the material with *lowest* water-vapor permeance directly under the membrane.

4. Do not use a highly water-permeable insulation alone.

5. Limit maximum thickness of highly absorptive insulating materials to 3 or 4 in.

6. Provide edge venting and/or membrane venting (to avoid rapid vapor-pressure buildup).

7. Limit maximum indoor winter dew point to 60°F (corresponding to 80°F, 50% RH).[1]

Before proceeding to the mathematical criteria for designing a self-drying roof, let us first discuss these general rules and their derivation from the experimental results.

1. Omission of the vapor retarder is of course central to the self-drying concept, which recognizes the high probability of periodic moisture invasion into the roofing system. For insulations installed dry and never wetted from a subsequent leak, a good, properly installed vapor retarder helps maintain dry insulation. But for roof insulations that are initially wet, or later wetted by a leak, the vapor retarder impedes moisture escape from the underside of the roof system.

2. A self-drying roof's insulation must have moisture-absorptive capacity to prevent leak water from dripping into the occupied space. Moisture-absorptive capacity also accommodates the much smaller quantity of condensed water vapor collected during the winter, when the vapor-pressure gradient perpetually promotes upward migration of water vapor from the occupied space.

3. When two insulating materials are used, the benefits of locating the insulation with the lesser permeance directly under the membrane are demonstrated by the NBS experimental results. Two test specimens comprised a 3-in.-thick lightweight perlite concrete and 1-in. polystyrene foam: specimen A with the 3-in. lightweight concrete (permeance = 4.5) on top; specimen B with the 1-in. polystyrene formboard (permeance = 1.2) on top. Specimen B successfully dried out within the 9 weeks allotted for self-drying after the simulated roof leak, whereas specimen A failed to dry within this allotted time. Specimen B also retained a much greater proportion of its dry thermal resistance: an average 89 vs. 62 percent for specimen A.

   Specimen B dried more efficiently than specimen A because the more permeable material's bottom location enabled it to vent accumulated moisture to the interior. With its polystyrene-board soffit acting like a partial vapor retarder, specimen A retained a greater proportion of its

---

[1]Powell and Robinson, op. cit., pp. 42ff.

accumulated moisture, 1.4 lb/ft², after the same exposure had reduced specimen B to a moisture equilibrium of 0.4 lb/ft², 70 percent less than specimen A's.

One especially well-performing specimen had 3-in. perlite concrete over a 1½-in. fiberglass formboard, an extremely permeable material, rated at 28 perm vs. 4.5 for the perlite concrete. This specimen rapidly regained its insulating value after its simulated roof leak. In addition to the highly permeable fiberglass formboard, the perlite concrete's low thermal resistance created a steep summer temperature gradient and a consequently steep downward vapor-pressure gradient that accelerated expulsion of absorbed moisture.

4. The recommendation against using a single, water-absorptive insulating material is a corollary of rule 3. An absorptive insulation used alone permits water concentration directly under the membrane, where it might expand voids at the membrane-insulation interface on freezing in cold weather.

5. The recommendation to limit the thickness of highly water-absorptive insulations to 3 or 4 in. springs from the poor drying exhibited by thicker, water-absorptive materials. A 6-in. layer of lightweight insulating concrete with high moisture content at the beginning of the test remained wet in its central region throughout all its exposures, including summer drying. Doubling the thickness of insulating concrete fill does not double its thermal resistance, substantiating proof of higher moisture retention in thicker sections.[1]

6. Edge and membrane (top-surface) venting offers no significant benefits for removing moisture. It also has little significant effect in restoring the insulation's thermal resistance. Breather-type vents installed on 11 specimens had little effect on moisture content and R values. This conclusion was substantiated by a calculation of the probable drying rate of a cellular concrete specimen from vent "breathing." This calculation assumed daily expulsion of air saturated at the higher mean insulation temperature and its replacement by dry air as the specimen cooled during simulated summer exposure. According to the calculation, it would take years to reduce moisture content of cellular concrete by only a few percent.[2]

[1]Ibid., p. 35.
[2]Ibid., pp. 12ff.

7. The recommendation to limit indoor air dew point to 60°F (corresponding to 80°F, 50% RH) should satisfy all normal occupancies in either winter or summer. For special occupancy—swimming pools, high-humidity manufacturing processes, and the like—the self-drying roof concept may be inappropriate.

The foregoing general rules provide chiefly qualitative guidance for designing self-drying roofs. In a further extension of these rules, the NBS researchers promulgated more quantitative criteria for self-drying design. Because of the many complex variables, however, they failed to formulate simple, practical engineering expressions that would predict thermal and self-drying roof performance from the physical properties and thicknesses of various roof components. Laboratory tests are required for new designs (defined as other than those listed in the Powell-Robinson study). And unfortunately, the NBS study omits from its tested specimen listings two popular deck materials: lightgage steel and plywood.

As previously indicated, the NBS self-drying tests yielded the following data:

- Time required for specimen to expel initial construction moisture, to approximate moisture equilibrium under simulated summer conditions

- Rate of moisture gain during winter exposure immediately following summer exposure

- Duration of second summer-exposure condition required to expel winter-gained moisture

- Self-drying performance (lb/ft^2 moisture removed) during second summer exposure following insulation wetting simulating roof leak (see Fig. 6-7).

Over the 38-week test period, the criteria for qualifying a test roof assembly as "self-drying" were

1. Average $R_t/R_d = 0.6$

   where R_t = average specimen thermal resistance over total

   38-week test period

   R_d = specimen's dry thermal resistance

   (For materials with only normal hygroscopic moisture, R_t/R_d should exceed 0.8.)

2. $M_t/M_c \leq 2$,

> where M_t = average evaporable moisture content (lb/ft²) after 16-week summer drying
>
> M_c = hygroscopic moisture content (lb/ft²) at 75°F, 90% RH

For any component, you compute M_c by the following formula:

$$M_c = d \frac{W_h - W_d}{W_d} \times \text{vol. (ft}^3)$$

> where d = material's dry density (lb/ft³)
>
> W_h = constant weight at 75°F, 90% RH
>
> W_d = constant weight after oven drying (215°F)

3. The maximum permissible average rate of regaining moisture under simulated winter test conditions = 0.05 lb/(ft²·week) (for a 6-week exposure period immediately following the first summer-exposure period). The specimen is considered at approximate moisture-content equilibrium during exposure testing when its weight change does not exceed 0.05 lb/(ft²·week) for two consecutive weeks.[1]

4. After the second summer exposure, the specimen should dry to a moisture content M_t no greater than that at the end of the first summer exposure. (This requirement is designed to prevent progressive, perennial moisture accumulation.)

5. After the leak-simulating water (10 percent by volume) is added, during the second summer, the self-drying time to achieve moisture-content equilibrium should not exceed 9 weeks. And during this 9-week period the specimen should regain its "preleak" R value.

6. This formula should hold:

$$2 \leq \frac{p_t}{M_c} \leq 200$$

> where p_t = permeance (perm) of roomside component
>
> M_c = hygroscopic moisture content (lb/ft²) at 75°F, 90% RH

The low limit of 2 in the previous equation is designed to ensure adequate permeance to expel water vapor from the insulation during

---

[1] Ibid., p. 44.

summer drying. On the other hand, the high limit of 200 is designed to limit moisture accumulation during the winter moisture-gaining season, when vapor migrates upward from the warm interior.

Examples of tested specimens falling below the lower limit included a 7-in. vermiculite concrete fill placed over cement asbestos board (p_t/M_c = 0.024) and 2½-in. wood fiberboard over expanded shale concrete (p_t/M_c = 1.4). No tested specimen failed to satisfy the higher limit. A 2-in. gypsum concrete over 1½-in. fiberglass formboard came fairly close, at 153. The porous fiberglass formboard would admit large quantities of water vapor into the roof system.

To derive the full benefits of a self-drying roof requires attention to field conditions that can cause excessive moisture in even a well-designed, self-drying roof system. Powell cites a case history illustrating field moisture problems. The roof system comprised an aggregate-surfaced built-up membrane on poured gypsum deck over fiberglass formboard supported by open web steel joists. In the laboratory, this system ranked as one of the best tested assemblies, requiring only 5 weeks drying time to dissipate initial moisture, compared with 16 or more weeks required by half the tested assemblies.[1] During this roof's first spring, however, a layer of water appeared near the lower surface of the fiberglass formboard. The explanation was as follows: After pouring the gypsum concrete during the previous winter, the contractor used portable, fuel-burning heaters within a weather enclosure during completion of the building's finishing operations. Initial moisture, plus moisture generated within the weather-enclosed building, accumulated within the built-up roof system. With the advent of spring, solar radiation, reversing the upward vapor-pressure gradient maintained during the winter by the heating of the humid enclosure, caused a downward migration of moisture to the lower surface of the fiberglass formboard. Areas with unventilated suspended ceilings suffered water damage.

Ventilation of the interior in combination with solar heating soon dissipated the accumulated moisture. But all moisture damage could have been completely prevented by simply ventilating the enclosed space during winter construction operations.[2]

The tremendous quantities of moisture generated by hydrating construction materials cured by fuel-burning heaters are often overlooked. Combustion in an oil or propane heater produces more water as a by-product than the weight of the consumed fuel—1 gal water for each gallon of heating oil and 30 gal water for a 200-lb tank of propane. Cementitious materials release vast quantities of water vapor—roughly 1 ton water for each 1000 ft² wall ceiling plaster or 4-in.-thick structural

[1]Ibid., table 4, p. 31.
[2]Ibid., p. 40.

concrete slab.[1] Ventilation of enclosures where such materials are curing is absolutely essential to prevent excessive wetting of the roof system.

Unfortunately, as noted, the NBS researchers found the problem of self-drying roof design too complex to be solved by formula. Laboratory tests of unlisted systems are required to assess self-drying performance during the roof's service life. Slight perennially accumulating moisture gains could build up to hazardous levels, destroying the roof's thermal performance and ending its useful service life far short of the generally demanded 20 years. These sensitive indexes cannot be reliably computed; they require experimental determination. Thus there is an obvious need for further research testing of self-drying roof systems. These tests should include steel, plywood, and precast concrete decks, all omitted from the NBS test program. Despite these omissions, Building Science Series 37 can be profitably studied for the principles of self-drying roof design.

THEORY OF VAPOR MIGRATION

Water vapor infiltrates a roof through two mechanisms: *diffusion* and *air leakage*. Compared with heat transmission, diffusion corresponds roughly to conduction; air leakage corresponds precisely to convection.

DIFFUSION

Diffusion can be explained by the kinetic theory of gases—a theory whose basic principles account for changes of state from liquids (or solids) to gases and the accompanying energy changes. In a sealed container of air with water in the bottom, at any temperature above absolute zero, some liquid (or solid) molecules tear loose from the surface and escape into the air above. If constant pressure is maintained within the sealed container, then for each temperature a different quantity of water vapor (gaseous water molecules) escapes from the liquid and diffuses through the air.

So long as the water supply holds out, the liquid surface reaches equilibrium soon after the mixture reaches a stable temperature, with as many water molecules plunging back into the liquid as escaping from its surface. When equilibrium is reached, the atmosphere is "saturated"; that is, the air-vapor mixture contains as much water vapor as it can retain at that temperature. The graph in Fig. 6-2 displays this capacity of air at constant pressure to "hold" varying quantities of water

[1]Tyler S. Rogers, *Thermal Design of Buildings*, John Wiley & Sons, Inc., New York, 1964, p. 42.

vapor at different temperatures. At normal atmospheric pressure of 14.7 psi and temperature of $-10°F$, a cubic foot of air can hold 0.3 grain of water vapor; at 70°F, it can hold 27 times that amount, or 8.1 grain.

So long as the liquid water holds out, the air-vapor mixture remains saturated. However, if all the water evaporates and the temperature rises, the water-vapor content of the air mixture represents something less than the saturation limit. The ratio of the volume of water vapor actually diffused through the mixture to the volume of water vapor in saturated condition is the *relative humidity* (RH). But because vapor pressure is of greatest interest to the roof-system designer, the most meaningful definition for our purposes is: Relative humidity is the ratio of actual vapor pressure to the vapor pressure of a vapor-saturated-air-vapor mixture at constant temperature and pressure.

When the air-vapor mixture is saturated with vapor, the air temperature and the dew point are the same. But whenever the relative humidity is less than 100%, the dew point is of course lower than the air temperature. For example, at 70°F, 10% RH, the dew point is 13°F.

Besides dew-point temperature, another condition is necessary to produce condensation. Water vapor flowing through permeable materials or air spaces passes through the theoretical dew-point plane into colder regions without condensing. (Condensation may actually occur, but it is followed by instantaneous reevaporation.) Migrating water vapor normally condenses only when it reaches an obstructive surface that stops or drastically retards its movement. This phenomenon is attributable to the kinetic energy of flow, which maintains the water vapor in a gaseous state at a super-cooled temperature until it collides with a sub-dew-point obstruction.[1]

Another phenomenon, associated with the diffusion of water vapor throughout air, explains the penetration of water vapor into built-up roofing. The diffused water vapor is actually low-pressure steam, possessing its high latent heat energy. Like any other diffused gas, water vapor mixed with air exerts a pressure independent of the pressure exerted by the gases constituting the air (oxygen, nitrogen, carbon dioxide, and so on). In accordance with Dalton's law of partial pressure, the partial vapor pressure is directly proportional to the volume occupied by the water vapor.

If water vapor is supplied to a space, the vapor pressure rises rapidly with increasing temperature. Absolutely dry air (i.e., air with no suspended water vapor whatever) confined in a space of constant volume gains 2.2 psi (from standard atmospheric pressure of 14.7 to 16.9 psi) under a temperature rise from 70 to 150°F. (In accordance with Charles' law, the pressure increases in direct proportion with the rise in *absolute*

[1]Ibid., p. 52.

temperature—from 530 to 610°R.) If to each cubic foot of dry air we add 76 grain of water vapor, the total pressure increases 5.6 psi under the same 80°F temperature rise. The water vapor exerts a partial pressure of 3.4 psi, or 490 psf, more than half again as much as the partial-pressure increase in the heated dry air. Thus warm, humid inside air exerts unbalanced vapor pressure impelling vapor migration *from* the warm *toward* the cold side of the roof (see Fig. 6-7).

A curious consequence of the law of partial pressures is the water vapor's tendency to flow from a region of high vapor pressure toward a region of low vapor pressure, *regardless of the total atmospheric pressure.* Thus even if the outside air pressure exceeds the inside air pressure, the water vapor tends to diffuse through the built-up roofing system from a warm, humid interior toward a cold, dry exterior. The roof merely impedes the natural tendency of both the air and water vapor to diffuse throughout the atmosphere in equal proportions. In our vast atmospheric system, buildings are a mere local accident, and in seeking to achieve a perfectly proportioned mixture, the gaseous molecules in the atmosphere may move through walls and roofs in opposite directions—with inward-bound air molecules passing outward-bound water-vapor molecules—as they seek their own partial-pressure level.

As an example of the pressure differentials encountered, consider a heated building interior within the normal range of heating and humidification—say, at 70°F and 40% RH. Outside temperature is −10°F, 100% RH (winter design conditions for Chicago). Under these conditions, 1 lb indoor air contains 0.0063 lb (44 grain) water vapor and 1 lb outdoor air contains 0.0005 lb (about 3 grain) water vapor. (These data are obtained from a psychrometric table or chart.) According to Dalton's law, the partial pressure of the water vapor is directly proportional to the percent by *volume* (not weight) of vapor. Because the air–water-vapor mixture closely approximates an "ideal" gas, the vapor volume varies inversely with the ratio of the molecular weights of water vapor (18.02) and the air (28.96). Assuming the pressure of the air-vapor mixture, p_m = standard atmospheric pressure of 14.70 psi, we compute the vapor pressure p_v as a percentage of the air pressure p_a as follows:

$$p_v = \frac{28.96 \text{ (molecular weight of air)}}{18.02 \text{ (molecular weight of vapor)}} \times 0.0063 = 0.01 p_a$$

$$p_m = p_v + p_a = 14.70 \text{ psi}$$

$$= 1.01 p_a$$

$$p_a = \frac{14.70}{1.01} = 14.5 \text{ psi}$$

$$p_v = 0.01 \times 14.5 = 0.145 \text{ psi} \times 144 = 21.1 \text{ psf}$$

We can thus assume for this example a vapor migration attributable to diffusion from the warm interior toward the cold exterior—a migration impelled under a pressure imbalance of 19.4 psf (obtained by subtracting the exterior vapor pressure of 1.7 psf from the 21.1-psf interior vapor pressure). As explained earlier, this vapor-pressure imbalance induces an inside-out diffusion regardless of any possible atmospheric pressure imbalance forcing dry air toward the interior. In addition to the partial-pressure difference, the rate of diffusion depends on the length of the flow path and of course the permeance of the media through which it passes (deck, vapor retarder, insulation, and so forth).

Air Leakage

Recent research demonstrates that air leakage, which conveys water vapor with it, can be a much greater factor than diffusion in accounting for vapor migration. If the components of the built-up roofing system—deck, vapor barrier, insulation, and built-up membrane—were perfect unpunctured surfaces, then diffusion would be the sole means of vapor migration. But cracks, pinholes, and joint openings in deck, insulation, and flashing let air pass from the high- to the low-pressure side of the roof.

Under the normally small pressure differences existing above and below the roof, the volume of air moved through these small openings is insignificant so far as it affects building heating and ventilation. But a relatively small volume of air can transport a troublesome volume of water vapor. At any point in the path where this migrating water vapor is stopped by a surface at or below the dew point, it condenses.

Several factors make the existence of atmospheric pressure differentials between inside and outside almost inevitable. The major factors are temperature differences and wind. Pressure differentials created by mechanical ventilation or exhaust systems are normally less important.

Chimney Effect In a heated building, the chimney effect produces higher pressures and consequent exfiltration in the upper part of the building and lower pressures and consequent infiltration in the lower part of the building. (Like a chimney, the heated building sets up a convection current, with cold air moving into the low-pressure space constantly vacated by the expanding, heated air, which rises and raises the pressure in the upper parts.) In a cooled building, the opposite occurs, with infiltration occurring at the upper levels and exfiltration at lower levels as the cooled, dense lower air presses out and sets up a downward convection current.

Wind Suction   Wind creates a positive static pressure on a windward wall in proportion to the wind's kinetic energy (velocity squared). In conformance with the Bernoulli principle, the wind across a level or slightly pitched roof produces an uplift varying from a factor of 3.0 p_s or more at the leading edge to 0.2 p_s in some other region of the roof (p_s is the static pressure against a vertical surface).

As a hypothetical example of how pressure difference is created above and below the roof, consider a 400-ft-high building with a steady 20-mph wind blowing across the roof and a temperature differential of 60°F (70°F inside, 10°F outside).

At any level in the building, the pressure differential accompanying temperature difference can be computed from the following formula:

$$p_c = 7.6h \left(\frac{1}{T_i + 460} - \frac{1}{T_o + 460} \right)$$

$$= 7.6 \times 200 \left(\frac{1}{70 + 460} - \frac{1}{10 + 460} \right) = -0.365 \text{ in. water}$$

where p_c = theoretical pressure (in. water) attributable to chimney effect

h = distance (ft) from "neutral zone" (level of equal inside-outside pressure, estimated at midheight for multistory building without air-sealed floors)

T_o = outside temperature (°F)
T_i = inside temperature (°F)

Because static pressure for a 20-mph wind equals −0.193 in. water, an estimate of 0.5 p_s (for average roof suction) yields a static pressure of about −0.1 in. water. Adding the two components of the pressure differential and multiplying by a conversion factor of 5.2 yields about 2.4 psf exfiltration pressure *upward* through the roof.

The practice of mechanically pressurizing airconditioned buildings through a 10 to 20 percent excess of fresh supply air over exhaust air can add another small increment of pressure imbalance to that produced by chimney effect and wind, further aggravating the problem of vapor migration. In arriving at a decision, the designer should weigh the benefits of mechanical pressurization against its liabilities. In humidified buildings the designer may even want to provide a small interior suction to reduce air leakage instead of deliberately increasing it.

Sample Vapor-Retarder Calculation

Winter Condition   To illustrate a vapor-migration problem for the normal (winter) condition, assume a roof construction as shown in the cross section and the following data: outside temperature T_o = 0°F, 90% RH, and inside temperature T_i = 70°F, 35% RH (see Fig. 6-8).

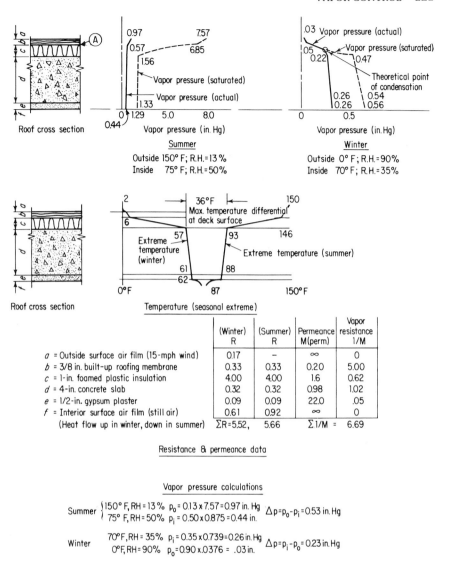

Fig. 6-8 Vapor-retarder calculations, although imprecise, indicate the difficulty of solving the moisture problem with a vapor retarder.

Because air-leakage computations can be made only for known, measured cracks and other openings—a clearly impossible task—the following computations consider only diffusion. It is naïve to assume great accuracy in such computations, so pressures (in in.Hg) are carried only to the second decimal place. Due to many imponderables, vapor-migration computations are far less accurate than ordinary structural or mechanical design computations for, say, axial column stresses or heat losses. Thus vapor-migration computations are more an index to a problem than a precise design technique. In general, the designer must assume that at its worst vapor infiltration is greater than the computations indicate. As a mitigating factor, the worst condition is normally temporary.

The designer first tabulates thermal resistance (R values), perm values (M), and vapor resistances (perm-value reciprocals) for the various roof components and air layers (see Fig. 6-8). For a thermal steady state, the temperature loss upward through the roof cross section is proportional to the sum of the R values of the components between the interior and a given plane, divided by the total sum of the R values. Thus

$$T_x = T_i - \frac{\Sigma R_x}{\Sigma R}(T_i - T_o)$$

where T_x = temperature at plane X
$\quad \Sigma R_x$ = sum of thermal resistances from interior to plane X
$\quad \Sigma R$ = total thermal resistances

The vapor-pressure gradient through the roof cross section can be computed in similar fashion. (It is analogous to drop in hydrostatic head in a series of canal locks.) Thus

$$p_x = p_i - \frac{\Sigma(1/M)_x}{\Sigma(1/M)}(p_i - p_o)$$

where p_x = vapor pressure at plane X
$\quad p_i$ = interior vapor pressure
$\quad p_o$ = exterior vapor pressure
$\quad \Sigma(1/M_x)$ = sum of vapor resistances from interior to plane X
$\quad \Sigma(1/M)$ = total vapor resistances

The critical plane is obviously A, the interface between the insulation and the built-up membrane (see Fig. 4-6). Because it lies on the far side of the relatively permeable insulation, it represents the worst possible combination of low temperature and high humidity and thus poses the greatest risk for condensation.

$$T_a = 70 - \left(\frac{5.02}{5.52} \times 70\right) = 6°F$$

$$p_a = 0.26 - \left(\frac{1.69}{6.69} \times 0.027\right) = 0.19 \text{ in.Hg}$$

We can now calculate the vapor flow to and from plane A. However, when we check the saturated vapor pressure at A, we find that it is only 0.05 in.Hg, which is far below the calculated 0.19 in.Hg. Thus we know that condensation must take place at (or possibly below) plane A.

To calculate the vapor flow to plane A, we divide the pressure differential by the vapor resistance:

$$\text{Vapor flow } to \text{ plane } A = \frac{0.26 - 0.05}{1.69} = 0.12 \text{ grain/(ft}^2 \cdot \text{h)}$$

$$\text{Vapor flow } from \text{ plane } A = \frac{0.05 - 0.03}{5} = 0.004 \text{ grain/(ft}^2 \cdot \text{h)}$$

These grains of water vapor are in transit through the roof, representing a more or less "steady state" of vapor migration, which varies with changing temperature and humidity. Because the air-vapor mixture at plane A is saturated, all the migrating water vapor must condense. Thus we can compute the condensation rate merely as the difference between the vapor flow to plane A and the vapor flow from plane A— 0.12 grain/(f$^2 \cdot$h).

To prevent condensation at plane A (the insulation-membrane interface), a vapor retarder must add sufficient vapor resistance on the high-pressure (warm side) of plane A to reduce the rate of vapor migration at plane A by at least 0.12 grain/(ft$^2 \cdot$h). Because the condensation rate (the excess of vapor flow to plane A over the vapor flow $from$ plane A) must equal 0, we set the vapor flow to plane A equal to the vapor flow from plane A.

$$\text{Vapor flow to plane } A = \frac{0.26 - 0.05}{1.69 + X} = 0.004 \text{ grain/(ft}^2 \cdot \text{h)}$$

where X = required vapor resistance added by vapor barrier
$$0.21 = 0.004 \ (1.69 + X)$$
$$X = 53.5$$

$$\text{Required perm rating} = \frac{1}{X} = 0.02$$

To ensure this virtually 0 perm rating is, to say the least, difficult.

Summer Condition   In summer the problem of vapor migration is normally much less serious than in winter. Consider the same roof construction with outside temperature 100°F, 50% RH; airconditioned interior at 75°F, 50% RH (see Fig. 6-8). Assume further that the roof surface heats up to 150°F, reducing the relative humidity to 14%.

Computation of temperatures at the various planes between adjacent roof components indicates no planes at which condensation will occur, at least because of vapor migration resulting from diffusion. At plane A, the critical membrane-insulation junction under winter conditions, there is no problem if the built-up membrane is in sufficiently good condition to prevent extensive air leakage. The membrane's low insulating value combines with its low permeance (good enough to qualify the membrane as a good vapor barrier), to produce a high temperature (146°F) with a low relative humidity (8%) at plane *A*. Because the saturation vapor pressure remains well above the computed vapor pressure through the entire roof cross section—from the built-up membrane through the deck and the ceiling plaster—there is no problem of vapor-migration diffusion.

ALERTS

Design

1. As a basis for deciding whether to use a vapor retarder, calculate the location of the dew point and the rate of vapor migration under the worst winter condition (see example at end of chapter). If vapor-retarder location is below dew point, do not specify it.

2. A vapor retarder on a roof destined for future penetrations is highly vulnerable to damage and likely to fail.

3. If in doubt, do not specify a vapor retarder. Use a vapor retarder only if a study of your conditions indicates a positive need for it. Normally, it should not be necessary unless the interior relative humidity exceeds 40% and January temperatures average less than 35°F. (For typical winter relative humidities under different occupancies, see Table 6-3.)

4. If unhumidified, exterior air is drawn into the building, a vapor retarder is normally not required, unless the interior relative humidity is 60% or more.

5. In conjunction with a vapor retarder, specify an insulation that under prolonged exposure to moisture will (a) permit lateral transfer of vapor for venting pressure relief and (b) retain sufficient strength despite the absorbed water. Specify a water-resistant method of bonding the insulation to the vapor barrier.

6. Never specify a vapor retarder between a poured structural deck and poured insulating concrete fill. (It will seal in moisture.)

7. Do not specify a vapor retarder in the roof over an unheated interior.

8. Consider an impervious, closed-cell board insulation in place of a vapor barrier.

9. Require flashing and enveloping of vapor retarder at all roof penetrations.

10. Check the vapor retarder's ability to take nailing or other anchorage punctures and still maintain a satisfactory perm rating.

11. Check a vapor retarder for its effect on the roof assembly's fire rating (see Chapter 15, "Fire Resistance").

12. For a vapor retarder hot-mopped to a deck, specify ASTM Type II (flat) or III (steep) asphalt. Specify rosin-sized paper, felt, or tape over joints in wood sheathing, plywood, or other prefabricated units to be nailed.

Field

1. Do not use the vapor retarder as a temporary roof.

2. Where cellular foamed glass insulation is substituted for a vapor retarder, place board with tight joints.

3. Prohibit installation of light, flexible plastic vapor retarders on windy days.

4. Check steel decks for deflection, transverse dishing, or other irregularities that could prevent the vapor barrier from bonding to cold-applied adhesive.

5. Check spreader-applied ribbons of cold adhesive for proper consistency, thickness, and height.

SEVEN

Elements of the Built-Up Membrane

MEMBRANE PRINCIPLES

The built-up roof membrane is the roof system's weatherproofing component. It comprises two basic elements, felts and bitumen combined into a laminate, usually surfaced with a third element: a mineral aggregate surfacing embedded in a bituminous top or "flood" coat. Properly designed and applied, the felt-bitumen laminate forms a flexible roof cover with sufficient strength to resist normal expansion and contraction forces.

Bitumen is the waterproofing agent and thus the most important membrane element. If it had sufficient fire resistance, strength, rigidity, and weathering durability, the membrane could theoretically (although not practicably) be fabricated entirely of bitumen.

Felts stabilize and reinforce the membrane, like steel reinforcement in a concrete slab, providing about 90 percent of its tensile strength. The felt fibers restrain the bitumen from flowing in hot weather and resist contraction stresses and cracking in winter (see Fig. 7-1). Felts also isolate the different layers of bituminous waterproofing, which helps the mopper apply the bitumen uniformly.

Aggregate surfacing protects the bitumen from damaging solar radiation. Through a combination of heat and photochemical oxidation, sun rays accelerate bitumen embrittlement and cracking. Mineral aggregate surfacing forms a fire-resistive skin that prevents flame spread and protects the membrane from abrasion caused by rain, wind, and foot traffic. It can help resist the corrosion from acid mists condensing on the roof in industrial areas. An aggregate surfacing also acts as ballast, offering some wind-uplift resistance, and as a shield against the impact of hailstones.

Aggregate surfacing also makes feasible the pouring of a heavy surface flood coat of bitumen. The closely massed aggregate forms tiny dams that retard the lateral flow of heated bitumen and allow it to con-

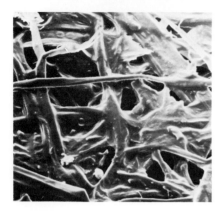

Fig. 7-1 Saturated, No. 15 asphalt-saturated organic felt, magnified 240 times by scanning electron photomicrograph, shows cellulose fibers only lightly coated with asphalt, with interfiber voids allowing easy water access when surface is exposed. (Tremco.)

geal, to a depth about three times that of a mopped interply layer, 60 lb/square of asphalt for the flood coat vs. 20 lb/square of interply mopping.

When properly designed and constructed, an *aggregate-surfaced membrane* is *water-resistant*; it can tolerate minor local ponding lasting 1 or 2 days. A *smooth-surfaced membrane* or *mineral-surfaced cap sheet* (or roll roofing) membrane is essentially *water-shedding*, requiring steeper slope for faster drainage (see Table 7-1). Lacking equivalently surfaced protection, these sloped roofs normally require recoating of their initial protective coating or replacement much sooner than aggregate-surfaced roofs. (For more detailed discussion, see the "Surfacing" section later in the chapter.)

As with surfacing, the number of felt plies affects membrane durability. Each felt ply adds a layer of waterproofing bitumen adhesive to the multi-ply membrane, providing an additional line of defense. A large number of felt plies also tends to equalize membrane properties throughout the membrane and increase tensile strength proportionately. The roofing industry has traditionally assigned 5 years' anticipated service life to each felt ply, that is, 20-year life for a four-ply membrane, 15-year life for a three-ply membrane, and so on.

This crude rule has some validity, judged by a statistical study conducted by Simpson, Gumpertz & Heger in 1977–1978 and reported by Cash.[1] Calculated life expectancies for four-ply membranes were double those calculated for two-ply membranes, with a three-ply membrane in between, according to an expert group of 104 respondents comprising roofers, consultants, materials manufacturers' representatives, government researchers, and so forth. Incidentally, no type of

[1]Carl G. Cash, "Durability of Bituminous Builtup Membranes," *Durability of Building Materials and Components*, ASTM STP691, 1980, pp. 741–755.

built-up membrane was collectively rated at more than 50 percent probability of lasting 20 years.

MEMBRANE MATERIALS

Roofing Bitumens

Bitumens are basically heavy, black or very dark brown hydrocarbons, divided into three major classes: natural asphalt, petroleum asphalt, and coal tar pitch. Only petroleum asphalt and coal tar pitch concern the designer of built-up roof systems.

Despite several significant physical and chemical differences, both asphalt and coal tar pitch have the following physical properties:

- Excellent resistance to water penetration and extremely low water absorptivity

- Durability under prolonged exposure to weather

- Good internal cohesion and adhesion (e.g., to roofing felts and insulation)

- Thermoplasticity (i.e., reversible, temperature-produced changes from semielastic solid to viscous fluid)

These four properties contribute to the long histories of successful roofing performance of both roof bitumens—well over a century for coal pitch and nearly a century for petroleum asphalt.

The most basic functional distinction between the two roofing bitumens is that asphalt comes in a much greater range of viscosities, which make it suitable for slopes up to 6 in. (see Table 7-2). On the other hand,

Table 7-1 Minimum Recommended Roof Slopes

Membrane Type	Minimum slope, in./ft	Notes
Aggregate-surfaced	¼	
Smooth-surfaced	½	¼-in. slope approved by manufacturers of some glass-fiber felt, smooth-surfaced membranes
Mineral-surfaced roll roofing	1	With concealed nails, minimum 3-in. top lap, 19-in. selvage, double-coverage roll
Mineral-surfaced roll roofing	2	With exposed nails

Table 7-2 Required Roofing Bitumen Properties

Property	Asphalt* Type I		Type II		Type III		Type IV		Coal Tar Pitch† Type I		Type III	
	Min.	Max.	Min.	Max.	Min.	Max.	Min.	Max.	Min.	Max.	Min.	Max.
Roof slope‡ (in./ft)	¼	½	½	2	1	3	1	3⁺	¼	¼	¼	¼
Softening point⁴, °F (°C)	135(57)	151(66)	158(70)	176(80)	185(85)	205(96)	210(99)	225(107)	126(52)	140(60)	133(56)	147(64)
Flash point, °F (°C)	437(225)		437(225)		437(225)		437(225)		248(120)		248(120)	
Penetration (tenths of mm):												
32°F (0°C)	3		6		6		6					
77°F (25°C)	18	60	18	40	15	35	12	25				
115°F (46°C)	90	180		100		90		75				
Ductility, 77°F, cm	10		3		2.5		1.5					
Specific gravity¶, 25/25°C									1.22	1.34	1.22	1.34

*Per ASTM D312-78.
†Per ASTM 450-78.
‡Roof-slope limits range upward from generally recommended lower limit of ¼ in./ft.
⁴Asphalt softening point per ASTM D2398; coal tar pitch softening point per ASTM D36. If ASTM D2398 test is used instead of D36, softening-point values for coal tar pitch are increased by about 2°C (4°F).
¶Asphalt specific gravity, although not limited by ASTM, generally runs around 1.03 for roofing asphalt.

coal tar pitch comes in only two roofing grades (plus a waterproofing grade), with maximum viscosity about the same as that of so-called "dead-level" (ASTM Type I) asphalt. It is generally limited to ½-in. maximum slope.

Asphalt is the dense, "bottom-of-the-barrel" residue left from petroleum distillation. The asphalt content of crude petroleum varies from zero to more than half. Moreover, asphalts from different sources may vary greatly, so an asphalt-producing refinery must carefully select its crude oils to ensure a sufficient quantity of satisfactory asphalt.

The heavy asphalt residue left after distillation of the crude's more volatile constituents requires further processing to qualify as roofing asphalt. The raw asphalt "flux" is heated to about 500°F. Air (or pure oxygen), blown through a perforated, hollow pipe pin wheel, bubbles through the hot liquid asphalt (see Fig. 7-2). The longer the blowing continues, the tougher and more viscous (i.e., the less fluid, or "steeper") the asphalt becomes, and the more suitable it is for roofs of steepening slope. The same flux can produce the whole range of roofing asphalts—from Type I (dead-level) to Type IV (special steep), a softening-point range of 135°F (minimum for dead-level asphalt) to 225°F (maximum for special steep).

Chemically, the accelerated oxidation process in the blowstill lengthens the long, chainlike hydrocarbon molecules by dehydrogenation, driving off hydrogen atoms in gaseous water molecules, length-

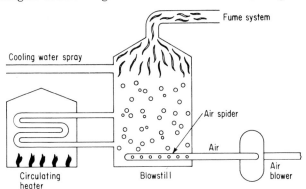

Fig. 7-2 A single charge of asphalt in the blowstill can produce all four types of roofing asphalt, ASTM Types I to IV, with constantly rising viscosity and softening points, from 135°F (minimum for Type I) to 225°F (maximum for Type IV), determined by time in the blowstill at temperatures of 325°F and higher. Oxygen, bubbled through the liquid asphalt by the revolving air spider, reacts with hydrogen in the asphalt to form water vapor. This dehydrogenation process links smaller units into longer, carbon chain molecules. (National Roofing Contractors Association.)

ening the carbon chains, and also distilling some lighter hydrocarbon molecules (see Fig. 7-3). Continued blowing creates a gel structure in the asphalt, and the growing number of lengthened hydrocarbon molecules (asphaltenes) accompanying continued blowing makes the asphalt stronger, tougher, and more elastic, all desirable qualities in a roofing asphalt. Compared with the less oxidized asphalts with a lower softening point, the steeper-blown asphalts are less temperature-susceptible; it takes a greater temperature range to change them from brittle solids to viscous fluids, or vice versa.

But in roofing asphalts, the good qualities become inextricably associated with some bad. Compared with asphalt of a lower softening point, steep asphalt has the following liabilities:

- Less durability

- Less water repellence

- Higher probability of overweight interply moppings with blister-forming voids because of the greater difficulty in applying the moppings properly

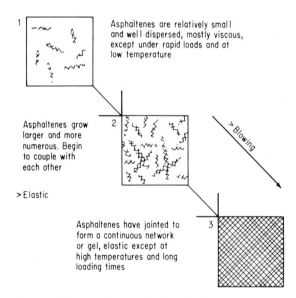

1 Asphaltenes are relatively small and well dispersed, mostly viscous, except under rapid loads and at low temperature

Asphaltenes grow larger and more numerous. Begin to couple with each other

2

> Blowing

> Elastic

Asphaltenes have jointed to form a continuous network or gel, elastic except at high temperatures and long loading times

3

Fig. 7-3   Hydrocarbon molecules lengthen into so-called asphaltenes as blowing process continues, producing a more viscous, elastic asphalt. (National Roofing Contractors Association.)

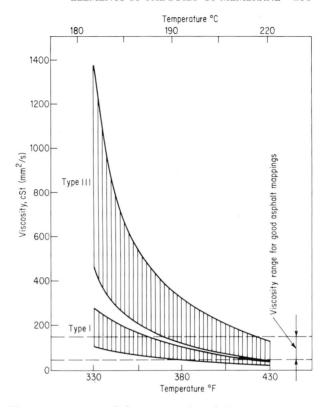

Fig. 7-4 Type III asphalt requires substantially higher temperatures (roughly 50° higher) than Type I asphalt to attain viscosities with the desired 50- to 150 cSt range. And as shown by the curves, Type III asphalts from different sources exhibit a greater range of viscosities at the same temperature than Type I asphalt. (National Bureau of Standards.)

This greater difficulty in getting continuous, void-free interply moppings is the major drawback of steep (Type III) asphalt compared with dead-level (Type I) asphalt. Here is why steep asphalt poses a more difficult field-application problem. National Bureau of Standards (NBS) researchers have associated a 50- to 150-cSt viscosity range with proper mopping fluidity to produce a uniform, void-free interply asphalt film of 15 to 20 lb/square weight.[1] To maintain hot asphalt within this viscosity range requires a minimum 390°F *application* temperature for Type III asphalt, minimum 345°F for Type I asphalt (see Fig. 7-4).

[1]W. J. Rossiter and R. G. Mathey, "The Viscosities of Roofing Asphalts at Application Temperatures," *Build. Sci. Ser.* 92, NBS, December 1976, p. 14.

Because hot asphalt generally cools from 500 to 390°F in about half the time it takes to cool from 500 to 345°F, the timing of the asphalt-mopping, felt-laying operation is much more critical for Type III than for Type I asphalt. At lower initial temperature—say, 450°F—the asphalt cools to 395°F in less than half the time it takes to cool to 345°F (see Fig. 7-5). Even if Type III asphalt is heated 50°F hotter than Type I asphalt, it normally cools to the equiviscous temperature (EVT—the temperature required to produce the theoretically ideal 125-cSt viscosity for mopping asphalt) in far less time than Type I asphalt because cooling rate increases exponentially with temperature difference between hot asphalt and ambient air.

Like minimum temperature for proper mopping viscosity, minimum temperature for good adhesion of felts is lower for Type I asphalt than for Type III, apparently by the same 45 to 50°F range that differentiates its softening point and EVT range from Type III asphalt's range. Minimum surface temperature (as detected by a hand-held, infrared thermal detector) should not fall below 250°F for Type I asphalt or coal tar pitch or below 300°F for Type III asphalt, according to roofing specialist consultant Gerald B. Curtis, Westmont, New Jersey. Based on extensive field experience, these temperatures are absolute minima. NBS is currently conducting field-based research to establish safe minimum temperatures for interply adhesion for different bitumens. Until such definitive research is available, the foregoing temperatures seem reasonable.

An even greater temperature difference than the 50°F figure just cited is indicated by Johns-Manville research into the effect of temperature on the mopping characteristics of different asphalts. Testing six different Type I and III asphalts, the Johns-Manville researchers required 450°F to get "good" mopping characteristics for all six Type III tested specimens, but only 375°F, 75°F less, to get similar good results with Type I asphalt. (The average weight of these good moppings of both asphalt types was 20 lb/square.) Moreover, Type I asphalt moppings were rated generally better at 350°F than Type III asphalt moppings at 425°F.[1] The Johns-Manville researchers also noted the greater safety factor, or spread, between the EVT and flash points for Type I asphalts compared with the narrower safety factor with Type III asphalts: an average 195°F spread for 12 tested Type I asphalts vs. 124°F for 13 tested Type III asphalts.

Overweight moppings with blister-originating voids and greater vulnerability to thermal-contraction stress are thus more likely to occur

---

[1]Richard Ducy, "Equiviscous Temperature of Roofing Asphalts," *RSI Magazine*, February 1978, pp. 66, 67 (Tables 4 and 5).

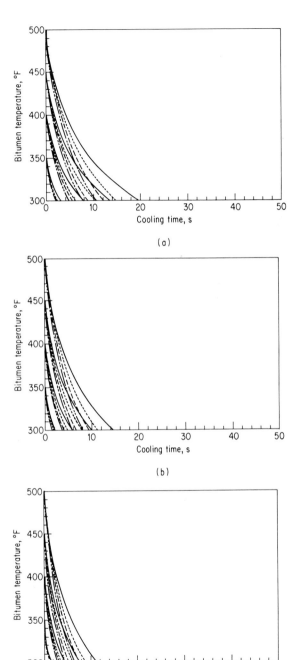

Fig. 7-5 The computer-calculated curves, developed by NBS researchers, show cooling rates for asphalt mopped to felt with an underlying plywood substrate. Air, felt, and plywood temperatures are (a) 120°F, (b) 70°F, and (c) 20°F. Wind speeds are shown as 0 mph for solid lines, 10 mph for dotted lines, 20 mph for dashed lines, and 30 mph for dashed/dotted lines. Note the tremendous impact of wind speed, a more important factor than ambient temperature.

with Type III than with Type I asphalt. Corroborating field evidence from a Johns-Manville research report indicates that 85 percent of the blistered low-slope roofs and 90 percent of the split low-slope roofs analyzed by laboratory test cuts had Type III asphalt interply moppings. The validity of this statistical evidence depends, of course, on the use of Type III asphalt in significantly lower percentages than the 85 to 90 percent range on the problem roofs. It appears to be a reasonable assumption.

Thus, as a general rule, specify the softest bitumen consistent with roof slope. The disadvantage of steeper asphalts normally outweighs their benefits. Only in special circumstances—for example, in a severely cold climate where asphalt embrittlement is a problem, or in an extremely hot climate like Phoenix's, where liquification can disrupt roofing application—should you consider a steeper asphalt than required by slope. And when it is specified for interply moppings, steeper asphalt requires tighter field controls to prevent premature congealing before the felts are rolled in.

Type II asphalt has been proposed by Johns-Manville researchers as a satisfactory compromise, multipurpose asphalt that solves the "two-asphalt" problem: the need for two kettles or tankers to supply Type III asphalt for adhering insulation, base sheet, and flashing and Type I asphalt for membrane moppings and flood coat. Some roofers solve this practical problem by using Type III asphalt throughout the entire roof system, even for a dead-level roof's flood coat. As noted, such roofs run higher risks of failure.

Where this practical obstacle looms large, Type II asphalt can be used for everything except flood coats on poorly drained roofs and possibly for adhering flashing. With its 158 to 176°F softening-point range, Type II asphalt can serve as an insulation adhesive without Type I asphalt's threat of bleeding through deck joints. And with its lower EVT, Type II asphalt reduces the threat of interply voids and generally poor adhesion posed by the faster congealing Type III asphalt. These conclusions spring from field investigations reported by the Johns-Manville researchers.[1]

Coal tar pitch is a jet black substance, denser and more uniform than asphalt. Normally recommended only for slopes of ½ in. or less, coal tar pitch comes in an acceptable softening-point range of 126 to 147°F, compared with asphalt's 135 to 225°F, minimum for Type I to maximum for Type IV.

Coal tar pitch is a by-product of the so-called "destructive distillation" of bituminous coal in the manufacture of coke, the carbon used

---

[1]C. Slahetka and J. McCorkle, "Type II Asphalt: The Way to Go on Low-Slope Roofs," *RSI Magazine*, October 1979, pp. 82ff.

to alloy steel. Subjected to temperatures well over 2000°F in the coke ovens, the coal is heated without air, and this destructive distillation process drives off gases and vapors, leaving the coke as a more or less pure carbon for charging the blast furnace. After the evaporated coal gases are cooled and condensed, crude tar is separated from the other compounds in the condensate. Coal tar pitch is 1 of some 200,000 products made from this crude coal tar.

Coal tar pitch is a good waterproofing agent because of a physical property known as *cold flow*. Even at moderate ambient temperatures—say, around 60°F—coal tar pitch slowly heals cracks formed at lower temperatures. This cold-flow property stems from a chemical peculiarity of coal tar pitch. Although its molecules have very strong *intra*molecular attractions between linked benzene rings, they have relatively weak *inter*molecular bonds; the molecules readily flow over one another, in slow response to gravity. This desirable waterproofing property is obviously what limits coal tar pitch's use to low-slope roofs. Used even on a moderately sloped roof, coal tar pitch, over years, migrates from higher to lower areas, leaving the roof's high points with a thinned layer of waterproofing bitumen.

Type I (dead-level) asphalt also has this cold-flow property, but usually to a lesser degree than coal tar pitch. More viscous asphalts, Types II to IV, lose this property in inverse proportion to their viscosities.

Revised ANSI-ASTM Standard 450-78, emended to reflect the changed properties of coal tar pitch, provides for two roofing grades: Type I (softening-point range 126 to 140°F) and the slightly steeper Type III (softening-point range 133 to 147°F) for use on slightly higher slopes (½-in. absolute maximum). (Type II coal tar pitch is a waterproofing grade.)

Incompatibility of Asphalt and Coal Tar Pitch

Mixing asphalt and coal tar pitch violates a well-publicized roofing industry taboo because the two bitumens are chemically incompatible. The worst mixing occurs when asphalt melts in a kettle coated with the remnants of coal tar pitch (or vice versa). Such mixing can produce a mongrel bitumen with unpredictable properties.

Applying one bitumen over the other can produce similarly deleterious consequences:

- Asphalt applied over coal tar pitch may soften and flow off, leaving exposed coal tar pitch, which weathers rapidly.

- Coal tar pitch applied over asphalt may harden and crack.[1]

[1]See P. M. Jones, "Bituminous Materials," *Can. Build. Dig.*, no. 38, May 1968, p. 38-3.

In mixtures of the two materials, coal tar pitch, the donor or "exudative" bitumen, becomes hardened and embrittled through loss of its lighter or more actively solvating molecules. Asphalt, the recipient or "insudative" bitumen, becomes softened and more fluid.

Tests conducted at Owens-Corning Fiberglas Corporation's Granville, Ohio, laboratory, indicate that the incompatibility between asphalt and coal tar pitch is greatest when the two bitumens have similar softening points, i.e., when Type I asphalt is mixed with roofing-grade coal tar pitch. But when the asphalt has a high softening point—such as the 220 to 260°F coating-grade asphalt film on coated felts—the mixing hazard abates. This principle explains the successful combination of flashings mopped with steep asphalt on coal-tar-pitch roofs and even the compatibility of coal-tar-pitch flood coats on asphalt-coated-felt membranes or glass-fiber felts, which are coated with a similarly hard asphalt.

Coal Tar Pitch vs. Asphalt

Commercial passions flare when the subject of coal tar pitch vs. asphalt arises; discussing this topic is much like refereeing the Hatfield-McCoy feud. To establish as cool an atmosphere as possible, I shall preface this discussion by reiterating a previously stated thesis: Both coal tar pitch and asphalt are intrinsically excellent roofing materials, with long records of proven performance and thousands of roofs still performing satisfactorily, some after service lives of 30 to 50 years. For the majority of built-up roof projects—perhaps the vast majority—both asphalt and coal-tar-pitch roof systems will perform satisfactorily, *if properly designed and constructed.*

Nonetheless, there are significant differences between the two materials. As noted, asphalt is a more varied, more versatile bitumen than coal tar pitch. Properly processed asphalt is suitable for slopes up to 6 in., whereas coal tar pitch requires special precautions on slopes over ½ in. Under the right conditions, asphalt can be left exposed on the roof surface, but coal tar pitch always requires a protective aggregate surfacing because of its excessive fluidity at high temperature. Asphalt also has markedly less temperature susceptibility than coal tar pitch; i.e., it takes a greater temperature drop to change asphalt from a viscous fluid to a brittle solid. Coal tar pitch makes this transition over a narrower temperature range, whereas asphalt retains its plasticity at considerably lower temperature than coal tar pitch and at higher temperature retains higher viscosity.

As a countervailing advantage, coal tar pitch has a generally better cold-flow, self-healing quality and somewhat less vulnerability to deteriorative oxidation than asphalt. This self-healing property, shared in a less dependable degree by Type I asphalt, is required to protect the underlying felts from destructive moisture and ultraviolet radiation.

Here, in conclusion, are several broad generalizations about the relative merits of coal tar pitch and asphalt:

Because of its lesser temperature susceptibility, asphalt will probably perform better than coal tar pitch under the following conditions:

- In climates characterized by prolonged, severe winter cold, but not by prolonged, intense summer heat

- In climates subjected to great daily temperature extremes

Because of its greater chemical stability and waterproofing quality, coal tar pitch will probably perform better than asphalt under the following conditions:

- In climates characterized by prolonged, intense summer heat, but not by prolonged, severe winter cold

- In locations subjected to intense sunlight

- On roofs where ponding of water occurs frequently

Felts

Roofing felts are nonwoven fabrics classified as either *organic* or *inorganic* (asbestos or glass-fiber). The paper-manufacturing process used to produce organic and asbestos felts orients the fibers in their longitudinal (machine or roll) direction, making them roughly twice as strong in that direction as in the transverse (cross-machine or cross-roll) direction. Some glass-fiber felts, however, are virtually isotropic (they have nearly equivalent longitudinal and transverse tensile strength).

Today's roofing felts have evolved, step by step, from a crude process originated in the 1840s, when square sheets of ship's sheathing paper, treated with a mixture of pine tar and pine pitch, became the felts for the prototypical built-up roofs.[1] Coal tar was next substituted for pine tar, as a more fluid saturant, but the square sheets were still dipped manually into the melted saturant and the excess pressed out. Then came the substitution of paper or felt rolls, running through continuously operating saturators. And finally came the substitution of distilled coal tar pitch for the more expensive pine-pitch-coal tar mixture. Asphalt-saturated felts appeared before the end of the century.

[1]Herbert Abraham, *Asphalts and Allied Substances: Their Occurrence, Modes of Production, Uses in the Arts, and Methods of Testing,* 6th ed., Van Nostrand Rheinhold Company, New York, 1966, p. 48.

Organic felts contain cellulose fibers—shredded wood and felted papers. (Although sometimes erroneously called rag felts, they now have little rag content.) Organic felts are impregnated with coal tar pitch or asphalt (see Table 7-3). Asphalt-saturated organic felt costs least and remains (in 1980) the most popular roofing felt on the market.

Asbestos felts are manufactured from natural, hollow, threadlike fibers of magnesium and calcium silicates obtained from the crushing and refining of chrysotile ore. These fibers are fireproof and highly resistant to aging. They resist deterioration from humidity, solar radiation, and other destructive agents better than do organic fibers.

Asbestos felts contain organic fibers, limited to 15 percent (by weight) by ASTM Standard D250. Organic fibers add "bulk" to the asbestos felts, enabling them to absorb more asphalt saturant. Asbestos fibers do not absorb asphalt saturant.

Glass-fiber mats, the newest of the three basic roofing felts, and obviously destined to outsell asbestos felts within a few years because of their intrinsic advantages, entered the United States market in the late 1940s. Glass-fiber filaments are drawn from molten-glass streams through tiny orifices made of precious metals. The molten glass comes from batches of sand, limestone, and soda ash, continuously fed into a furnace fired above 2500°F. The long, fiberglass filaments, 12 to 15 per inch cross section, are usually bound with a thermosetting binder, phenol-formaldehyde or urea-formaldehyde.

The nearly isotropic glass-fiber mats owe this property to the more random pattern of the reinforcing filaments, in contrast with the longitudinal orientation of cellulosic and mineral fibers in organic and asbestos felts and in some glass-fiber mats. In the strongest glass-fiber mats (ASTM D2178, Type IV), the filaments are extremely long. These mats have three times the required transverse tensile strength of a 15-lb (nominal) saturated organic or asbestos felt (see Table 7-3). (Note also in Table 7-3 that ASTM D2178 requires the same minimum longitudinal and transverse strength for all glass-fiber mats, whereas the longitudinal/transverse strength ratio is 2/1 for organic and asbestos felts.)

Glass-fiber mats also differ from organic and asbestos felts in that they require a more viscous impregnating asphalt. For saturated organic and asbestos felts, asphalt saturant varies from 100 to 160°F softening point. But because the glass fibers do not absorb the impregnating asphalt, a much harder asphalt is required to provide a stable coating for the reinforcing glass-fiber mat.

Felts are manufactured into three kinds of sheets:

- Saturated felts

- Coated felts

- Mineral-surfaced sheets

Each type represents a progressive stage in the manufacturing process.

Saturated felts, as previously indicated, are saturated with bitumen (asphalt for organic, asbestos, and glass-fiber felts; coal tar pitch is restricted almost exclusively to organic felt because cellulosic fibers are the only satisfactory medium for maintaining stable uniformity in the distribution of this relatively fluid bitumen).

Coated felts are saturated felts subjected to an additional manufacturing stage—coating with blown asphalt, generally around 220°F softening point, but ranging up to 260°F, and stabilized with finely ground minerals (silica, slate dust, talc, dolomite, trap rock, or mica) to improve their durability and resist cracking in cold weather. Finely ground minerals—commonly, talc, mica, or silica—are also dusted on the inside surface of coated felts as a releasing agent or parting agent to prevent contacting surfaces from adhering when the felt is rolled. Coating a saturated felt greatly improves its moisture resistance.

Mineral-surfaced sheets are coated felts taken a step beyond the coating process: Mineral granules (colored slate, rock granules) are embedded for the weather-exposed surface. The heaviest mineral-surfaced roll roofing with organic felt weighs a nominal 90 lb/square. (It is accordingly known as a 90-lb mineral cap sheet.)

Saturated felts generally serve as ply or base felts; coated felts serve as base sheets, rarely as ply felts; mineral-surfaced sheets serve as surfacing (cap) sheets. Grid-grooved mineral-surfaced sheets serve as venting base sheets. Both saturated and coated felts are also used in vapor retarders, and mineral-surfaced sheets are sometimes used as exposed base flashing.

Organic felts and, to lesser degree, asbestos felts, absorb water, with threatening consequences to membrane performance because water absorption can promote such major premature failures as blistering, ridging, and splitting.

Surfacing

Hot-mopped, built-up roof membranes have four basic types of surfacing:

- Mineral aggregate (embedded in hot bituminous flood coat)

- Asphalt (hot- or cold-applied)

- Mineral-surfaced cap sheet

- Heat-reflective coatings

Embedded aggregate surfacing has this basic advantage: It makes possible the pouring of a heavyweight flood coat, average ⅛ in. thick,

Table 7-3 Typical Built-Up Membrane Roofing Felts

Designation	ASTM specification	Type	Minimum total weight, lb/sq	Minimum dry felt weight, lb/sq	Minimum saturant weight, lb/sq	Minimum tensile strength, psi, 77°F, 25°C	
						Longitudinal	Transverse
Asphalt-saturated organic felt	D226	I, 15	13	5.2	7.3	30	15
		II, 30	26	10.0	15.0	40	20
		III, 20	17	6.8	9.6	35	17
Coal tar-saturated organic felt	D227	15	13	5.2	7.3	30	15
Asphalt-saturated asbestos felt	D250	I, 15	13	8.5	3.4	20	10
		II, 30	28	17.5	7.0	40	20
Asphalt-impregnated glass mat	D2178	I	7.3*	1.0	6.1	15	15
		III	9.7*	1.5	7.8	22	22
		IV	7.0*	1.7	4.7	44	44
		V	14.6*	1.0	10.8	30	30

	ASTM	Type					
Asphalt-saturated and coated organic base sheet	D2626	I	37	5.2	7.3†	35	20
		II	39	7.0	9.8†	45	20
Asphalt-saturated and coated organic base sheet	D3158·		29	5.2	7.3‡	35	20
Asphalt-saturated and coated asbestos base sheet	D3378	I	37	8.5	3.4†	30	15
		II	39	10.0	4.4†	40	20
Venting asphalt-saturated and coated inorganic base sheet	D3672	I	60	9.0	3.6	30	15
		II	50	0.8	....	25	20

*Average of all rolls. Total weight also includes weight of comminuted (i.e., pulverized) mineral antistick surfacing, limited to maximum 2.5 lb/square for Type IV, 1.5 lb/square for Types I, III, and V.

†Plus 18 lb/square minimum coating asphalt and mineral surfacing, with minimum 3.5 lb/square asphalt coating on each side of felt, exclusive of mineral stabilizer and surfacing.

‡Plus 12 lb/square minimum coating asphalt and mineral surfacing, with minimum 3.5 lb/square asphalt on each side, exclusive of mineral stabilizer and surfacing.

three times the thickness of a standard interply mopping (see Fig. 7-6). Individual aggregate pieces, applied simultaneously with the hot, fluid bitumen, dam the flood coat to its thickened depth; this thickened flood coat enhances the membrane's waterproofing quality.

Aggregate surfacing provides other important benefits:

- Shielding from damaging solar radiation, which accelerates photochemical oxidation of bitumen by a factor of 200 and ultimately threatens the felts.

- Improved resistance to bitumen erosion from scouring action by wind and rain.

- Improved impact resistance (from hailstones, foot traffic, dropped tools, and falling tree limbs).

- Improved fire resistance (preventing flame spread or ignition by burning brands).

- Superior wind-uplift resistance. (A rough surface disrupts the laminar airflow patterns associated with uplift pressure, the 3 to 4-psf aggregate weight acts as partial ballast, and the thick flood coat stiffens the membrane, making it less vulnerable to "ballooning" and undulation.)

- Reduced roof-surface temperature compared with a black, smooth-surfaced membrane.

The last benefit has several advantages. Most obvious, it reduces cooling-energy consumption (or produces cooler temperatures in unair-conditioned buildings). It also tends to reduce the deteriorative oxida-

Fig. 7-6 The flood coat of an aggregate-surfaced membrane is poured (manually, as shown, or mechanically), not mopped like interply bitumen. (GAF Corporation.)

tion that embrittles bitumen. According to figures cited by R. L. Fricklas, gravel surfacing can reduce peak roof-surface temperature by 20°F or more.[1] A rise of 18°F roughly doubles an organic material's oxidation rate.[2] Reduction of roof-surface temperature can be especially important in climates characterized by prolonged heat, humidity, and rainfall because water dissolves oxidized asphaltic compounds, exposing fresh asphalt to the photo-oxidative attack of sunlight and direct attack by atmospheric oxygen. Aggregate surfacing thus shields the bituminous flood coat from harmful ultraviolet radiation and excessive heat, agents that combine to greatly accelerate the chemical deterioration of the bituminous flood coat.

Aggregate surfacing does, however, have some disadvantages compared with smooth-surfaced roofs. Because it obscures the underlying felts, aggregate surfacing makes it more difficult to spot membrane defects, for example, small, growing blisters; ridges; fishmouths; curled felts; even splits. Repairing an aggregate-surfaced membrane is also more difficult than repairing a smooth-surfaced membrane, partially offsetting its advantage in generally requiring fewer repairs. And if an aggregate-surfaced membrane requires tearoff-replacement, it presents a much tougher problem than a smooth-surfaced roof. Disposing of the old aggregate can be extremely difficult and expensive. Slag or gravel sticks to workers' bitumen-covered shoes and equipment wheels; carried back to the new work area, it can get trapped between new felts, forming voids (the origin of blisters) and puncturing the felts. Entrapped aggregate within a new membrane almost certainly portends shortened service life, increased problems, and repair bills.

The maximum 3-in. slope limit generally recommended by roofing manufacturers for aggregate-surfaced membranes is designed to prevent both the heavy flood coat and the aggregate from sliding down the slope during application or during hot weather.

In addition to strength, hardness, and durability, aggregates require the following two basic properties:

- Opacity
- Proper sizing and grading

Opacity is an important property of surfacing aggregates. Translucent aggregates that permit passage of solar radiation lack a major fea-

---

[1] R. L. Fricklas, "Technical Aspects of Retrofitting," *Proc. 5th Conf. Roofing Technol.*, NBS-NRCA, April 1979, p. 32.

[2] P. G. Campbell, J. R. Wright, and P. B. Bowman, "The Effect of Temperature and Humidity on the Oxidation of Air-Blown Asphalts," *Mater. Res. Stand. ASTM*, vol. 2, no. 12, 1962, p. 988.

ture of good surfacing aggregate. Carl G. Cash's research on roofing aggregates demonstrates the potential hazards of translucent aggregate. Comparing a highly opaque aggregate with a highly translucent quartz aggregate in test panels exposed for 3 months (September through November, 1978), Cash discovered 100 percent light-stained area on felts covered by the translucent aggregate, compared with only 3 percent light-stained area for the felts surfaced with opaque aggregate.[1] After 3 months' exposure, felts surfaced with translucent quartz aggregate exhibited the same stain intensity as unsurfaced, totally exposed "control" felts after 1 month exposure. As a shield against solar radiation, translucent aggregates thus appear virtually useless.

Proper sizing and grading of aggregates are needed to ensure proper nesting, which is vital to continuity of surface protection and continuous flood coat. ASTM Standard D1863, "Standard Specification for Mineral Aggregate Used on Built-Up Roofs," sets a maximum size of ¾ in., with a desired minimum of ⁵⁄₁₆ in., although permitting a small percentage (by weight) to pass through a No. 8, 2.36-mm (³⁄₃₂-in.) sieve.

Undersizing of aggregate is more serious than a comparable degree of oversizing. Excess fines in undersized aggregate spread into hot, fluid bitumen sink into the flood coat. Submerged, these fine aggregate particles perform no protective function whatever, only the negative function of interrupting the continuous waterproofing film. On the surface, loose, undersized aggregate is less stable than larger, properly graded aggregate under the action of wind, flowing water, and roof traffic. This reduced stability also detracts from the aggregate's protective function as a shield against ultraviolet radiation.

Undersizing also reduces aggregate embedment, a fact established by Cash's cited research. The quantity of adhered aggregate for laboratory test samples was a direct linear function of mean aggregate diameter between the limiting sizes of 2 mm (0.08 in.) and 12 mm (½ in.). Adhered aggregate dropped from 300 lb/square embedment for the 12-mm mean aggregate diameter to 150 lb/square for the 2-mm aggregate (see Fig. 7-7). Moreover, because field-test samples generally had less adhered aggregate than laboratory test samples (up to 60 lb/square flood-coat weight), field-applied aggregate may suffer even greater embedment loss from undersizing than laboratory-prepared specimens.

Oversized aggregate, a less common, less severe problem than undersized aggregate, can nonetheless cause trouble. Oversized interstices between pieces break the continuity of the surface protection. In the vertical plane, these larger interstices may leave voids that can entrap silt and promote plant growth.

---

[1]Carl G. Cash, "On Builtup Roofing Aggregates," presented to ASTM D8 Main Committee, Dec. 6, 1978, p. 5.

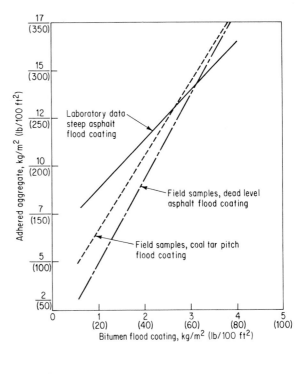

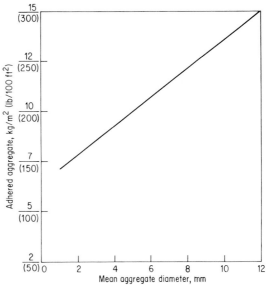

Fig. 7-7 Tests reported by Carl G. Cash indicate linear relationship between flood-coat thickness and weight of adhered aggregate (top). Adhered aggregate also varies linearly with mean aggregate diameter (bottom). (From Carl G. Cash, "On Builtup Roofing Aggregates," presented to ASTM D8 Main Committee, Dec. 6, 1978.)

147

The most common roof aggregates are river-washed gravel, crushed stone, and blast-furnace slag, a fused, porous substance separated in the reduction of iron ore, chemically comprising silicates and alumino-silicates of lime.

Slag is an excellent aggregate. Its opacity provides superb shielding from solar radiation. Its pitted, porous surface provides excellent adhesion with the cooled bitumen, which, as a hot fluid, flows into the slag's surface cavities and forms keys that lock the aggregate in place. The slag's alkaline surfaces also improve adhesion because of their greater affinity for bitumen than for water.[1] Although gravel is a good aggregate, its smooth, often rounded surfaces provide less surface area for bonding, thus making it inferior to slag.

Apart from gravel, slag, and crushed stone, roof designers should approach other roofing aggregates with intense suspicion. Georgia roofing-grade white marble chips may satisfy ASTM requirements, but a host of occasionally used aggregate materials do not. Dolomite (marble chips) can provide a good heat-reflective surface, but it has several flaws. Small translucent chips may admit damaging solar radiation that can degrade the flood-coat bitumen. The resulting embrittlement can loosen the aggregate, exposing more bitumen to photo-oxidative reactions and accelerating the membrane's deterioration. Moreover, dolomite chips are often coated with dust, which may weaken the aggregate bond to the bitumen. Excessive quantities of salt or free alkaline material also disqualify aggregates as suitable membrane surfacings.

Other occasionally used, but highly dubious, surfacing aggregates include limestone, volcanic rock, crushed oyster and clam shells, crushed brick, tile, or cinders. The crushed coral rock used extensively in southern Florida is a similarly dubious surfacing aggregate, consistently undersized. Canadian researchers D. C. Tibbetts and M. C. Baker warn against the temptation to use what is *available* instead of what is *suitable*.[2] The first-cost saving may be repaid many times over in reduced membrane service life.

Full-weight bitumen flood coat is required to bond an adequate quantity of aggregate to the membrane surface. Cash's cited research, paralleling earlier research by Jim Walter researchers, demonstrates this requirement. For asphalt field samples, adhered aggregate ranged linearly downward from about 270 lb/square embedded aggregate for a 60-lb asphalt flood coat to less than 90 lb/square embedded aggregate for a 20-lb flood coat (see Fig. 7-7).

---

[1]D. C. Tibbetts and M. C. Baker, "Mineral Aggregate Roof Surfacing," *Can. Build. Dig.*, no. 65, May 1968, p. 65-3.

[2]Ibid.

Skimping on the flood coat thus compounds the problem of constructing a good, weathertight membrane. Not only does it reduce the thickness of the water-resistant bitumen film, it also impairs the shielding of this film by sufficient aggregate coverage, exposing the bitumen to damaging solar radiation and general weather erosion.

A double-pour aggregate surfacing, featuring two flood-coat applications, can solve several problems with aggregate surfacing. Blast-furnace slag as the lower surfacing exploits the superior embedment and opacity attainable through using this porous material, thus providing basic protection. In the top flood coat, where exposed bitumen is a lesser threat to the membrane's integrity than in the bottom flood coat, white marble chips, or similarly light-colored aggregate, can provide superior heat reflectance. Additional cost for a double-pour might range around $25/square at 1980 prices. It could more than pay for its initial cost in cooling-energy savings (from reduced surface temperature) and superior weatherproofing. (Double-pour membranes are required for deliberately ponded, water-cooled built-up membranes.)

Double-pour aggregate surfacing is credited as the chief reason for the excellent performance of built-up roof membranes at the U.S. Naval Base in Guam, where these specially designed roof systems have given 20 years or more of excellent service. Coral-gravel aggregate is washed with water, dried, and then sprayed lightly with diesel fuel (as a primer) before it is applied. This unique policy was instituted after typhoon winds had persistently swept aggregate from single-pour aggregate-surfaced roofs. The unwashed coral gravel used in these single-pour roofs is dusty and thus difficult to embed solidly in a bituminous flood coat.

An insufficient quantity of embedded aggregate, below 50 percent of weight total, often results from tardy spreading of aggregate, after the flood-coat bitumen has congealed, thereby losing the required fluidity for adhesion with the aggregate. Excessive moisture and dust can also reduce the quantity of embedded aggregate. Acceptable limits for these aggregate contaminants are not, however, as severe as formerly believed, according to research by Cash.[1] Up to 2 percent dust (defined as aggregate passing a No. 200 mesh sieve) and 2 percent moisture caused no detectable loss of aggregate adhesion on the laboratory test samples. Mean adhered aggregate was roughly the same for 0, 0.5, 1, and 2 percent dust or water content. Furthermore 2 percent water content makes the aggregate surface palpably wet. ASTM Standard D1863-80 accordingly permits maximum 2 percent moisture for crushed stone and gravel, 5 percent for slag, and 2 percent dust content.

---

[1]Cash, "On Builtup Roofing Aggregates," op. cit.

Is the flood coat of an aggregate-surfaced built-up membrane the membrane's key waterproofing element? "No," says E. H. Rissmiller, Senior Research Associate at Jim Walter Research Corporation, as his comment on this old roofing industry controversy. According to Rissmiller, the flood coat's major purpose is to bond the aggregate to the membrane. It does provide some additional waterproofing, but it cannot function as the principal waterproofing agent because its film is broken by the following factors:

- Embedded aggregate projecting through the flood-coat film
- Air and moisture bubbles entrapped during application of the aggregate
- Alligatoring and cracking due to weathering and temperature changes
- Thin spots where the bitumen flows off ridges
- Roof traffic and impact from falling objects

This theory highlights the importance of uniform, continuous interply moppings which, by general consensus, form the foundation of a good built-up roof membrane. The poured top coat can be corrected if it is deficient. But repairing deficient interply moppings, which can also cause premature membrane failure, is much more difficult.

Regarding whether the flood coat or the interply moppings is *the* waterproofing element, the safest answer is to assume the indispensability of both. The answer can have important practical consequences. I have heard a roofing contractor argue that penetration of the interply moppings by existing aggregate entrapped between the insulation and the bottom felt ply did not matter because, after all, the flood coat provides the membrane's waterproofing. An owner or architect who believes such nonsense may pay for such credulity. Built-up membranes require good flood coats *and* good interply moppings to ensure their watertight integrity.

Smooth-Surfaced Membrane

A *smooth-surfaced membrane* is generally topped with 15 to 25 lb/ square of hot steep asphalt or cold-applied cutback or emulsion applied to an asphalt-saturated inorganic felt at a spread of 25 to 50 ft²/gal. A smooth-surfaced roof offers several positive advantages:

- Easier inspection, maintenance, and repair than an aggregated-surfaced membrane

- Easier installation of new openings (for vents, exhaust fans, heating, ventilating, airconditioned (HVAC) units, and so on) in buildings subject to changing uses

- Easier cleaning than aggregate-surfaced roofs (important for food-processing, fertilizer, paper mill, and other industrial plants whose stack exhausts settle on the roof)

- Easier reroofing or replacement at end of a roof's service life

- Slight reduction in dead load (by 3 to 4 lb/ft^2 aggregate weight)

As discussed, however, the disadvantages usually outweigh the advantages. Foremost of these disadvantages, a smooth-surfaced roof is a *water-shedding* rather than a *water-resistant* roof. All membranes, even those designed for deliberate ponding of water, should drain. But whereas aggregate-surfaced roofs can tolerate some intermittent ponding, a smooth-surfaced roof should be designed for rapid runoff—generally with a slope of at least ½ in./ft, preferably 1 in./ft.

Because of their extreme vulnerability to moisture, organic felts are unsuitable for smooth-surfaced roofs. Glass-fiber felts make the best smooth-surfaced membranes. Asbestos-felted, smooth-surfaced roofs sometimes betray their vulnerability to water by the characteristic "finger-wrinkling" that indicates swelling from moisture absorption, generally evidence of exposure to moisture during phased construction. (Remember that asbestos felts are permitted to contain 15 percent organic fibers and starch binder.)

Mineral-Surfaced Roll Roofing

Mineral-surfaced cap sheets (roll roofing) are also suitable for roofs with minimum 1-in. slope.[1] This minimum 1-in. slope requirement should be considered an absolute for organic roll roofing felts. In the military services' "roofing bible," NBS researcher W. C. Cullen recommends minimum 3-in. slope for asphalt roll roofing (application parallel to eave, with concealed nailing), with 2-in. minimum for emergency construction.

Minimum ¼-in. slope is, however, approved by at least one commercial manufacturer for glass-fiber–mineral-surfaced sheets. Note however that this recommendation does not apply to either organic or asbestos felts.

---

[1]Asphalt Roofing Manufacturers Association, *Manufacture, Selection and Application of Asphalt Roofing and Siding Products,* 12th ed., 1974, p. 20.

Heat-Reflective Surfacings

Heat-reflective surfacings perform three vital functions:

- Reduce cooling energy consumption (especially important in warm climates with long cooling seasons)[1]

- Protect membrane bitumens from deteriorating photo-oxidative chemical reactions that accelerate exponentially with increasing temperature

- Reduce the rate of blister growth

This last benefit depends on two factors associated with reduced membrane temperature: (1) reduced internal pressure inside the blister void and (2) increased bitumen cohesive strength and cleavage resistance at lower temperature. Rising temperature produces both an exponential growth of pressure within a vapor-saturated blister chamber and an exponential decline in bitumen strength. Thus it is conceivable that heat-reflective coatings can even prevent blister formation in some critical instances; i.e., the reduced internal pressure attributable to the heat-reflective coating may be too slight to pry the adhered felts apart around the perimeter of a potential blister-forming void.

Color is the key to heat-reflective cooling. Under a sunny, summer sky, with air temperature 90°F, a black roof surface might climb to 170°F, compared with 135°F for an aluminum-coated surface, still lower for a white surface.

Heat-reflective surfacings are available for all three types of built-up surfacing. Heat-reflective aggregates include white marble chips and other light-colored aggregates. Gray slag cuts the peak roof temperature by about 15°F from a black-surfaced membrane's temperature. Heat-reflective cap sheets are surfaced with light mineral granules. For smooth-surfaced roofs, aluminum or other light, pigmented coatings can be brushed, rolled, or sprayed.

BUILT-UP MEMBRANE SPECIFICATIONS

Today's variety of membrane specifications evolved from two basic ancestors:

- The older five-ply, wood-deck specification, with the two lower plies nailed to the deck and the top three mopped

---

[1]C. W. Griffin, "Cost Savings with Heat-Reflective Roof Coatings," *Plant Eng.*, July 10, 1980.

• The four-ply concrete-deck specification, with the base ply mopped to the deck and all four plies mopped

Subsequent modifications ultimately reduced the number of felt plies from four or five saturated felts to two coated felts, in some instances. With the widespread failure of the two-ply membranes, the trend has swung back toward the more traditional three- and four-ply systems.

The typical modern built-up roof specification over rigid insulation board calls for all shingled plies—preferably a minimum of four. Over a lightweight concrete insulating fill or a gypsum deck, the typical specification calls for a nailed coated base sheet plus a minimum of three shingled plies.

Application of the felt starts at an edge, or low point (so the shingled joints will not "buck" water), with a felt cut into strips whose widths are an even factor of the 36-in. felt width divided by the number of plied felts. Thus for three-plied shingled felts (a four-ply membrane, including base sheet), the edge strips are 12, 24, and 36 in. wide (see Fig. 7-8).

Next come the regular 36-in.-wide shingled felts. The overlap dimension is computed by dividing the felt width minus 2 by the number of shingled plies. Thus for three shingled plies, the overlap = (36

Fig. 7-8 Shingling of roofing felts requires their simultaneous application. With three shingled plies, the exposure of each 36-in.-wide felt is 11⅓ in., computed by dividing the 36-in. felt width minus 2 in. by the number of plies.

— 2)/3 = 11⅛ in., which is also the distance of the first sheet from the edge.

This overlap ensures that any vertical cross section always has at least the minimum number of required plies.

Shingling of Felts

Shingling of felts is a universal practice in built-up roofing membranes. It originated as a means of facilitating and accelerating application of the membrane. Overlapping the felts enables the roofer to complete the membrane as the roofing crew proceeds across the roof, instead of applying alternating layers of felt and bitumen across large areas, or even over the entire roof, in separate stages.

Shingling of felts may reduce the membrane's resistance to hydrostatic pressure. Waterproofing membranes for foundations, walls, and concrete-topped basement roofs are often applied with alternating laminations, with an entire ply placed and mopped before the next ply is applied. These parallel planes of mopped interply bitumen probably provide better waterproofing defense than the moppings between shingled-felt plies. Shingled plies can slowly wick moisture from an exposed felt edge diagonally down through the membrane to the base sheet—or, in the case of a totally shingled built-up membrane—to the insulation substrate. Defective application, e.g., a fishmouth at a felt lap, provides a direct path to the substrate. Waterproofing membranes laid ply-on-ply are not vulnerable to this diagonal moisture penetration through the entire membrane cross section.

Accompanying this singular liability of shingling are two important advantages:

- Improved slippage resistance
- Reduced risk of poor interply mopping adhesion

Improved slippage resistance results from the shingled felts' structurally integrated construction. Phased, or laminar, construction promotes slippage because it provides a continuous, unbroken film of bitumen between felts. When a phased built-up membrane slips, the slippage invariably occurs at the phased plane. Shingling provides slightly more frictional resistance (equal to the slippage force times the sine of the acute angle between the shingled felts and the membrane substrate plane) in addition to the pure parallel shearing resistance (see Fig. 7-9). Because the bitumen's tensile resistance exceeds its horizontal shearing resistance, a shingled-felt membrane resists slippage better than a totally laminated ply membrane or even a membrane with one

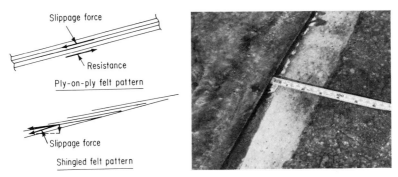

Fig. 7-9 With an unshingled, ply-on-ply felt pattern (top drawing), slippage force parallel to the slope is resisted only by the horizontal shearing strength of the asphalt mopping film (unless the felts are nailed). A shingled felt pattern (bottom drawing) provides additional friction resistance, represented by the coefficient of friction multiplied by the slippage force and the sine of the tiny acute angle between the shingled felts and the roof-deck plane. A 3-in. slippage occurred on the 1-in. sloped, smooth-surfaced roof (photo) because the roofer omitted the specified nails to anchor the felts. Shrinkage of wet felts and overweight moppings evidently aggravated the problem.

laminated ply formed by phased application, usually a base plus two or three shingled plies.

Reduced risk of poor interply mopping adhesion results from the slower cooling rate of several simultaneously applied bituminous moppings compared with the cooling rate of individually applied moppings. With three or four shingled felts, there is three or four times as much hot bitumen as with application of merely a single felt, and this greater mass of hot bitumen cools at a significantly slower rate than a single mopping.[1]

Coated Base Sheet

Using a coated base sheet as the membrane's bottom felt ply became an industry standard for organic and asbestos felt membranes in response to widespread ridging, or "picture-framing," which afflicted the industry two decades ago. Water vapor migrating upward through board-insulation joints can penetrate into saturated felts, condense, and cause the ridging phenomenon discussed in Chapter 12, "Premature Membrane Failures." Coated felts, with their added waterproofing film, can better resist this moisture penetration than saturated felts. As

[1]See R. M. Dupuis, J. W. Lee, and J. E. Johnson, "Field Measurement of Asphalt Temperatures During Cold Weather Construction of BUR Systems," *Proc. Symp. Roofing Technol.*, NBS-NRCA, September 1977, pp. 272ff.

a highly impermeable asphalt film on the membrane's underside, the base sheet is applied as a single ply, not shingled with the ply felts. (Shingling the base sheet with the other felt plies would obviously defeat its purpose because it would no longer form a continuous coated-felt bottom surface.)

For some years the coated base sheet remained an industry standard for asphalt organic and asbestos felt membranes (but not for glass-fiber-felt membranes, which are less vulnerable to ridging). In the basic nailable wood-deck specification, one coated base sheet replaces two saturated felts.

In recent years this trend has swung away from use of the coated base sheet toward the old specification with all saturated, shingled felts, simultaneously applied. Because of their lower permeability and their superior water-absorptive resistance, coated base sheets still predominate in membranes applied over wet decks—lightweight insulating concrete, poured gypsum, or structural concrete—and also over nailable wood decks, where they have better nailing-holding strength. But many specifiers have returned to the old saturated, all-shingled ply system for membranes over board insulation, for several reasons:

- Coated felts applied in a single operation (i.e., unshingled) are subjected to more rapid cooling of hot-mopped asphalt, with consequently higher risk of membrane thermal-contraction splitting, slippage, or blistering resulting from mopping voids.

- The high softening point of the coated-film asphalt (up to 260°F compared with the 100 to 160°F softening-point saturant in saturated felts) may shorten the time allowed for bonding the felt to the substrate with the hot asphalt. This shortened adhesion time also heightens the probability of mopping voids and consequent blistering.

- Phased application, in which the various membrane felt plies are installed in separate operations (often on different days) as opposed to the simultaneous shingling operation, heightens the risks of slippage and blistering. (See the following section, "Phased Application," for further discussion of this problem.)

Phased Application

In phased application, the felts are applied in two (or more) operations, with a delay between operations. At the felting break plane, the felts are necessarily unshingled. This break plane usually occurs between application of a base felt (usually coated, as indicated in the preceding section) and the top three plies of a four-ply membrane.

To some roofing experts, phased application refers to *any* delay in the field fabrication of the membrane—including the application of aggregate surfacing and a delay in placing the top felt plies. But to others, phased application refers only to a break in the felt application. In this manual, "phased application" (or "phased construction") refers to a break in felt application; "delayed surfacing" refers to a delay in flood-coat and graveling-in operations.

Phased application offers convenience to a roofing contractor; while the coated base sheet serves as a temporary roof, the contractor can spread the workers around many roofing projects on good working days and get buildings "in the dry." But if the delay in completing the felt application lasts even overnight, it threatens the membrane's integrity, and the longer the felts remain exposed, the greater the risk. Although less for coated felts than for saturated felts, this risk nonetheless applies to all felts.

Phased felt application poses the following hazards:

- Heightened risk of slippage, which, in a phased membrane, always occurs at the bitumen plane between base sheet and shingled felts

- Heightened risk of interply blistering at the base sheet's top surface, from mopping voids caused by condensation (dew) or rain-deposited water on this exposed felt surface (particularly at a fish-mouthed lap joint, where water can wick)

Some roofers have a cavalier attitude toward this latter hazard, arguing that the hot-mopped bitumen will evaporate all surface moisture, or even felt-absorbed moisture, when it hits the felt. Like most wishful thinking, this belief is totally false. You can count on a practical maximum of 50 percent moisture evaporation, according to research by T. A. Schwartz and Carl G. Cash, of Simpson, Gumpertz & Heger.[1] Because the Schwartz and Cash tests were performed in a laboratory at room temperature, the percentage of dissipated moisture doubtless declines in winter, when hot-mopped bitumen cools rapidly. Entrapment of undissipated moisture in moppings' voids creates the classic condition for blister formation and subsequent growth (see Chapter 12, "Premature Membrane Failures").

Other hazards from phased application include the following:

- Exposing organic or asbestos felts to condensation, rain, or even high relative humidity can cause the felts to expand from

---

[1]T. A. Schwartz and Carl G. Cash, "Equilibrium Moisture Content of Roofing and Roof Insulation Materials and the Effect of Moisture on the Tensile Strength of Roofing Felts," *Proc. Symp. Roofing Technol.*, NBS-NRCA, September 1977, p. 242.

moisture. Moisture absorption can curl the felts' edges as the exposed side dries and contracts (see Fig. 7-10). These curled edges form open conduits for water entry by wicking action. They sometimes protrude through the flood coat and aggregate surfacing. Although saturated felts are most vulnerable, even coated felts eventually absorb moisture if exposed long enough.

- Uncompleted membranes are more vulnerable to traffic damage than completed, surfaced membranes.

- Felts left unsurfaced for phased application accumulate dust that can weaken adhesion of the next interply mopping.

To avert the foregoing problems with phased application, the roofer often applies a glaze coat, a light bituminous mopping designed to protect the felts until they get their final surfacing. This expedient helps, but the glaze coat presents an easily damaged traffic surface, tacky in summer and brittle in winter (see Fig. 7-11).

This glaze coat, whether applied to the base sheet or the top felt surface awaiting flood coat and surfacing aggregate, should be kept as thin as practicable, to prevent alligatoring. One major manufacturer recommends squeegeeing the glaze-coat surfacing to maintain it at 10 lb/ square, roughly 0.02 in. thick.

In contrast with their unanimity in disapproving of phased felt application (with a day or more delay in membrane completion), roofing experts are divided on their opinion of phasing flood coat and aggregate

Fig. 7-10 Saturated (uncoated) felts left exposed even for short intervals absorb moisture and then curl at the edges as the top side dries and contracts. Curled felts also indicate poor adhesion and inadequate mopping along the felt edge.

Fig. 7-11 Glaze coating (left) although not a perfect remedy, does maintain a plane surface for uncompleted membranes. Exposed felt (right) curls. (National Bureau of Standards.)

surfacing. Fiberglass membranes can safely remain unsurfaced for up to 6 months, according to David Richards, of Owens-Corning Fiberglas Corporation. Even for organic and asbestos felt membranes, there are benefits in delaying flood-coat surfacing operations until the completion of felt-laying instead of completing each membrane segment—from felt-laying through graveling-in—in the same day's work. According to New York City architect Justin A. Henshell, delaying the graveling-in operations has two important advantages:

- Less chance of tracking aggregate into new areas being felted, thereby avoiding the blistering hazard from voids created by entrapped aggregate pieces

- Less chance (or temptation) to use the wrong flood-coat asphalt (Type III, instead of the normally correct Type I for low-slope roofs) because the roofer can then order the correct Type I asphalt for one continuous flood-coating operation

A compromise, allowing 4 days' maximum delay between felt application and flood-coat surfacing application, is recommended by R. J. Moore, of ARMM Consultants, Gloucester City, New Jersey. This policy avoids several hazards of same-day completion and delaying flood-coat-surfacing until the total completion of the felt application.

Like so many other aspects of built-up roofing design and application, this problem requires compromise and trade-off, regardless of which policy is chosen. There are many matters with overwhelming consensus. But this problem of when to schedule graveling-in operations is definitely not one of them. As a key requirement whenever surfacing aggregate application is delayed, the exposed felts must be glazed for temporary protection against moisture invasion. This glaze coat should be limited to 10 lb/square.

Temporary Roofs

A temporary roof, comprising built-up layers of felt or a heavy, coated base sheet, is advisable when the job is hampered by any of the following conditions:

- Prolonged rainy, snowy, or cold weather

- Necessity of storing building materials on the roof deck

- Mandatory in-the-dry work within the building before weather permits safe application of permanent roofing system

- Large volume of work on roof deck by tradespeople other than roofer

A minimum thickness of insulation must be applied over fluted steel decks as a substrate for the temporary roof.

Temporary materials, including insulation, should generally be removed before the permanent roofing is installed. The risk of damaging the temporary roof, subjected to a heavy volume of construction traffic, is too great to warrant its use as a vapor retarder or bottom plies of the permanent roof. Owners who save the costs of replacing temporary roofing materials are gambling small current gains against potentially large future losses.

Some experienced roofing contractors, however—notably J. Roy Martin, of Columbia, South Carolina—report satisfactory results repairing a felt-asphalt vapor retarder and leaving it in place after it served its stint as a temporary roof. Provision for drainage at the vapor-retarder plane both drains the temporary roof and then, after insulation and membrane are installed, disposes of water that invades the roofing sandwich.

Some manufacturers similarly allow for repair of temporary roofs left in place as vapor retarders. To justify this practice, however, requires *extraordinary*—in the literal sense of *more-than-ordinary*—care in the repair operations, with rigorous inspection and identification of punctures and other defects, removal of dust and general cleanup, and scrupulous follow-through on repair work.

Joints in Built-Up Membranes

Roof expansion-contraction joints accommodate movement from thermal expansion or contraction, from drying shrinkage of poured decks, or from structural movement—e.g., deck bearing-seat rotation—without overstressing and cracking, buckling, or otherwise impairing the roof system.

Expansion-contraction joints should allow for movement in three directions: perpendicular and parallel to the joint in the horizontal roof plane, and perpendicular to the roof in the vertical plane. Roof expansion-contraction joints are recommended at the following locations:

- Junctures between changes in deck material (e.g., from steel to concrete)

- Junctures between changes between span direction of the same deck material

- Junctures between an existing building and a later addition

- Deck intersections with nonbearing walls or other surfaces where the deck can move relative to the abutting wall, curb, or other building component

- Maximum distances of about 200 ft

- Junctures in an H-, L-, E-, U-, or T-shaped building

Expansion-contraction joints are mandatory at changes in deck material or span direction because of the need to avoid relative movement in both the vertical and horizontal planes, where movement can occur parallel as well as perpendicular to the joint line. An expansion-contraction joint at these critical lines protects the membrane from a complex combination of concentrated stresses: splitting or cracking from joint rotation at deck or joist bearings, from differential vertical deflection (i.e., shearing forces), from axial separating forces, or from horizontal shearing forces that may develop along the intersection of two different deck materials or along the line where the same deck material changes span direction.

In lightweight insulating concrete decks designed for composite structural action between corrugated steel-deck centering and concrete fill, a line of changing deck direction presents a drastically weakened cross section. A 26-gage (0.02-in.) corrugated high-strength steel sheet provides nearly 100 times as much tensile strength (roughly 2000 lb/in.) as 3-in.-thick lightweight insulating concrete. A line where this steel is omitted is consequently highly vulnerable to concrete tensile cracking. And wherever the concrete cracks, the membrane is also likely to crack along this line of stress concentration.

The expansion-contraction joints just discussed are building, not merely roof expansion-contraction, joints. They generally continue through the entire building superstructure, from rooftop to foundation, with a minimum 1-in. width. Double, parallel columns and beams are required to ensure independent movement of the integral structural segments on each side of the expansion joint.

The building designer should consider the roof when designing building expansion-contraction joints. Most recommendations for joint spacing set a higher figure than the approximate 200 ft recommended, but these recommendations generally ignore the special requirements of the roof. Contraction, not expansion, is the major problem of the roof system. (For this reason, expansion joints are appropriately termed "expansion-contraction" joints.) Closer spacing, especially in colder climates, is warranted by the roof's extreme exposure. In insulated air-conditioned buildings, the structure experiences perhaps 30°F seasonal and 10°F maximum daily temperature differential, but the roof membrane may experience nearly 200°F seasonal and 100°F daily temperature differential. If the roof system had perfect attachment at the deck-insulation, insulation-membrane interfaces, there would still be no problem with a healthy built-up membrane, regardless of expansion-contraction joint spacing. With uniform thermal-stress distribution, membrane stress would be the same for a 10-ft as for a 1000-ft expansion-contraction joint spacing.

However, the chances of membrane-stress concentration increase with the distance between expansion-contraction joints. At 0°F, a built-up membrane's ultimate strain drops below 1 percent (except for some glass-fiber-mat membranes), and its coefficient of thermal expansion-contraction increases rapidly in the $+30°$ to $-30°$F temperature range. With increasing distance between joints comes a corresponding expanse of roof for any loose boards to move. This greater potential for lateral movement increases the risk of stress concentration—for example, where a curbed opening restrains an insulation board from accommodating itself to movement in other nearby boards, thus widening a joint and producing a membrane-stress concentration.

Rotation of roof beams from live-load deflection can split the built-up membrane if the members are relatively deep. For a uniformly loaded span with $L/360$ live-load deflection, the end rotation at a simple span (i.e., noncontinuous) support is 0.51°. If the beam—say, a prestressed T—is 18 in. deep, that rotation offsets the top of the beam by tan 0.51° $\times$ 18 = 0.16 in. If another similarly loaded beam bears on the opposite side of the support, the total movement equals $\frac{5}{16}$ in. In cold weather, that movement could split most built-up membranes, with their ultimate elongations of less than 1 percent. As a consequence of end rotation, a roof-system expansion-contraction joint may be required along a line where simple-span roof beams sit on a bearing wall or girder.

Membrane control joints are not recommended. A control joint is essentially a roof expansion-contraction joint over a continuous structural deck, i.e., a location where no building expansion-contraction joint occurs. The entire concept of a control joint is based on an erroneous theory of membrane behavior: that the built-up bituminous membrane behaves elastically, like steel, with completely reversible thermal movement, expansion as well as contraction.

A built-up membrane is not elastic, but *viscoelastic*. As time passes, tensile stresses at service loadings disappear at constant strain. Cold-weather membrane contraction has no counterpart hot-weather expansion, except in minor thermal ridging of the flexible membrane. Consequently, an aging membrane tends to experience cumulative shrinkage.[1] Because of the membrane's long-term shrinkage, a control joint's ultimate position is open. It is obviously highly vulnerable to leakage because it is in the roof plane.

In place of a control joint, use a *roof area divider* or *relief joint*. Unlike a control joint, a roof area divider anchors the membrane. It is especially useful for eliminating stress concentration at reentrant (270°) corners (see Fig. 7-12). Locate these joints at high points, with drainage away from the joint in both directions.[2]

MEMBRANE PERFORMANCE STANDARDS

The 1974 publication "Preliminary Performance Criteria for Bituminous Membrane Roofing" (*Build. Sci. Ser.* 55, NBS), briefly described in Chapter 2, promulgates 20 performance "attributes" required by built-up membranes:

- Tensile strength
- Notch tensile strength
- Tensile fatigue strength
- Limited creep
- Flexural strength
- Flexural fatigue strength
- Pliability
- Shear strength
- Impact resistance
- Ply adhesion

[1] For an excellent discussion of this phenomenon, see R. G. Turenne, "Shrinkage of Bituminous Roofing Membranes," *Can. Build. Dig.*, no. 181, April 1979.

[2] For another excellent discussion on this general topic, see R. G. Turenne, "Joints in Conventional Bituminous Roofing Systems," *Can. Build. Dig.*, no. 202, January 1979.

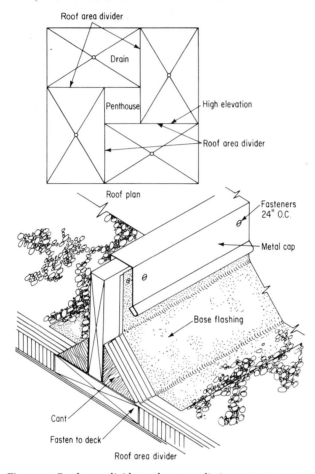

Fig. 7-12 Roof area divider reduces or eliminates stress concentrations at 270° corners. It also breaks up long stretches of roof, with a stable area anchoring all components. (National Roofing Contractors Association.)

- Wind-uplift resistance
- Limited thermal expansion-contraction
- Limited moisture expansion
- Limited moisture effects on strength
- Abrasion resistance
- Tear resistance
- Impermeability
- Weather resistance

- Fire resistance

- Fungus-attack resistance

From this list of required attributes, Cullen and Mathey developed a "performance format" of 10 criteria, each with a stated requirement, criterion, test method, and commentary. A major goal of these preliminary performance criteria is to aid manufacturers developing new built-up membranes with scientific means of evaluating the membrane's prospective performance during its service life, without waiting 20 years for retrospective evaluation. But inevitably these criteria, preliminary though they are, have been used by consultants and some designers, on the sound premise that something, although admittedly imperfect, is still better than nothing but past experience, the traditional guide for roofing technology.

Let us consider several criteria to demonstrate the NBS modus operandi. First, take the simple tensile-strength requirement of 200 lb/in. in the membrane's weakest (i.e., transverse) direction. From three different sources—field observations and large- and small-scale tests of many commercially marketed membranes—Cullen and Mathey established 200 lb/in. as the minimum tensile strength required by a *new* membrane to resist external and internal stresses imposed during a typical membrane's service life. For this particular criterion there was a standard ASTM test available: ASTM D2523, for testing built-up membranes' load-strain properties.

For the related requirement of "tensile fatigue strength," however, there was no standard ASTM test, so the NBS researchers devised their own criteria (100,000 cycles of repeated 20 lb/in. at 73°F and 100,000 cycles of repeated 100 lb/in. force at 0°F). Hailstone-impact resistance is another performance requirement for which NBS developed its own standard test, reported in *Hail-Resistance Test, Build. Sci. Ser. 23*, NBS.

The performance criterion generating the most confusion and controversy is doubtless thermal shock factor (TSF), discussed in several other sections of this manual. TSF is a *calculated* criterion, set at 100°F, for the temperature drop that must be theoretically resisted by a membrane test sample clamped at constant length. The formula for calculating TSF is

$$\text{TSF} = \frac{P}{M\alpha}$$

where TSF = temperature drop (°F)
P = tensile strength (lb/in., 0°F)
M = load/strain modulus (lb/in., 0°F)
α = coefficient of thermal expansion-contraction from 0 to −30°F, in/(in. · °F)

Now no one claims that this TSF precisely parallels the complex thermal stresses experienced by a roof membrane in service. The field membrane is normally restrained by its deck or insulation substrate. Moreover, a field membrane is seldom, if ever, *totally* restrained during a large temperature drop.

Objections that the TSF concept fails to simulate the field membrane's thermal behavior miss the point. There are numerous situations on actual roofs when the membrane's substrate moves. Additional contraction stress may act in combination with thermal stress. The TSF is merely an index for comparing different membranes. With other factors assumed equal, a membrane with a higher TSF resists thermal-contractive stress better than a membrane with a lower TSF.

There is another point to be made about TSF. To objections that a 100°F temperature drop is too severe, remember that this criterion applies to a *new* membrane. NBS studies of existing membranes, discussed elsewhere in this manual, indicate that aging deterioration of the membrane—strength loss, increase in thermal coefficient and stiffness as the bitumen age hardens—reduces even a good membrane's TSF by as much as 70 percent. The performance criterion for roofing, like a load factor for a structural beam, column, or truss, contains a safety factor. Moreover, for mild climates, free of temperature drops into subfreezing or subzero range, the 100°F TSF criterion can be waived, according to Cullen and Mathey.

Flexural and tensile fatigue testing of typical four-ply built-up membranes, reported in a subsequent NBS research publication, is designed to demonstrate the membrane's durability under repeated cyclical loading.[1] Flexural fatigue loading spans a spectrum from low-cycle, large-amplitude flexing from foot traffic to wind-induced, high-frequency oscillations of low amplitude. Thermal cycling exemplifies one kind of tensile fatigue, but there are others: e.g., long term tensile cycling from alternating contraction and expansion of membrane felts, a process that tends, however, toward permanent shrinkage.

ALERTS

General

1. Check with the manufacturers of built-up roofing and composition flashing to make sure that proposed roof system and materials are compatible with roof deck, vapor retarder, and insulation.

2. On major roofing projects, require either inspection by the manufacturer or inspection service provided by the owner.

[1]G. F. Sushinsky and R. G. Mathey, "Fatigue Tests of Bituminous Membrane Roofing Specimens," NBS Tech. Note 863, April 1975.

Design

1. Use a nailed, asphalt-coated base sheet for first ply in built-up roofing membranes over "wet" decks—e.g., poured gypsum, lightweight insulating concrete fill.

2. Do not specify coal tar pitch on roofs with slope exceeding ½ in.

3. Specify asphalt with the lowest practicable softening point suited to slope and climate as follows:

Slope (in./ft)*	ASTM designation	Softening point, °F
½ in. or less	D312, Type I	135–151
½–3	D312, Type II	158–176
½–6	D312, Type III	185–205
3–6	D312, Type IV	205–225

*Warm climates may limit the slopes for asphalts of lower softening point to less than maximum slopes specified above.

4. When hot-mopped bitumen is used to bond a base sheet to insulation, specify Type III (or in some instances Type II) asphalt.

5. Apply felts *perpendicular* to the longitudinal, continuous joints of board insulation, if practicable.

6. Provide trafficways or complete traffic surfacing for roofs subjected to more than occasional foot traffic (see Figs. 7-13 and 7-14).

7. In roofs over cold-storage space, do not place membrane in direct contact with cold-storage insulation. (For discussion of cold-storage insulation, see Chap. 5, "Thermal Insulation.")

8. Avoid contact between built-up membrane and metal flashings and other metal accessories, especially below the level where water collects.

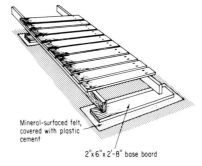

Fig. 7-13 Construction of wood walkway starts with embedment of mineral-surfaced, coated felt strips in hot bitumen. Next comes a layer of plastic cement, into which the base boards are set. Aggregate surfacing follows between and around the walkway bases. (Owens-Corning Fiberglas Corporation.)

Mineral-surfaced felt, covered with plastic cement

2"x6"x2'-8" base board

Fig. 7-14   Granule-coated treads, made with an asphalt filler sandwiched between coated felts, are placed in flood coat to form walkway. (Philip Carey Corp.)

Field

1. Prohibit storage of felts over new concrete floors. Require pallets covered with kraft paper. Stack felts on ends. Avoid prolonged storage of felts on site. Felts should be stored in an environment with RH ≤ 40%.

2. Require a smooth, dry, clean substrate, free of projections that might puncture the felts. Take precautions on wood and precast decks to prevent bitumen drippage [see Chap. 4, "Structural Deck" (Design Alert No. 4)].

3. Make sure insulation is firmly attached to substrate.

4. Prohibit use of heavy mechanical roof-construction equipment that may puncture the membrane or deflect the deck excessively.

5. Prohibit use of asphalt-saturated felt with coal tar pitch and vice versa.

6. Require a visible thermometer and thermostatic controls on all kettles, set to manufacturer's recommended limits. Require rejection of bitumen heated above the specified maximum and reheating of bitumen too viscous for mopping. Avoid prolonged storage of bitumen in heated container. (It can lower the softening point by 10 to 20°F.)

7. Hot bitumen temperatures at time of application:

Type III asphalt	375°F–450°F
Type I asphalt	350°F–425°F
Coal tar pitch	250°F–325°F

When EVT is available, require hot bitumen temperature at application within $\pm 25°F$ of the EVT.

8. When an asphalt tanker is on the site, maintain asphalt temperature below the *finished blowing temperature* (FBT) and feed asphalt into a kettle for further heating before pumping it to the roof.

9. In cold weather, require double or insulated lines and insulated asphalt carriers and mop buckets. (Also store felts in warm enclosure.)

10. Check for continuous, uniform interply hot moppings, with no contact between adhered felts, with average interply mopping-weight tolerance = ± 15 percent.

11. Require manual brooming following 6-ft maximum distance behind unrolling felt.

12. Require immediate repair of felt-laying defects: fishmouths, blisters, ridges, splits.

13. Terminate day's work with complete glaze-coated seal stripping, preferably coated with plastic cement.

14. Aggregate-surfaced membrane may be glaze-coated with flood coat and aggregate surfacing delayed until later, if necessary. Limit delay between felt application and flood-coat aggregate surfacing to a maximum 4 days.

15. Prohibit phased application, i.e., leaving felts exposed overnight or longer before top felts are applied.

EIGHT

Flashings

Flashings are the most common source of roof leaks, according to expert industry consensus. Back in the 1960s, a study of 163 roofs in southern California indicated 68 percent flashing leaks compared with only 13 percent membrane leaks. Building owners' experience generally tends to confirm flashings as a primary source of leaks. Whenever you encounter leaks of unknown origin, check the flashings first as the main suspect.

Flashing failures (analyzed near the end of this chapter) result from faulty design, poor field application, and defective materials. Flashings generally demand more of the roof designer's attention than other roof components. Do not leave the design of flashing details to the roofing contractor.

The need for good flashing design grows more important because of modern buildings' ever-increasing number of roof penetrations—for rooftop heating, ventilating, and airconditioning (HVAC) equipment, solar collectors, ventilators, electrical and mechanical equipment serving industrial processes of growing complexity. Reduce penetrations through the roof to the lowest practicable number. Then, as another preliminary step in good flashing design, *coordinate the location of flashed components with the drainage plan. Locate flashed joints at high points.*

FLASHING FUNCTIONS AND REQUIREMENTS

Flashings seal the joints at gravel stops, curbs, chimneys, vents, parapets, walls, expansion joints, skylights, scuttles, drains, built-in gutters, and other places where the membrane is interrupted or terminated. *Base flashings* are essentially a continuation of the built-up roof membrane at the upturned edges of the watertight tray. They are normally made of bituminous, plastic, or other nonmetallic materials applied in an operation separate from the application of the membrane itself.

171

Counterflashings (or *cap flashings*), usually made of sheet metal, shield or seal the exposed joints of base flashings (see Figs. 8-1 to 8-3).

There are two basic types of base and counterflashing combinations:

- Vertical terminations (at walls, parapets, or curbs)

- Horizontal terminations (e.g., at roof edges, drains, or vents)

To serve its varying functions, flashing requires the following qualities:

- Impermeability to water penetration.

- Flexibility, for molding to supports and accommodating thermal, wind, and structural movement.

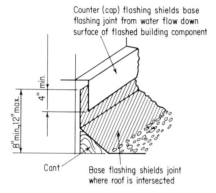

Counter (cap) flashing shields base flashing joint from water flow down surface of flashed building component

Base flashing shields joint where roof is intersected

Fig. 8-1 Base and counterflashing.

Fig. 8-2 Counterflashing, erroneously omitted on left side of brick wall control joint, shields joint in composition base flashing.

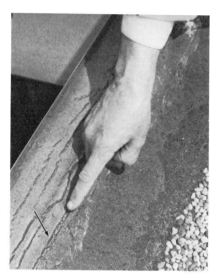

Fig. 8-3 Plastic cement surfacing displays random shrinkage cracking plus indicated longitudinal crack where fabric base flashing terminates. This is a poor edge detail. Counterflashing should shield base flashing joint (see NRCA Detail G toward end of chapter). Random shrinkage cracking can result from excessively thick flashing cement.

- Compatibility with the roof membrane and other adjoining surfaces, notably in coefficient of thermal expansion-contraction and chemical constitution.

- Stability (resistance to slipping and sagging).

- Durability, notably weather and corrosion resistance (flashings should last at least as long as the built-up membrane).

MATERIALS

Flashing materials span a broader spectrum than built-up membrane materials. They include a wide variety of metals plus the same felt and bituminous laminations used in fabricating built-up membranes and new plastic or elastomeric sheets. Because repair of leaking and deteriorating flashings is so common and costly, first-cost economy should rank last among the factors governing material selection and design. Quality and durability are paramount.

Using nonmetallic materials for base flashings and metals for counterflashings exploits the best qualities of each. Because it has a similar coefficient of thermal expansion, plied felt or fabric-based flashing works best with the built-up membrane as base flashing. And because of their good weather and corrosion resistance, metal counterflashings provide superior surface protection. As long as the base and counterflashings are designed to allow relative movement, they can function together despite their different thermal-expansion coefficients (see Table 8-1).

Cold-applied and bituminous plastic cements constitute an important class of flashing materials. A crucial distinction among these materials is between ordinary asphalt (*plastic*) roofing cement and the stiffer *flashing* cement. Both are basically so-called asphaltic cutbacks—Type I asphalt (Type II for some manufacturers) dissolved in mineral spirits of 300 to 400°F boiling range to produce 55 to 65 percent asphalt-solution plastic at ordinary temperatures. They can be troweled at temperatures down to 20°F.

Flashing cement differs from plastic cement in both the quantity and the quality of the mineral fibers added to the cutback to give it special characteristics. Plastic cement contains 15 to 45 percent asbestos or glass fibers and mineral fillers designed to maintain some plasticity in the ordinary range of roof-surface temperatures after the solvent thinner evaporates. Flashing cement, however, contains 30 to 45 percent longer asbestos or glass fibers and mineral fillers designed to produce a stiffer, sag-resistant cement.

These differing properties set their proper uses:

- Specify flashing cement for vertically flashed and canted surfaces where sag resistance is important.

- Specify plastic cement for horizontal or slightly sloped flashing joints that must accommodate relative movement (e.g., the joint between a gravel-stop flange and the membrane felts underneath).

Heat-reflective, aluminum-fibrated coatings normally provide a better surface coating for base flashings than flashing cement, especially

Table 8-1   Thermal Expansion of Counterflashing Metals

Metal	Thermal expansion coefficient $\times 10^{-6}$ in./(in.)·(°F)	Expansion of 10-ft length for 100°F temperature rise, in 64ths in.
Galvanized steel	6.7	5
Monel	7.8	6
Copper	9.4	8
Stainless steel (300 series)	9.6	8
Aluminum	12.9	10
Lead	15.0	12
Zinc, rolled	17.4	13

in hot climates and locations exposed to direct sunlight. Compared with a black coating, an aluminum coating can drop flashing surface temperature by 20°F or more, even after it ages. This temperature reduction not only reduces the risk of sagging, it enhances the asphalt's durability because its chemical degradation rises exponentially with increasing temperature.

Base-Flashing Materials

Base flashings embrace a whole spectrum of roofing materials: conventional saturated and coated felts; composition base flashing made from reinforced/laminated asbestos felt and scrim; fabrics woven from fiberglass or cotton and impregnated with a bituminous or mastic material; synthetic materials, including vinyls, neoprenes, butyl rubber.

Most base flashings used with conventional bituminous built-up membranes are composition flashings, which is a generic term denoting any nonmetallic, bituminous flashing, including fabric or felt, applied with hot-mopped asphalt or cold-process bituminous flashing cement.

Flashing fabrics generally provide greater strength per ply than felts. Close-woven fabrics are also more puncture-resistant. As still another advantage over felt flashings, they can be more easily molded to flashed surfaces.

Laminated composition flashings, with felt on the exposed surface and fabric backing, exploit the complementary qualities of these materials. For exposed base flashings, an inorganic felt (asbestos or fiberglass) provides durability. Stronger, lighter, but less durable fabric (coarse cotton or fiberglass scrims) provides strength. Laminated asbestos felt composition sheets weigh from 50 to 65 lb/square. Glass-fiber flashing sheets weigh 36 to 80 lb/square. Surfacing with mineral granules (like roll roofing) is also provided for the exposed surface on some of these composition flashing sheets. Only inorganic (asbestos, glass-fiber), reinforced, granular-surfaced sheets provide acceptable weathering qualities.

The heavier, stiffer asbestos felt–cotton fabric flashing sheets may require hot-mopped steep asphalt to make them workable enough to conform with the flashed surface. Lighter sheets can be installed with a cold-troweled flashing cement.

Although unsatisfactory for the surfacing sheet, organic felts are suitable for interior, unexposed base flashing plies. But even here inorganic felts are preferable because flashing felts are more likely than membrane felts to be exposed to weathering attack.

For walls and parapet base flashings, the choice between hot-mopped asphalt or cold-applied flashing cement as the bonding and interply adhesive may be dictated by anchorage provisions. Because hot-mopped flashings must be nailed, a precast concrete parapet or similarly nonnailable wall section requires cold-applied flashing cement.

With nailable walls, e.g., concrete with wood nailing strips, the choice involves important trade-offs. Hot-mopped steep asphalt offers speed, convenience, and economy, but it presents two major problems:

1. The need for quickly setting the hot-mopped flashing against the base before the asphalt congeals and fails to bond

2. The possibility of mistakenly substituting Type I for Type III asphalt, with consequent risk of sagging

Because of its great sag resistance, cold-applied flashing cement is better than hot-mopped asphalt for vertically flashed surfaces. In long runs, cold-applied flashing cement may raise labor costs, but the added expense may be well worth it when there is much doubt about application of proper materials. It is especially advantageous in cold weather, which makes it extremely difficult to get good adhesion with hot-mopped asphalt. Unlike hot-mopped asphalt, flashing cement can serve as both an interply waterproofing and adhesive agent between plies and as a surfacing material. Asbestos fibers make flashing and plastic cements weather-resistant as well as stable. Fabric and organic felt flashings must be surfaced with flashing cement, plastic cement, or clay-type asphalt emulsion. Mineral-surfaced roll roofing, often placed over the lower plies of hot-mopped flashing felt, requires no field-applied surfacing.

Because of the superior waterproofing provided by properly troweled layers of flashing cement, cold-applied flashing systems with fabric can sometimes have fewer plies than the membrane. (Hot-mopped felt flashings generally require, as a minimum, the same number of plies as the membrane.) A "three-course," cold-applied flashing system has one reinforcing ply sandwiched between two trowelings of flashing cement; a "five-course" flashing system has two reinforcing plies with three trowelings. A one-ply, three-course flashing system relies too heavily on good-quality work and good maintenance to be generally reliable, according to some roofing experts, who advocate a minimum two-ply, five-course base flashing.

Metal is generally unsuitable as base flashing material. It normally lacks the required flexibility for molding to supports; it is difficult to connect to the built-up membrane; and metals have incompatible coefficients of expansion-contraction. At hot summer temperatures, when metal expands, bitumen-saturated felts experience little or no expansion. But at the coldest winter temperatures, bitumen-saturated felts contract at rates varying from about 30 to 500 percent greater than the various metals used as counterflashings.

Metal base flashings are often detailed at skylights, ventilators, HVAC equipment, scuttles, and similar roof-penetrating components (see Fig. 8-4). These are generally poor details. The roof-penetrating components should be set on curbs flashed with bituminous materials, as shown in NRCA Detail R.

A notable exception to the rule against metal base flashings is at roof drains, where the need for stability and a good bolted connection to the metal drain frame favors lead or copper. A drain flashing seldom measures more than 2 ft, at most. Thermal contraction of a lead flashing of that width amounts to $\frac{1}{32}$ in. maximum; the bituminous materials bonded to the lead flashing can readily accommodate movements of that tiny magnitude. Moreover, areas around drains normally exhibit more stable temperatures because of uninsulated sumps or thinner insulation.

As a class of materials, single-ply synthetic flashings—vinyl, neoprene, butyl rubber, nonplasticized chlorinated polyethylene reinforced with polyester mesh—still lack adequate performance history for general approval. Among the troubles encountered with some of these materials are deterioration in sunlight and gradual loss of plasticizer. They have sometimes cracked under cold-weather shrinkage and embrittlement. As a consequence, at least one elastomeric flashing material, polyvinylidene chloride, was withdrawn from the market. However, some single-ply flashing materials are performing well and promising durability.

Uncured neoprene, a synthetic rubber sheet, offers a superior combination of qualities for flashing: ease of installation (because it readily conforms to corners and other flashed surfaces), excellent ozone and ultraviolet resistance, and flexibility at extreme low temperature.

Fig. 8-4 Metal base flashing shown is bad design in several important respects. (a) It requires a difficult connection between membrane and metal. Different rates of thermal contraction, especially at subfreezing temperatures, can split the membrane. (b) The cap flashing has inadequate lap over the metal base flashing. (c) Standing water reveals that this vulnerable, badly designed joint, far from being elevated above general roof level, is at a low point of the roof.

Aluminum-foil-faced, plastic-core bituminous single-ply flashing offers similarly good performance plus high heat reflectivity, which maintains lower temperature during hot summer weather.

Before specifying these new materials, designers should investigate their past performance, demanding technical data from laboratory and field tests. Costly failures have resulted from using unsuitable flashing materials.

Plastic—notably vinyl—may be incompatible with asphaltic materials. An oily film may appear between the plastic flashing and the bituminous materials, and after 2 or 3 years the flashing may delaminate. Leaks in these delaminated places are usually impossible to reseal because dirt and dust can collect in the delamination planes.

Counterflashing Materials

Counterflashings (or cap flashings) shield the exposed joints of base flashings and shed water from vertical surfaces onto the roof. Because of their exposed locations, counterflashings must be rigid and durable; thus metal generally proves the best material. The metals used for counterflashings include copper, aluminum, galvanized steel, stainless steel, and lead.

Where no cap flashing is provided, the top edge of a composition base flashing is sometimes counterflashed with similar bituminous material. When counterflashing covers a large area, it is a "stripping" or "seal" (see Fig. 8-5).

Whenever there is a possibility of delay by the sheet metal contractor installing a metal counterflashing, the top edge of all base flashing requires a temporary bituminous counterflashing to protect the vulnerable base flashing joint. This temporary counterflashing can consist of a three-course seal of asbestos felt and flashing cement. At a fascia-counterflashed edge detail, the temporarily exposed base flashing joint

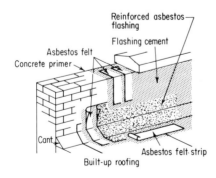

Reinforced asbestos flashing

Flashing cement

Asbestos felt

Concrete primer

Cant

Built-up roofing

Asbestos felt strip

Fig. 8-5 Bituminous and counterflashing on masonry parapet is known as a "waterproofing wall system." It is generally limited to a 2-ft height above top of cant. Omission of a wood backer board, as in this detail, is permissible only when the roof deck's structural framing bears on the masonry wall.

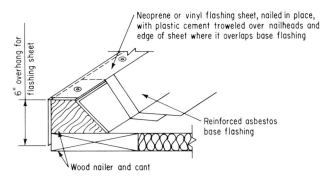

6" overhang for flashing sheet

Neoprene or vinyl flashing sheet, nailed in place, with plastic cement troweled over nailheads and edge of sheet where it overlaps base flashing

Reinforced asbestos base flashing

Wood nailer and cant

Fig. 8-6 The detail shows temporary seal provided at base-flashing edge strip if installation of metal fascia/counterflashing detail cannot proceed immediately following application of the base flashing.

can be protected with a single neoprene, vinyl, or felt flashing sheet (see Fig. 8-6).

PRINCIPLES OF FLASHING DESIGN

Flashing design requires observance of the following four basic principles:

- Locate flashed joints above the highest water level on the roof, providing positive drainage away from flashed joints.
- Allow for differential movement between base and cap flashings.
- Contour flashed surfaces to avoid sharp bends (45° maximum) in bituminous base flashings.
- Connect flashings solidly to firm supports.

Flashing Elevation

As the most basic principle of flashing design, often ignored, locate flashed joints above the roof water line. Because of the numerous penetrations required through a modern building's roof, it is often difficult to follow this rule. It requires coordination of the drainage design with the flashing layout. Locate gravel stops, wall intersections, skylights, equipment bases, and vent stacks at high points.

When it is impracticable to follow this rule, at least keep flashings out of low areas where ponding might occur. Even well-flashed joints are more vulnerable than the membrane to penetration by standing water, and poorly flashed joints are extremely vulnerable. Failure to keep flashed joints out of standing water is one of the most common and costly errors in roof design. It is also one of the most difficult and expensive to correct.

Base flashings at vertical surfaces should extend at least 8 in. above the highest anticipated waterline (see Fig. 8-1) for two reasons. First, there is the necessity of protecting the flashed joint from being penetrated by rain driven against the vertical surface or snow piled against it. The second reason concerns the field-application procedure. A vertical dimension of 8 in. (absolute minimum of 6 in.) provides working room for application of the base flashing, i.e., for nailing or otherwise anchoring the flashing felts above the cant strip. Maximum elevation for base flashing on a vertical surface is 12 in. above membrane grade line. This maximum-height limit is a safeguard against sagging, which becomes more probable as the weight of the flashing increases with its vertical dimension. A vertical flashing must resist the force of gravity throughout every second of its service life. Even perfectly formulated flashing cement or steep asphalt tends to flow a little at the highest range of roof temperatures reached even in the northern United States and Canada, and this plastic flow is always downward. Limiting the base flashing's height above the membrane level, especially hot-mopped flashing, thus limits the sagging stress applied to nails, reglet friction wedges, or other flashing anchorage devices.

Limiting flashing height also reduces the vulnerability to differential movement at the intersection of vertical and horizontal surfaces. And, as a practical matter, this flashing-height limitation works well with economical use of felt rolls. Cutting a 36-in.-wide felt roll along its longitudinal centerline yields an 18-in. width, with 4-in. horizontal extension along the membrane, about 5½ in. along a 4 × 4 in. cant, and 8½ in. up the vertical surface.

Differential Movement

Flashing details must provide for differential movement among the different parts of the building. Anchor base flashings to the structural roof deck, free of the walls or other intersecting elements, and anchor the counterflashing to the wall, column, pipe, or other flashed building component.

This need to accommodate relative movement between structurally independent building elements is often overlooked—notably in flashing details that incorrectly connect base flashing directly to the walls (see NRCA Detail H toward the end of this chapter for the correct method of isolating base and counterflashing).

Differential movement is the basic reason why metal counterflashings, edge strips, gravel stops, and equipment curbs should be kept from contact with bituminous materials. As indicated earlier in this chapter, the coefficients of expansion-contraction of metals and bituminous materials vary greatly, especially at low temperatures. At the coldest winter temperatures ($0°$ to $-30°F$), built-up roof membranes have expansion-contraction coefficients ranging around 35×10^{-6} in. lin./ °F, roughly three times the coefficient for aluminum, five times the coefficient for galvanized steel. At subzero temperatures, the contracting membrane can pull away from the metal. Wherever it breaks the bond becomes a potential source of leaks (see Fig. 8-7).

Note also that, contrary to prevailing opinion, it is *membrane* contraction, not *metal* contraction, that causes separation at subzero temperatures. Whenever contraction stresses exceed the adhesive forces bonding the membrane to the metal and also exceed the membrane's ultimate tensile strength, the membrane splits.

Thus, wherever practicable, keep metal from contact with bituminous materials. Where this is impracticable, notably at some perimeter gravel-stop details, put the primed flange on top of the membrane, coat the joint between the two dissimilar materials with plastic cement, and provide two stripping seals (also set in plastic cement) as shown in NRCA Detail O.

Flashing Contours and Cants

A third basic rule of flashing design is to avoid right angle corner bends in bituminous base flashing. To reduce the risk of cracking the felts, use cants of 45° slope.

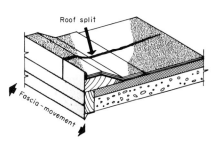

Fig. 8-7 Bonding the membrane directly to the metal flange of a gravel stop can split the membrane when it contracts at low temperature because its thermal coefficient is greater than that of any commonly used roofing metal.

Wood cants (southern yellow pine, Douglas fir, or equivalent), pressure-treated with water-base preservative, are generally preferable to fiberboard cants, especially where they abut vertical nailers at equipment or wall curbs. Wood cants help brace these vertical nailers, whereas fiberboard nailers perform no structural function. Wood cants are also more durable. When flashing leaks wet a fiberboard cant, it tends to deteriorate and soften. However, a treated wood cant continues to provide firm, solid support for the flashing. Wood cants should be securely anchored, to prevent warping.

Flashings should be supported continuously, firmly adhered to their backing, and not left to bridge openings.

Flashing Connections

Anchorage for flashings is generally the same as for the built-up membrane and subject to the same limitations. Secure nailing for flashings is even more important than for the membrane because perimeter flashings—e.g., gravel stops—are more vulnerable to wind uplift and other physical damage. On vertical surfaces, flashings exert a constant gravity force against flashing nails. Flashing nail spacing should not exceed 8 in.

Perimeter flashing failures initiate most wind-uplift roofing failures, according to Factory Mutual. Flashing nails driven into wood (or lag screws used for the same purpose) should satisfy the following FM requirements:

- Twisted or threaded shanks

- Corrosion-resistant

- 1½-in. minimum penetration into wood

- 100-lb withdrawal for each nail holding flashing or cleats, 150-lb withdrawal resistance for nails anchoring wood cants, top nailers, and fascias

- 3-in. minimum spacing for lighter metal flashings (24-gage steel, 0.032-in. aluminum, or 20-oz copper)

- Staggered nailing patterns for wind perimeter cant-strip surfaces, with spacing *in any one row* not exceeding the foregoing 3-in. requirement

Steel nails driven through dissimilar flashing metals—e.g., aluminum or copper—could cause galvanic corrosion if water provides an electrolytic circuit linking the two metals. Neoprene washers, besides sealing nailing holes, help isolate incompatible materials from corrosion-generating contact. And the underlying hot-mopped asphalt or flashing cement should press up around the nail head at the penetration point, for further isolation and water protection.

For nailing into masonry mortar joints, specify 1½-in.-long barbed, hardened, simplex-type roofing nails; for concrete anchorage, specify case-hardened steel driven through steel disks.

SPECIFIC FLASHING CONDITIONS

Flashings are required for the following conditions:

1. Edge details, e.g., gravel stops, eaves

2. Walls, parapets, and other vertical surfaces

3. Roof-penetration connections—for vents, skylights, roof drains, scuttles, airconditioning equipment, columns, and so forth

4. Expansion-contraction joints

5. Water conductors, e.g., built-in gutters, valleys, scuppers

Each kind of flashing shares in the general problems previously discussed and also has its own peculiar problems, discussed in the following text.

Edge Details

Gravel stops perform four functions:

- As a barrier preventing loose aggregate from rolling off the roof

- As edge termination for the membrane

- As rain shield

- As top surface for anchorage of a fascia strip

To serve these functions, gravel stops require a rigid material: metal or premolded plastic. End laps, expansion joints, and membrane connections must all be watertight yet flexible enough to accommodate thermal movement and differential movement of dissimilar materials. Flange widths should be a minimum 4 in., but not greater than the horizontal width of the underlying nailer. The flange should be primed before installation.

Gravel stops, like other flashed joints, must be raised above the general roof elevation. This can be accomplished by simply sloping the roof down from the gravel stop or via a tapered edge strip (see NRCA Detail Q). On roofs with perimeter drainage systems, the tapered edge strips should form a cricket surface.

Edge strips are subjected to especially high wind-uplift forces; they form the leading edge of the roof airfoil. Most blowoffs start with wind loosening the roof-edge detail, bending and twisting the metal fascia strip, then peeling the membrane off the insulation. To ensure adequate stiffness in the fascia, the roof designer should consult Factory Mutual's *Loss Prevention Data Sheet 1-49* for recommended metal thickness and anchorage details. For metal-deck roofs, edge flashing details require proper blocking as a shield against wind access to flutes under the insulation at the upper deck surface.

To accommodate thermal expansion and contraction stresses in a gravel stop, the roof designer has two alternative strategies:

1. Closely spaced (3-in.) nailing, designed to restrain the gravel stop from expansion and contraction

2. Sleeved joints for 10-ft lengths of fascia units, designed to accommodate linear expansion and contraction

The first strategy, nailing, is suitable for thin-gage metal, in which expansion-contraction forces are light enough to be resisted by the closely spaced nails (see NRCA Detail O). The second strategy is for thicker sections—e.g., extruded aluminum sections—whose thermal forces might distort closely spaced nails. Here the 10-ft sections of fascia are anchored at midlength, with ends left free to expand and contract, inserted into sleeves anchored to the cant (see NRCA Details A and G).

For stark examples of how *not* to detail a gravel stop, see Figs. 8-8 to 8-10. These details violate the two most basic rules of flashing design:

• They are *not* located above the highest water level anticipated on the roof. (They are in fact both used on dead-level roofs, on which ponding occurred over some segments of flashed joints.)

• They do *not* allow for differential movement.

Where metal gravel stops are stripped in with the membrane, splits often occur, extending from the edge perpendicular to the perimeter

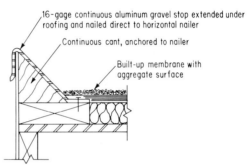

16-gage continuous aluminum gravel stop extended under roofing and nailed direct to horizontal nailer

Continuous cant, anchored to nailer

Built-up membrane with aggregate surface

Fig. 8-8 This gravel-stop detail violates several basic principles of flashing design. (1) The gravel stop serves a dual function as base flashing, thus violating the rule against metal base flashing, which has a different coefficient of thermal expansion-contraction from the membrane. (2) The metal flange is placed *under* rather than on top of the membrane, thereby compounding the problems resulting from differential movement of the two materials. (3) This vulnerable joint was not raised above the general roof slope; it occurred on a dead-level roof, subject to ponding at random places all over the roof. The correct design for this detail is shown in NRCA Detail G.

line toward the roof interior (see Fig. 8-7). Such splits will most likely occur at subzero temperatures. In the 0 to −30°F temperature range, bitumen is a brittle solid, and the membrane coefficient of expansion-contraction becomes extremely high (three to five times that of the metal gravel stop). In climates subject to such extremely cold temperatures, notably the northern and midwestern states, it is especially important to isolate gravel stops from the membrane with a layer of plastic cement.

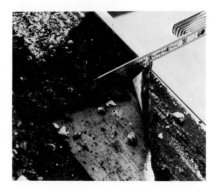

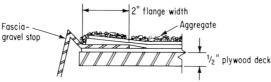

Fig. 8-9 The gravel stop depicted here violates the same basic rules as the gravel stop in Fig. 8-8. As the only significant difference, the gravel-stop flange is inserted between the membrane base and the two shingled plies. The gravel-stop–fascia flange should have been placed on top of the membrane, and the flashed joint should be elevated, as shown in NRCA Detail O.

Because of flashed joints' vulnerability to ponded or running water, edge cant strips are preferable to gravel-stop–fascia details. (See NRCA Detail G.)

Vertical Flashings

Flashings at masonry walls or parapets come in two types, depending on the need to accommodate differential movement. Where the roof deck and the wall are structurally connected—e.g., where the wall is supported on a monolithically cast concrete frame that incorporates the roof deck as a concrete slab, or roof beams bear on the wall—the flashing detail may not need to allow for differential movement (see NRCA Details J and V). Where the roof structure is not tied to the wall—e.g., where the roof's structural members span parallel to the wall, or where the wall is a nonbearing, curtain wall—an expansion joint is required to accommodate differential movement (see NRCA Detail H). For this detail, a vertical nailer behind the cant forms a blocking backer for the base flashing. Like the cant, this vertical nailer must be anchored to the

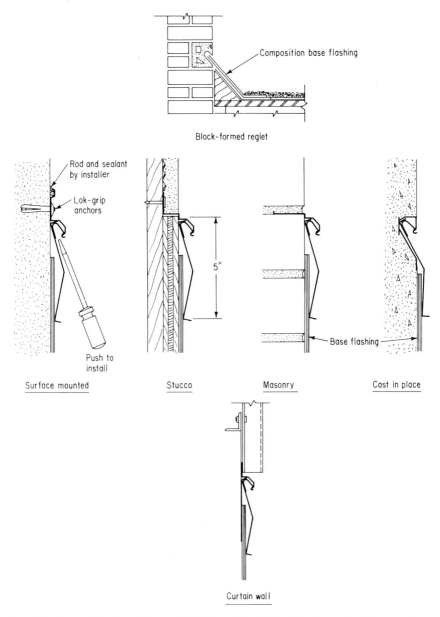

Composition base flashing

Block-formed reglet

Rod and sealant by installer

Lok-grip anchors

Push to install

Surface mounted

Stucco

5"

Masonry

Base flashing

Cast in place

Curtain wall

Fig. 8-10 Conventional continuous block-formed reglet (top) is deservedly disappearing, extremely difficult to caulk and maintain, and highly vulnerable to water penetration behind base flashing. Stainless steel clip-spring assembly (bottom details) eliminates reglet for base flashing and shields it with metal counterflashing. (IMETCO.)

187

structural deck. (Even when an expansion joint is not required, a vertical backer board against the wall may be generally advisable.)

Compressible insulation inserted in the 1-in. gap between the back of the vertical nailer and the wall blocks convection currents that waste heating energy and carry moisture to the underside of the counterflashing through joints in the wood blocking. Because the conveyed air would come from the warm interior, water vapor would condense on the cold surfaces, leaking back into the building as the water vapor thawed. In cold climates, freezing of this condensed migrating water vapor can build up ice layers, possibly damaging the flashing. And upon thawing, this condensed moisture could leak back into the building.

A convenient insulation for sealing off this convective vapor flow is 1-in.-thick fiberglass batt insulation with a vapor retarder (two layers of kraft paper adhered with bitumen) on one face. The roofer folds this insulation double, with the vapor retarder on the outside, squeezes the envelope (nailing tabs up) into the gap behind the vertical nailer, bends the tabs over, and nails or staples them to the nailer's top surface. NRCA Detail H shows a flexible vapor retarder, required to obstruct vapor flow up through the insulation filler.

Two-piece, through-wall metal counterflashing has several advantages:

- Delayed installation of the vertical shielding segment permits installation of the base flashing without bending up the projecting metal and possibly deforming it from proper alignment when it is rebent. Turning up a one-piece metal counterflashing to repair or replace base flashings work hardens the metal and often forms a concave lip where the metal enters the wall. Water collecting on this lip usually works its way into the wall, and in time this invading water causes spalling.

- The horizontal receiver segment shields against water filtering through the core of the masonry wall and then laterally into the rear of the base flashing.

- Metal flashings can sometimes be salvaged when reroofing is required.

Note that cleats are generally required to brace the bottom of the exposed counterflashing strip.

Because of the threat of water penetrating into the back of the base flashing, through-wall flashing rather than a reglet should be used. Avoid using a prefabricated cant and reglet block system into which the base flashing is inserted. (See Fig. 8-10.) Gravity stress on the base flashing and cracking and loosening of the reglet caulking make this an extremely vulnerable joint requiring constant surveillance. However, if for one reason or another through-wall flashing is impracticable— e.g., in a concrete wall—use a prefabricated cap-flashing system and vertical waterproofing detail or NRCA Detail U. The best elastomeric sealants for NRCA Detail U are silicone or polyurethane. Flashing cement or oil-based caulking are short-lived.

Roof Penetrations

Curbs are required around most kinds of roof openings. It is necessary to allow for differential movement by isolating the base flashing from the counterflashing and the roof-penetrating component from the built-up membrane.

It is also necessary to provide structural framing, or stiffeners, around openings, especially in metal or plywood decks. Flexible decks can be wracked by wind or other lateral forces on the roof-penetrating component, and this wracking can damage the flashing and promote leakage at the vulnerable flashed joint.

Some roof-penetrating components—notably skylights, scuttles, and HVAC rooftop units—come with metal curbs and flanges that act as base flashings, with direct connection to the membrane (see Fig. 8-11). These details violate the general flashing rule to isolate bituminous materials from metal. In cold weather, the membrane contracts much more than the metal, promoting splits and delamination at points of stress concentration.

Rooftop airconditioning units require a decision as to what kind of curbed opening to detail: support on the roof or an opening through the roof (see NRCA Details N1, N2, N3, and R). The table in Detail N1 shows minimum leg heights required for varying equipment widths.

The National Roofing Contractors Association (NRCA) has published a helpful leaflet, "NRCA Roof Curb Criteria," with a listing of NRCA-approved curb details furnished by manufacturers of rooftop equipment. Here are the major requirements:

- A continuous metal-framed curb (16-gage minimum) furnished by the manufacturer of the roof-penetrating equipment, with a 2-in. wood nailer mounted at the curb top and minimum 8-in. vertical clearance (to top of nailer) above the finished roof surface

- No penetration of the curb for drains, electrical conduit, etc.

- Provision for metal counterflashing by sheet metal contractor

- 45° cant at the base of the curb

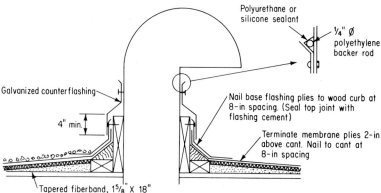

Fig. 8-11 This improperly flashed vent (top) leaked around the opening perimeter. It was corrected with the sketched detail (bottom), which (1) raised the flashing above the previously ponded roof elevation and (2) substituted a properly counterflashed bituminous base flashing for the defective metal base flashing in the original detail.

- Installation instructions requiring the supplier of rooftop equipment to provide a weathertight seal between unit and curb top

- Flashing of all piping and plumbing with roof flange-and-sleeve assembly extending a minimum 8 in. above the roof surface and the sleeve edge turned down inside the vent stack

To avoid the cost of providing separate support for base flashing at small pipes, vents, and similar small roof penetrations, designers usually omit these base flashing supports. Instead, the flanges of metal or plastic "pitch-pan" sleeves sit directly on the roof membrane, and the flanged sheet metal support carries the base flashing around it (see NRCA Detail E).

Such a detail is justified around a small roof-penetrating element because it is subjected to far less differential movement than a larger roof-penetrating element. But these details nonetheless require more frequent inspection and maintenance than basic flashing details with deck-supported base flashing. Support on the membrane is less stable than support on the structural deck. Flashed details so supported are thus more vulnerable to leakage from joint separation than conventionally flashed joints. The plastic cement may require frequent replenishment, with removal of the brittle, aged cement, to protect felt edges from exposure to water. Pyramidal roof vents, with their elevated base flashing, are generally more satisfactory than tubular stack vents.

For a roof-piercing vent stack, a curbed opening, counterflashed with a sheet metal umbrella, is recommended (see NRCA Detail F).

So-called "pitch pans" (filled with asphalt or coal tar pitch) are a cheap (in both connotations of low cost and low quality), generally unsatisfactory means of flashing roof penetrations. They pose a constant threat of leaks and require frequent inspection and maintenance, to replace bitumen embrittled by solar radiation (see Fig. 8-12). Whenever practicable, build pedestals for roof-penetrating components and flash them. Encase pipes and conduits in a properly flashed metal hood (see NRCA Detail M). A roof-penetrating column can be flashed as shown in NRCA Details L or N2.

Sometimes, however, pitch pans are the solution of last resort—for TV antennas, sign posts, and other components with irregular cross sections or diagonally angled entry into the roof. When forced to accept the pitch pan as a necessary evil, follow these rules:

- Raise pitch-pan elevation (with tapered insulation strips, if necessary) to roof-level high points.

- Use 16-oz copper, 4 in. high, with 4-in. flange.

- Pack any gap between the roof-penetrating element and deck with fiberglass insulation, mineral wool, or other compressible material to prevent drippage.

- Anchor a treated wood nailer (same thickness as insulation) around the pan opening, for nailing the flange.

- Set the flange in plastic cement on top of the membrane plies (bottom ply enveloped).

- Nail at 3-in. spacing, about 1 in. from outer edge of flange.

- Strip in with one ply before applying flood coat and aggregate.

Fig. 8-12 Fabricating a sheet metal box, detailed for lateral entry of pipes or conduit (top left), is a far better practice than allowing vertical entry into a pitch pocket (top right). The sleeper pitch-pan detail shown at the rooftop airconditioning units (middle and bottom) is a cheap, inferior substitute for the more expensive curbed or column-supported details approved by the NRCA (see NRCA Details N and R).

- Fill bottom stratum of pan with cement grout or plastic cement stiffened with portland cement before applying final surfacing stratum of bituminous sealer, sloped down to roof.

- If the roof system has a vapor retarder, flash it to preserve vapor-retarder continuity.

Drain flashings present a unique combination of hazards. Clogged drains may impound water for long intervals. Drains are also subjected to differential vertical movement between drainpipe and roof. A structural roof deck supported on shrinking 2 × 12 in. wood joists could lower the deck elevation ½ in., breaking or distorting the drain flashing at its connection. (An expansion joint or 45° offset in the drainpipe can accommodate most differential vertical movement.)

Lead or copper sheets generally provide the best connection to drain frames; they also provide the stability needed to maintain a watertight joint. As noted earlier, drain flashings constitute an exception to the general rule against metal base flashings because of severe erosive and corrosive forces at the drain. Differential movement of the metal drain flashing is a minor problem because of its small plan dimensions (see NRCA Details W1 and W2).

Expansion-Contraction Joints

Expansion joints, generally spaced from about 150 to 300 ft, relieve stresses that otherwise accompany thermal expansion and contraction and other building movement accompanying material shrinkage or foundation settlement. They should be provided at changes in span direction in the structural deck or framing, at changes in deck material, and at expansion joints extending through the structure. (See Chapter 7, "Elements of the Built-Up Membrane," for a fuller discussion of expansion-joint design.)

An expansion joint must accommodate tensile and contractive movement and shear movement in the roof plane and shearing movement perpendicular to the roof plane. It requires a curb for base flashing, and the membrane should slope down from this curb (see NRCA Details C1 and C2).

Like the gap between a vertical nailer and a wall, an expansion joint should be filled with compressible material to prevent condensation and icing from vapor migrating upward from the warm interior. Condensation on the underside of a metal bellows may drip back into the interior if the expansion-joint gap is not filled. (Some prefabricated expansion joints come with an insulated bellows.)

Uncurbed, unelevated expansion-joint details are a reckless attempt to economize. These very bad practices take several forms. Expansion joints featuring a fold in the membrane may behave like a roof ridge,

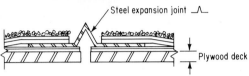

Fig. 8-13 The cheap, defective expansion-joint detail shown flouts the rules of good flashing design with a vengeance. This inverted V, galvanized steel expansion joint violates the rule against metal base flashing and the rule to elevate flashing, especially important for an expansion joint. Compounding these serious flaws, it aggravates the already serious leak hazard by inserting the metal flanges between membrane plies instead of placing them on top of the felts and stripping them in with two plies of felt.

with bitumen flowing down the sloping sides. Inverted V, galvanized steel, or other metal expansion joints in the plane of the membrane violate the rules against metal base flashing and elevated flashings (see Fig. 8-13). A designer who locates a joint designed for movement down at the general membrane level is inviting leakage.

Water Conductors

Built-in gutters are subject to surface abrasion. They are also difficult to connect to the membrane. For relatively flat gutters, which may retain water for long intervals, leakage in sliding expansion joints is also a problem.

Metal offers superior durability, but largely because of its differential thermal coefficient of expansion, it is difficult to connect permanently to the bituminous membrane. Gutters made of bituminous materials readily accommodate thermal movement, but they deteriorate so rapidly under wetting-drying cycles and the abrasive flow of water that they are limited to short valleys. NRCA Detail B shows a satisfactory method of flashing scuppers. Interior drainage is nonetheless preferable to perimeter drainage, for these reasons:

1. Interior drains are less vulnerable to ice-dam blockage than exterior drains.

2. Interior drains are generally less vulnerable to wear and damage (e.g., distortion from freeze-thaw cycling) than scuppers.

FLASHING FAILURES

Flashing failures, ultimately resulting in leaks, occur in several modes, and each failure mode has several possible causes, acting singly or in combination. Here are the basic flashing failure modes:

- Sagging

- Ponding leakage

- Leakage *around* the flashing

- Leakage directly *through* the flashing

- Separation of flashing materials

- Diagonal wrinkling (resulting in splits)

- Damage after construction (abuse)

Sagging

Sagging is a vertical (or steeply sloped) flashing's tendency to slide downward, ultimately tending to pull out from reglets, to tear, or to split. Sagging has numerous causes. It can result from omitting flashing nails, using an overly fluid adhesive—hot-applied asphalt with too low a softening point—or using plastic cement instead of the more viscous flashing cement. At summer roof temperatures, which may exceed 150°F, plastic bitumen gradually flows downward, stressing the flashing fabric and possibly tearing it. Flashings extended more than 12 in. up a vertical surface are also subject to sagging. The added weight of the flashing materials aggravates the general problem of perpetual gravity force. Membrane shrinkage, caused by inadequate insulation anchorage to the deck, can also exert sagging stress on base flashings.

Still another cause of sagging is omission of a primer on masonry walls. A primer of solvent-thinned (cutback) asphalt is required to prepare a masonry surface for either hot-applied steel asphalt or cold-applied flashing cement. If the masonry surface is not primed, mortar flakings, effluorescent precipitate, and other foreign particles may prevent good adhesion between the masonry surface and the bitumen.

Fig. 8-14 These base flashing gaps exhibit two flaws: (1) failure to bond the flashing to its vertical support and apply vertical sealer strip (left), and (2) incorrect, open-jointed corner (right). See Fig. 8-15 for correct corner detailing.

Ponding Leakage

Ponding is a greater threat to flashings than to the membrane. Flashing occurs at joints between different building components. In addition to the increased probability of leakage at any joint between materials, there is usually much greater probability of relative movement at a flashing than in the membrane. Thus arises the rule to keep flashed joints at high points, if practicable. If this is not practicable, at least keep flashing out of low points, with rapid drainage away from them. This policy requires active attention; the designer cannot simply let the elevation chips fall where they may. Yet many roof designers do just that.

Leakage can occur *around* poorly designed flashings. When vertically flashed joints are too low—i.e., less than 8 in. high, with less than a 4-in. cap-flashing overlap of base flashing—wind-driven rain, snow, or even deep ponded water can blow or rebound up and under the cap flashing, penetrating the masonry wall above the base flashing and descending inside the wall to invade the built-up roof system. If water ponds at such an inadequately detailed flashing, it is even more vulnerable to leakage from wind-driven rain and capillary rise when snow is banked against flashings. Even for a well-designed flashing of the correct height (between 8 to 12 in.), water can get behind the base flashing if the masonry is porous.

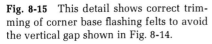

Fig. 8-15 This detail shows correct trimming of corner base flashing felts to avoid the vertical gap shown in Fig. 8-14.

Leakage directly *through* the flashing is usually the result of improper application—e.g., omission of a ply of fabric and an interply hot mopping or troweling of flashing cement or failure to embed flashing felts or fabrics into their adhesive. A loose flashing felt is especially vulnerable to puncture and the consequent admission of water. Hot-mopped flashings are especially liable to loose, faulty backing. (See Fig. 8-14.)

Separation of flashing materials often results from details that ignore metallic and bituminous flashings' different coefficients of expansion-contraction. As noted earlier, these failures often occur at gravel stops. They also occur at curbs around roof-penetrating components—scuttles, airconditioning equipment, and so forth. Such failures can be avoided by heeding the design principles set forth elsewhere in this chapter.

Diagonal Wrinkling

Diagonal wrinkling in base flashing along parapet walls follows a consistent pattern, with the wrinkles sloping down from the top toward the center of the wall (see Fig. 8-16). The most severe wrinkling occurs at building corners, diminishing in proportion to the distance from the corners and proximity to the center. Diagonal wrinkles obviously break the flashing felts' bond with their backing and thus make the flashing highly vulnerable to punctures and mechanical damage. And like membrane ridging, diagonal wrinkling of base flashing often sets in motion a degenerative process that ends in cracking and consequent leaking.

Diagonal wrinkling has several causes, acting independently or jointly. Membrane shrinkage both parallel to and perpendicular to (i.e., away from) the parapet produces these wrinkles. As discussed elsewhere in this manual, failure to anchor the roof system to the deck—notably at the deck-insulation interface—permits a contracting membrane to drag loose insulation boards across the deck surface. Because the flexible membrane cannot expand but only contracts under thermal or moisture-content cycling, diagonal wrinkling from this cause tends to worsen with time.

Fig. 8-16   Diagonal wrinkling in parapet flashing can result from (1) relative movement of membrane parallel to the wall from insecurely anchored insulation boards or (2) long term masonry wall expansion.

The worst diagonal wrinkling from membrane shrinkage usually occurs along a parapet perpendicular to the felt direction because built-up membranes shrink more in the transverse than in the longitudinal direction, in response to a temperature drop and drying of wet felts.

Another cause of diagonal wrinkling in base flashing, sometimes in combination with membrane shrinkage, is differential thermal expansion of the masonry parapet walls. Long term wall growth can produce the relative lateral movement that produces diagonal wrinkling in base flashing. This long term expansion results from the largely irreversible elongation caused by thermal cycling and moisture absorption in the inelastic masonry. In the northern hemisphere, sun-baked parapets may experience daily temperature differentials of 100°F and extreme annual differentials approaching 200°F, compared with 15°F daily and 25°F seasonal extreme temperature differentials for the roof deck and structural framing.

On parapets with stone or concrete copings, freeze-thaw cycles can also promote diagonal wrinkling of the base flashings. Thermal cycling cracks the mortar joints, letting water enter. Subsequent freezing expands the water, prying apart the coping stones and possibly dragging the top edge of the base flashing along with them.

This parapet-promoted diagonal wrinkling in base flashing can be avoided by isolating the base flashing from the parapet, as shown in NRCA Detail H.

Post-Construction Damage

Post-construction damage accounts for a significant proportion of flashing failures. Equipment and scuttle locations are the busiest locations on the roof, and a dropped tool or a carelessly swung worker's boot can puncture an inadequately supported flashing. Ladder locations, where a roof visitor climbs over a parapet or up a wall, with the lowest ladder rung starting over a foot above roof level, are likely spots for a careless kick that may puncture base flashing not solidly bonded to a firm backing. Maintenance crews working on ventilating or airconditioning equipment may even fail to reinstall counterflashing removed during repair operations.

NRCA DETAILS

Pages 200 through 212 illustrate NRCA Details A–Z, used by permission of the National Roofing Contractors Association.

SCUPPER THROUGH ROOF EDGE

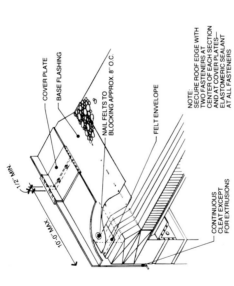

2" 4" MIN 4" MIN 4" MIN

2" 0" MAX.

2-PLY STRIPPING

¾" × 1" L RIVETED & SOLDERED TO APRON—NOTCHED TO PERMIT DRAINAGE

SET FLANGE IN MASTIC
-NAIL TO BLOCKING
-PRIME FLANGE
BEFORE STRIPPING

FELT ENVELOPE

TURN DOWN PLY
OF FELT TO BELOW
BLOCKING

1980

B

NOTES:

THIS DETAIL SHOULD BE USED ONLY WHERE THE DECK IS SUPPORTED BY THE OUTSIDE WALL.

THIS DETAIL CAN BE ADAPTED TO ROOF EDGES SHOWN IN DETAILS "G" AND "O" AND IS EASY TO INSTALL AFTER THE BUILDING IS COMPLETED. THIS DETAIL IS USED TO RELIEVE STANDING WATER IN AREAS ALONG THE ROOF EDGE. ALL ROOF SURFACES SHOULD BE SLOPED TO DRAIN.

ATTACH NAILER TO MASONRY WALL. REFER TO FACTORY MUTUAL DATA SHEET 1-49.

WOOD BLOCKING MAY BE SLOTTED FOR VENTING WHERE REQUIRED.

2

HEAVY METAL ROOF EDGE

1980

A

COVER PLATE

BASE FLASHING

NOTE:
SECURE ROOF EDGE WITH
TWO FASTENERS AT
CENTER OF EACH SECTION
AND AT COVER PLATES—
ELASTOMERIC SEALANT
AT ALL FASTENERS

NAIL FELTS TO
BLOCKING APPROX. 8" O.C.

FELT ENVELOPE

½" MIN

10'-0" MAX

CONTINUOUS
CLEAT EXCEPT
FOR EXTRUSIONS

NOTES:

THIS DETAIL SHOULD BE USED ONLY WHERE THE DECK IS SUPPORTED BY THE OUTSIDE WALL.

METALS OF 22 GAUGE STEEL, 0.050" ALUMINUM, 24 GAUGE STAINLESS STEEL OR HEAVIER ARE APPROPRIATE FOR THIS DETAIL. METALS OF THIS WEIGHT ARE VERY RIGID WHEN FORMED, AND FASTENING AT THE CENTER-LINE AND JOINT COVER WILL ALLOW EXPANSION AND CONTRACTION WITHOUT DAMAGING THE BASE FLASHING MATERIAL.

ATTACH NAILER TO MASONRY WALL. REFER TO FACTORY MUTUAL DATA SHEET 1-49.

WOOD BLOCKING MAY BE SLOTTED FOR VENTING WHERE REQUIRED.

1

CONSTRUCTION DETAILS/CONSTRUCTION DETAILS/CONSTRUCTION

EXPANSION JOINT

FASTENERS APPROX. 8" O.C.

FASTENERS APPROX. 8" O.C.—BOTH SIDES

BASE FLASHING—COVER TOP OF BASE FLASHING WITH VAPOR RETARDER

2" NOMINAL

NAIL TOP AND BOTTOM APPROX. 16" O.C.

FLEXIBLE VAPOR RETARDER TO SERVE AS INSULATION RETAINER—ATTACHED TO TOP OF CURB

CHAMFER EACH SIDE OF WOOD CURB TO DRAIN

8" MIN.

COMPRESSIBLE INSULATION

WOOD CANT TO PROVIDE STRUCTURAL STRENGTH

WOOD NAILER EACH SIDE SECURED TO DECK WITH APPROPRIATE FASTENERS APPROX. 24" O.C.

1980 C-2

4

EXPANSION JOINT

DRIVE CLEAT OR STANDING SEAM

FASTENERS APPROX. 8" O.C. BOTH SIDES

FASTENERS APPROX. 12" O.C.

BASE FLASHING—COVER TOP OF BASE FLASHING WITH VAPOR RETARDER

2" NOMINAL

NAIL TOP AND BOTTOM APPROX. 16" O.C.

FLEXIBLE VAPOR RETARDER TO SERVE AS INSULATION RETAINER—ATTACHED TO TOP OF CURB

CHAMFER TOP OF BOTH WOOD CURBS TO DRAIN TO ONE SIDE

DRAINAGE SLOPE

8" MIN.

COMPRESSIBLE INSULATION

WOOD CANT TO PROVIDE STRUCTURAL STRENGTH

WOOD NAILER EACH SIDE SECURED TO DECK WITH APPROPRIATE FASTENERS APPROX. 24" O.C.

1980 C-1

NOTE:
THIS DETAIL ALLOWS FOR BUILDING MOVEMENT IN BOTH DIRECTIONS. IT HAS PROVEN SUCCESSFUL WITH MANY CONTRACTORS FOR MANY YEARS.

3

EQUIPMENT OR SIGN SUPPORT

SET BOLTS IN ELASTOMERIC SEALANT
NEOPRENE PAD
FASTENERS APPROX. 24" O.C.
REMOVABLE COUNTERFLASHING
FASTENERS APPROX. 8" O.C.
BASE FLASHING
2" NOMINAL
FIBER CANT STRIP—SET IN BITUMEN

14" MIN TO BOTTOM OF EQUIPMENT

8" MIN

1980 · D-2

NOTE:
THIS DETAIL ALLOWS FOR ROOF MAINTENANCE AROUND THE EQUIPMENT SIGN. THE CONTINUOUS SUPPORT IS PREFERRED IN LIGHTWEIGHT ROOF SYSTEMS SINCE THE EQUIPMENT WEIGHT CAN BE SPREAD OVER MORE SUPPORTING MEMBERS, WHERE HEAVY STRUCTURAL SYSTEMS ARE USED OR WHERE THE LOAD CAN BE CONCENTRATED OVER A COLUMN, DETAIL "N" IS PREFERRED. CLEARANCE MUST BE PROVIDED FOR REMOVAL AND REPLACEMENT OF ROOFING AND FLASHING BETWEEN PARALLEL SUPPORTS.

6

AREA DIVIDER

FASTENERS APPROX. 24" O.C.
FASTENERS APPROX. 8" O.C.
2" NOMINAL
BASE FLASHING
FIBER CANT STRIP—SET IN BITUMEN
FASTEN WOOD BLOCKING TO METAL DECK WITH MECH. FASTENER

8" MIN

1980 · D-1

NOTE:
AN AREA DIVIDER IS DESIGNED SIMPLY AS A RAISED DOUBLE WOOD MEMBER ATTACHED TO A PROPERLY FLASHED WOOD BASE PLATE THAT IS ANCHORED TO THE ROOF DECK. AREA DIVIDERS SHOULD BE LOCATED BETWEEN THE ROOF'S EXPANSION JOINTS AT 100-200 FOOT INTERVALS, DEPENDING UPON CLIMATIC CONDITIONS AND AREA PRACTICES. THEY SHOULD NEVER RESTRICT THE FLOW OF WATER.

5

202

STACK FLASHING

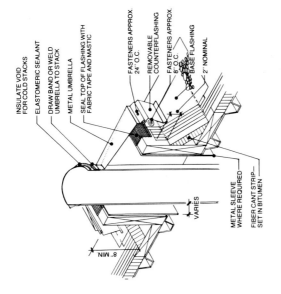

INSULATE VOID
FOR COLD STACKS

ELASTOMERIC SEALANT

DRAW BAND OR WELD
UMBRELLA TO STACK

METAL UMBRELLA

SEAL TOP OF FLASHING WITH
FABRIC TAPE AND MASTIC

FASTENERS APPROX.
24" O.C.

REMOVABLE
COUNTERFLASHING

FASTENERS APPROX.
8" O.C.

BASE FLASHING

2" NOMINAL

VARIES

8" MIN.

METAL SLEEVE
WHERE REQUIRED

FIBER CANT STRIP—
SET IN BITUMEN

1980

F

NOTE:
THIS DETAIL ALLOWS THE OPENING TO BE COMPLETED BEFORE THE
STACK IS PLACED. THE METAL SLEEVE AND THE CLEARANCE NECES-
SARY WILL DEPEND ON THE TEMPERATURE OF THE MATERIAL HANDLED
BY THE STACK.

8

ROOF RELIEF VENT

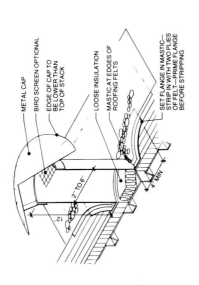

METAL CAP

BIRD SCREEN OPTIONAL

EDGE OF CAP TO
BE LOWER THAN
TOP OF STACK

LOOSE INSULATION

MASTIC AT EDGES OF
ROOFING FELTS

SET FLANGE IN MASTIC
STRIP IN WITH TWO PLIES
OF FELT—PRIME FLANGE
BEFORE STRIPPING

2" TO 6"

4" MIN

12

1980

E

NOTE:
THIS DETAIL IS USED TO RELIEVE MOISTURE VAPOR PRESSURE FROM
INSULATION. THE MOISTURE MAY HAVE ENTERED DUE TO LEAKS,
FAULTY VAPOR RETARDERS OR DURING CONSTRUCTION. THE SPAC-
ING OF RELIEF VENTS IS DETERMINED BY THE TYPE OF INSULATION
USED AND THE AMOUNT OF MOISTURE TO BE RELIEVED. THIS DETAIL IS
SOMETIMES USED FOR NEW ROOFS WHEN VAPOR RETARDERS ARE
USED AND A VENTING SYSTEM IS DESIRED.

7

BASE FLASHING FOR NON-WALL SUPPORTED DECK

- 3" LAP WITH SEALANT
- METAL REGLET
- FASTENERS APPROX. 24" O.C.
- LAP METAL AT JOINTS
- REMOVABLE COUNTERFLASHING
- FLEXIBLE VAPOR RETARDER TO SERVE AS INSULATION RETAINER
- 2" WIDE CLIP APPROX. 30' O.C.
- FASTENERS APPROX. 8" O.C.
- BASE FLASHING—COVER TOP OF BASE FLASHING WITH VAPOR RETARDER
- 2" NOMINAL
- WOOD CANT STRIP TO PROVIDE STRUCTURAL STRENGTH—NAIL TOP AND BOTTOM APPROX. 16" O.C.
- WOOD NAILER SECURED TO DECK WITH APPROPRIATE FASTENERS APPROX. 24" O.C.
- COMPRESSIBLE INSULATION

NOTES:
THIS DETAIL ALLOWS WALL AND DECK TO MOVE INDEPENDENTLY.
THIS DETAIL SHOULD BE USED WHERE THERE IS ANY POSSIBILITY THAT DIFFERENTIAL MOVEMENT WILL OCCUR BETWEEN THE DECK AND A VERTICAL SURFACE, SUCH AS AT A PENTHOUSE WALL. THE VERTICAL WOOD MEMBER SHOULD BE FASTENED TO THE DECK ONLY. THIS IS ONE SATISFACTORY METHOD OF JOINING THE TWO PIECE FLASHING SYSTEM. OTHER METHODS MAY BE USED.

1980 H

10

CONSTRUCTION DETAILS/CONSTRUCTION DETAILS/CONSTRUCTION

LIGHT METAL ROOF EDGE

- BASE FLASHING
- FASTENERS APPROX. 8" O.C.
- NOTE: SECURE ROOF EDGE WITH TWO FASTENERS AT CENTER OF EACH SECTION AND AT COVER PLATES— ELASTOMERIC SEALANT AT ALL FASTENERS
- WOOD CANT STRIP— NAIL TO BLOCKING
- CONTINUOUS CLEAT
- ½" MIN.
- 10'-0" MAX.
- 3½" MIN.
- 4" TO 6"

NOTES:
THIS DETAIL SHOULD BE USED ONLY WHERE THE DECK IS SUPPORTED BY THE OUTSIDE WALL.

THIS DETAIL IS SIMILAR TO DETAILS "A" AND "O". THE CANT STRIP, PLACED AS SHOWN, WILL RESULT IN A HIGHER FASCIA LINE. THE NO. 15 FELT SHOWN BEHIND THE FASCIA PROVIDES PROTECTION FOR THE FLASHING EDGE AND SEALS THE SYSTEM UNTIL THE METAL WORK IS INSTALLED.

ATTACH NAILER TO MASONRY WALL. REFER TO FACTORY MUTUAL DATA SHEET 1-49.

WOOD BLOCKING MAY BE SLOTTED FOR VENTING WHERE REQUIRED.

1980 G

9

BASE FLASHING FOR VENTED BASE SHEET

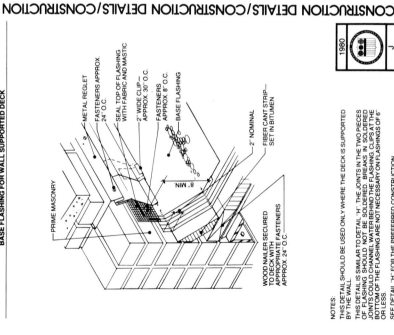

Labels:
- METAL REGLET
- FASTENERS APPROX. 24" O.C.
- 2" WIDE CLIP — APPROX. 30" O.C.
- FASTENERS APPROX. 8" O.C.
- BASE FLASHING
- 2" NOMINAL
- VENTED BASE SHEET
- FIBER CANT STRIP — SET IN BITUMEN
- APPROVED LOW DENSITY FASTENERS
- INSULATING CONCRETE
- WOOD NAILER SECURED TO DECK WITH APPROPRIATE FASTENERS APPROX. 24" O.C.

1980 K

NOTES:

THIS DETAIL TO BE USED OVER WET-FILL DECKS OR WHEN REROOFING OVER EXISTING INSULATION.

ALL PLIES AND FLASHING ARE TO BE SOLIDLY MOPPED TO THE BASE SHEET. CARE SHOULD BE USED NOT TO SEAL THE BASE SHEET TO THE PARAPET.

SEE DETAIL "H" FOR THE PREFERRED CONSTRUCTION.

12

BASE FLASHING FOR WALL SUPPORTED DECK

Labels:
- PRIME MASONRY
- METAL REGLET
- FASTENERS APPROX. 24" O.C.
- SEAL TOP OF FLASHING WITH FABRIC AND MASTIC
- 2" WIDE CLIP — APPROX. 30" O.C.
- FASTENERS APPROX. 8" O.C.
- BASE FLASHING
- 2" NOMINAL
- FIBER CANT STRIP — SET IN BITUMEN
- 8" MIN.
- WOOD NAILER SECURED TO DECK WITH APPROPRIATE FASTENERS APPROX. 24" O.C.

1980 J

NOTES:

THIS DETAIL SHOULD BE USED ONLY WHERE THE DECK IS SUPPORTED BY THE WALL.

THIS DETAIL IS SIMILAR TO DETAIL "H". THE JOINTS IN THE TWO PIECES OF FLASHING SHOULD NOT BE SOLDERED. BREAKS IN SOLDERED JOINTS COULD CHANNEL WATER BEHIND THE FLASHING. CLIPS AT THE BOTTOM OF THE FLASHING ARE NOT NECESSARY ON FLASHINGS OF 6" OR LESS.

SEE DETAIL "H" FOR THE PREFERRED CONSTRUCTION.

11

205

PIPING THROUGH ROOF DECK

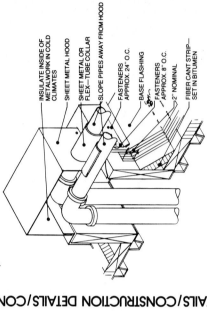

- INSULATE INSIDE OF METALWORK IN COLD CLIMATES
- SHEET METAL HOOD
- SHEET METAL OR FLEX—TUBE COLLAR
- SLOPE PIPES AWAY FROM HOOD
- FASTENERS APPROX. 24" O.C.
- BASE FLASHING
- FASTENERS APPROX. 8" O.C.
- 2" NOMINAL
- FIBER CANT STRIP— SET IN BITUMEN

NOTE:
THIS DETAIL ILLUSTRATES ANOTHER METHOD OF ELIMINATING PITCH POCKETS AND A SATISFACTORY METHOD OF GROUPING PIPING THAT MUST COME UP ABOVE THE ROOF SURFACE.

14

FLASHING STRUCTURAL MEMBER THROUGH ROOF DECK

- STRUCTURAL SECTION
- WELDED PLATE—WATERTIGHT
- INSULATE TO PREVENT CONDENSATION
- CAULK WITH ELASTOMERIC SEALANT
- FASTENERS AS NECESSARY
- COMPRESSIBLE ELASTOMERIC TAPE TO SPAN IRREGULARITIES
- SEAL TOP OF FLASHING WITH FABRIC TAPE AND MASTIC
- FASTENERS APPROX. 8" O.C.
- BASE FLASHING
- FIBER CANT STRIP— SET IN BITUMEN
- 8" MIN.

NOTE:
THIS DETAIL ILLUSTRATES ONE METHOD OF ELIMINATING PITCH POCKETS. THE CURBED SYSTEM ALLOWS FOR MOVEMENT IN THE STRUCTURAL MEMBER WITHOUT DISTURBING THE ROOF SYSTEM.

13

1980   M

1980   L

206

INSULATED DECK STEEL FRAME

STRUCTURAL FRAME

CAULK WITH ELASTOMERIC SEALANT

DRAW BAND

WATER TIGHT UMBRELLA

NAIL FLANGE TO WOOD NAILER – FLANGE SET IN MASTIC OVER ROOFING – PRIME FLANGE BEFORE STRIPPING

WELDED ANCHOR PLATE

4" MIN.

1980

N-2

16

MECHANICAL EQUIPMENT STAND

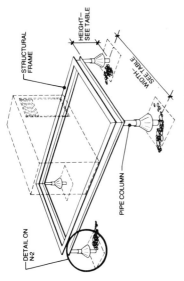

STRUCTURAL FRAME

HEIGHT – SEE TABLE

WIDTH – SEE TABLE

PIPE COLUMN

DETAIL ON N-2

WIDTH OF EQUIPMENT	HEIGHT OF LEGS
UP TO 24"	14"
25" TO 36"	18"
37" TO 48"	24"
49" TO 60"	30"
61" AND WIDER	48"

NOTE:
THIS DETAIL IS PREFERABLE TO DETAIL "D" WHEN THE CONCENTRATED LOAD CAN BE LOCATED DIRECTLY OVER COLUMNS OR HEAVY GIRDERS IN THE STRUCTURE OF THE BUILDING. THIS DETAIL CAN BE ADAPTED FOR OTHER USES, SUCH AS SIGN SUPPORTS.

1980

N-1

15

CONSTRUCTION DETAILS/CONSTRUCTION DETAILS/CONSTRUCTION

LIGHT METAL ROOF EDGE

JOINT COVER 4" TO 6" WIDE – SET IN MASTIC

METAL SET IN MASTIC – PRIME FLANGE BEFORE STRIPPING

STRIPPING

NAILS APPROX. 3" O.C. – STAGGERED

TURN DOWN ONE PLY OF FELT TO BELOW BLOCKING

1½"

12" TO 18" TAPERED EDGE STRIP

CONTINUOUS CLEAT

NOTES:

ENVELOPE SHOWN FOR COAL TAR PITCH AND LOW SLOPE ASPHALT.

ATTACH NAILER TO MASONRY WALL. REFER TO FACTORY MUTUAL DATA SHEET 1-49.

THIS DETAIL SHOULD BE USED ONLY WHERE DECK IS SUPPORTED BY THE OUTSIDE WALL.

THIS DETAIL SHOULD BE USED WITH LIGHT GAUGE METALS, SUCH AS 16 OZ. COPPER, 24 GAUGE GALVANIZED ALUMINUM OR 0.040" ALUMINUM. A TAPERED EDGE STRIP IS USED TO RAISE THE GRAVEL STOP. FREQUENT NAILING IS NECESSARY TO CONTROL THERMAL MOVEMENT.

WOOD BLOCKING MAY BE SLOTTED FOR VENTING WHERE REQUIRED.

18

N-3

CONCRETE DECK AND FRAME

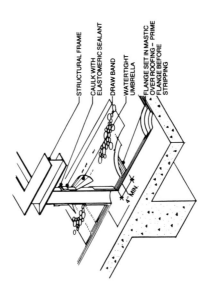

STRUCTURAL FRAME

CAULK WITH ELASTOMERIC SEALANT

DRAW BAND

WATERTIGHT UMBRELLA

FLANGE SET IN MASTIC OVER ROOFING – PRIME FLANGE BEFORE STRIPPING

4" MIN.

17

208

CURB DETAIL FOR ROOFTOP AIR HANDLING UNITS

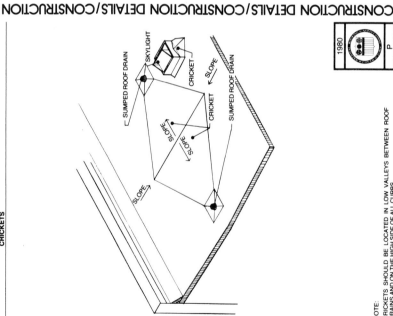

2" WOOD NAILER

16 GA. MIN. FRAME

INSULATION SUPPLIED BY CURB MANUFACTURER

SEAL STRIP

COUNTERFLASHING FASTENED APPROX. 18" O.C.-SUPPLIED AND INSTALLED BY SHEET METAL CONTRACTOR

FASTENERS APPROX. 8" O.C.

BASE FLASHING

2" NOMINAL

FIBER CANT STRIP SET IN BITUMEN

WOOD BLOCKING FASTENED TO DECK

14" NOMINAL MFGD. HEIGHT

ALTERNATE FRAME LOCATION FOR HEAVY UNITS

8" MIN.

1980

R

20

NOTE:
THE CURB, WOOD NAILER, INSULATION AND SEAL STRIP ARE TO BE SUPPLIED BY THE CURB MANUFACTURER. THE NOMINAL 14" CURB HEIGHT IS EFFECTIVE AS OF JANUARY 1, 1981.

CRICKETS

SUMPED ROOF DRAIN

SKYLIGHT

CRICKET

SLOPE

CRICKET

SUMPED ROOF DRAIN

SLOPE

SLOPE

SLOPE

1980

P

19

NOTE:
CRICKETS SHOULD BE LOCATED IN LOW VALLEYS BETWEEN ROOF DRAINS AND ON THE HIGH SIDE OF ALL CURBS.

SMOOTH CONCRETE—EXPOSED SURFACES MUST BE WATERPROOFED

CAULK WITH ELASTOMERIC SEALANT

ANGLE CLAMPING BAR WITH SLOTTED ANCHOR HOLES

FASTENERS IN EXPANSION SHIELDS

COMPRESSIBLE ELASTOMERIC TAPE TO SPAN IRREGULARITIES

SEAL TOP OF SYSTEM WITH FABRIC TAPE AND MASTIC

PRIME CONCRETE

FASTENERS APPROX. 8" O.C.

CHAMFER TOP TO DRAIN

2" NOMINAL

ATTACH WOOD NAILER TO CONCRETE WITH APPROVED METHOD

FIBER CANT STRIP—SET IN BITUMEN

1980
U

NOTE:
WHERE DECK IS SUPPORTED BY AND FASTENED TO THE CONCRETE WALL, VERTICAL WOOD-NAILERS SHOULD BE SECURED TO THE WALL WITH SUITABLE FASTENERS

22

CONSTRUCTION DETAILS/CONSTRUCTION DETAILS/CONSTRUCTION

THIS DETAIL ALLOWS FOR EXPANSION AND CONTRACTION OF PIPES WITHOUT ROOF DAMAGE

SET BOLTS IN ELASTOMERIC SEALANT

ADJUSTS VERTICALLY AND HORIZONTALLY

1980
S

NOTE:
NRCA REAFFIRMS ITS OPPOSITION TO PIPES AND CONDUITS BEING PLACED ON ROOFS. HOWEVER, WHERE THEY ARE NECESSARY, THIS TYPE OF PIPE ROLLER SUPPORT IS RECOMMENDED.

21

CONSTRUCTION DETAILS/CONSTRUCTION DETAILS/CONSTRUCTION

ROOF DRAIN

STRAINER

CLAMPING RING

METAL FLASHING*

STRIPPING FELTS

DECK CLAMP

TAPER INSULATION 24" TO DRAIN

1980

W-1

24

NOTES:

*MIN. 30" SQUARE, 2½ LB. TO 4 LB. LEAD OR 16 OZ. SOFT COPPER FLASHING, SET ON FINISHED ROOFING FELTS IN MASTIC. PRIME TOP SURFACE BEFORE STRIPPING.

MEMBRANE PLIES, METAL FLASHING, AND FLASH-IN PLIES EXTEND UNDER CLAMPING RING.

STRIPPING FELTS EXTEND 4" AND 6" BEYOND EDGE OF FLASHING SHEET.

**CLEARANCES FOR MULTIPLE PIPES—
BETWEEN PIPES AND FROM WALLS AND CURBS**

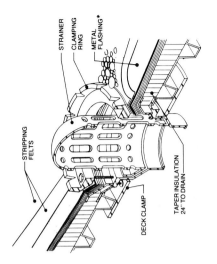

FIBER CANT STRIP— SET IN BITUMEN

12" MIN.

4" MIN.

4" MIN.

12" MIN.

12" MIN.

2" NOMINAL

12" MIN.

1980

V

23

211

GAUGE OR THICKNESS GUIDE

FOR METAL FASCIA EXPOSED TO VIEW

COPING

CAP FLASHING AND FASCIA

VARIATIONS

RECOMMENDED MINIMUM GAUGES FOR FASCIA SHOWN ABOVE

EXPOSED FACE WITHOUT BRAKES "A" DIMENSION	GALVANIZED IRON	COLD ROLLED COPPER	ALUMINUM 3003-H14
UP TO 6" FACE	26 GA.	16 OZ.	.040" (18 GA.)
6" TO 8" FACE	24 GA.	16 OZ.	.050" (16 GA.)
8" TO 10" FACE	22 GA.	20 OZ.	.064" (14 GA.)
10" TO 15" FACE	20 GA.	ADD BRAKES TO STIFFEN	.080" (12 GA.)

NOTE:

WHEN USING THE ABOVE TABLE, OTHER ITEMS SHOULD BE CONSIDERED, SUCH AS THE FASTENING PATTERN. FOR INSTANCE, IF THE METAL CAN ONLY BE FASTENED AT 10' INTERVALS, A HEAVIER GAUGE METAL WOULD BE REQUIRED.

1980

Z

26

ROOF DRAIN

STRIPPING FELTS

STRAINER

CLAMPING RING

METAL FLASHING*

OPTIONAL 1" × 4" SHEET METAL GRAVEL STOP—36" SQUARE MIN. SET IN FLASHING CEMENT

DECK CLAMP

TAPER INSULATION 24" DOWN TO DRAIN

NOTES:

*MIN. 30" SQUARE 2½ LB. TO 4 LB. LEAD OR 16 OZ. SOFT COPPER FLASHING, SET ON FINISHED ROOFING FELTS IN MASTIC. PRIME TOP SURFACE BEFORE STRIPPING.

MEMBRANE PLIES, METAL FLASHING, AND FLASH-IN PLIES EXTEND UNDER CLAMPING RING.

STRIPPING FELTS EXTEND 4" AND 6" BEYOND EDGE OF FLASHING SHEET.

1980

W-2

25

ALERTS

General

1. Check with roofing materials manufacturer for approval of all flashing details and materials, including primers, bitumens, fabrics, coated base sheets, surfaced cap sheets, and bituminous mastic for use with flashing details.

2. Check "NRCA Roof Curb Criteria" list for rooftop equipment to ensure that roof-penetrating components conform with good flashing practice.

3. Complement specifications with large-scale flashing details on drawings.

Design

1. Locate flashed joints at a roof's high points, if possible; in any event, above anticipated water level.

2. Keep HVAC units off the roof, if economically feasible. (This policy reduces the number of vulnerable flashed joints, keeps hazardous workers off the roof, and lengthens HVAC equipment service life.)

3. Minimize the number of penetrations through the roof. Group piping and roof-mounted equipment requiring flashing with pads, curbs, and so on into a smaller number of larger areas to reduce total quantity of flashing (see NRCA Detail N-1).

4. Locate stacks and other small roof-penetrating components at least 2 ft from walls or other vertical surfaces. (They are vulnerable to damage if placed too close to curbs requiring periodic maintenance and repair.)

5. Favor edge cant strip over gravel-strip edge details. If gravel strip is used, elevate the edge with tapered insulation board.

6. Wherever parapets or other walls intersect the roof, provide through-wall flashing rather than reglets.

7. Generally avoid metal base flashing; instead use composition flashing.

8. Avoid right angle bends in base flashing; use cant strips to provide an approximate 45° bend.

9. Favor treated, solid wood cant strips over fiberboard.

10. Specify water-based, salt-type preservative treatment, compatible with bitumen, for wood cant strips, nailers, and blocking used with roof assembly.

11. Never anchor base flashing directly to walls, structural members, pipes, or other building elements independent of the roof assembly. Always provide for differential movement.

12. Where metal cap flashing overlaps base flashing and at horizontal joints between cap-flashing segments, detail joint for differential movement.

13. Extend base flashing at vertical surfaces to a minimum 8-in. vertical dimension, maximum 12 in. above membrane elevation.

14. Avoid pitch pockets whenever practicable. (Use pedestals or curbed openings.)

15. Check Factory Mutual minimum gages for gravel stops (consult Factory Mutual's *Loss Prevention Data Sheet 1-49*).

16. Specify an expansion joint between drain and leader pipes, or offset the leader pipe 45°. (When these units are rigidly attached to columns, movement can break flashing connections and permit water entry into membrane.)

17. Specify sealer strips for exposed felted flashing vertical laps and horizontal toe sealer strips where base flashing overlaps membrane on smooth-surfaced roofs.

Field

1. When hot-mopped, felted flashings are specified, check to ensure use of steep asphalt (to prevent eventual sagging of base flashings adhered with low softening-point asphalt mistakenly applied). Back mopping of flashing felts is often preferable to application of hot-mopped asphalt on vertical surfaces.

2. Check for proper priming of walls and gravel stop's metal flanges before applying base flashing or stripping felts.

3. Check base flashing felts for maximum 12-ft length, staggered end laps (2-ft-minimum offset, 3-in.-minimum end laps with 4-in. asbestos sealer strips troweled with flashing cement).

4. Check for use of specified nails.

5. Check temperature for application of plastic or flashing cement (25°F minimum).

6. Install flashing, including counterflashing, as roof application progresses. If delay in installing counterflashing is unavoidable, apply a three-course stripping seal (one ply of fabric flashing between two trowelings of flashing cement) to seal joint and prevent water from entering behind base flashing until counterflashing goes in place.

7. Check for use of sealer strips at vertical flashing joints.

8. Establish a maintenance program with semiannual inspection followed by resurfacing of deteriorating surfaces, restripping of opened joints, renewal of all joint and reglet caulking, and repair of rips, tears, and other defects.

NINE

Cold-Process Built-Up Roof Membranes

Cold-process bituminous roof membranes differ from traditional hot-applied membranes in only one essential respect: The waterproofing, interply bituminous adhesive, and top coating is field-applied at ambient temperature, not heated to the 400 to 500°F temperature range specified for hot-mopped bitumens. There are two other limitations:

- Asphalt cutback is the only interply adhesive in general use.

- Felts are coated, not merely saturated, because coated felts fuse better with the cold cement and thus adhere better than saturated felts.

Glass felts have also entered the cold-process market, offering the advantages of greater dimensional stability and greater tensile strength.

Cold-process membranes, like hot-applied membranes, come in both smooth- and aggregate-surfaced types. In one popular cold-process membrane, aggregate surfacing consists of No. 11 mineral granules (the same granules used for asphalt shingles) conveyed from a ground-level hopper to the roof surface and sprayed by compressed air at a recommended rate of 50 lb/square into a fibrated asphalt cutback top coat weighing 18 lb/square (2¼ gal/square).

For the smooth-surfaced membranes, solvent-thinned asphalt cutback can serve as a coating as well as interply adhesive. However, it is inferior to an asphalt emulsion.

There are two types of asphalt-based emulsion surfacing: one with a mineral colloid (bentonite clay) emulsifying agent, the other with a chemical (or soap) emulsifying agent. The bentonite-clay emulsion resists weathering better than the cheaper chemical emulsions.

Both types of asphalt emulsion rely on the emulsifying agent to hold the two immiscible liquids (low-viscosity asphalt and water) in intimate, dispersed suspension. Asphalt particles are completely surrounded by water, and the molecules of the emulsifier surround each asphalt particle and separate it from its neighbors. Bentonite clay creates an ideal colloidal suspension, largely because of its swelling.

The bentonite-clay emulsion has superior durability largely because of its resistance to ultraviolet radiation. The platelike bentonite-clay particles stabilize the emulsion through a honeycomb structure left throughout the bitumen film as the water evaporates. Because of this stable honeycomb structure, a soft, and consequently more durable asphalt, can be used in the emulsion, still another factor promoting durability. The emulsion weathers through chalking, or slow erosion rather than drying out, like the asphalt cutbacks.

As disadvantages, the emulsions are less resistant to standing water and are also more vulnerable to cold weather during storage or application. They should be stored in an environment above 32°F and never applied when either subfreezing temperatures or rain is imminent. (A freeze-stabilized emulsion is available on special order.)

COLD-PROCESS ADVANTAGES

Cold-process roofing has expanded from its original use in maintenance to reroofing, a growing portion of the total roofing market, and then into new construction. The cold-process membrane's growing popularity stems from a long list of factors:

- A steadily narrowing price gap between the lower-priced, hot-applied and higher-priced cold-applied materials.

Fig. 9-1 Interply cutback asphalt adhesive is sprayed in advance of coated felt's unrolling in this three-ply, cold-process membrane.

Fig. 9-2 Rolling the longitudinal side laps of a cold-process membrane is recommended by some manufacturers to ensure good interply bond.

- Greater labor productivity for a four-member field crew compared with the normal five-member crew for hot-applied membranes (see Figs. 9-1 and 9-2).

- Reduced construction delays from tanker breakdowns and so on.

- Elimination of heavy equipment required for hot application (see Fig. 9-3).

- Reduced hazards—notably the elimination of fire risk to workers and the building.

- Reduction of air pollution hazards. (Some local governments ban open-air heating of bitumen under certain smog conditions.)

- Lighter weight (about 4 psf less than an aggregate-surfaced, hot-applied membrane).

- Fewer reported roofing problems.

The last benefit, manifested chiefly in reduced incidence of blistering, is more easily explained since the late 1970s' publication of research into hot bitumens' time-temperature curves. As explained elsewhere in this manual, the fast cooling of hot asphalt makes split-second timing essential to good interply mopping adhesion (as well as

Fig. 9-3   Cold-process membrane requires only light equipment—typically, a gasoline engine, an air compressor, an air-powered material-handling pump and a granule hopper—in place of heavy kettle or tanker, plus ancillary equipment required for hot-applied bituminous membrane.

for bonding hot-mopped insulation to the deck or a base sheet to the insulation). Hot-mopped steep (Type III) asphalt has an especially short "decay time," 5 s or so at 70°F, 10-mph outside conditions, before it cools from 500 to the 390°F minimum recommended temperature that will maintain its viscosity within the 50- to 150-cSt range required for forming a continuous mopping film. With a cold-process asphalt cutback, however, the time allowed for adhering the felt multiplies by a factor of 50 to 100. Evaporation of the solvent vehicle for the asphalt cutback is a slow process, giving the workers plenty of time to place the felt before adhesion is impaired.

Spray-applied, aggregate-surfaced, cold-process membranes have another field-application advantage over their hot-applied counterparts. Getting well-embedded aggregate in a hot built-up membrane flood coat requires precise field timing and coordination frequently not achieved. Flood coats are often too thin, well below the required 60 lb/square asphalt weight, and irregular. Aggregate embedment often falls below the minimum required for dependable protection against ultraviolet radiation. Good aggregate embedment literally requires split-second timing. The roofing mechanics have only a few seconds before hot bitumen congeals, making it too viscous to adhere to the spreading aggregate. Unless the aggregate-spreading and flood-coat-pouring operations are precisely timed, the flood coat flows too thin, and the tardily spread aggregate fails to adhere, thus jeopardizing the flood-coat bitumen's integrity.

The spraying process for both top coat and mineral granules of aggregate-surfaced, cold-process membranes requires far less coordination than the aggregate-spreading operation on a hot-applied membrane (see Fig. 9-4). There is no serious obstacle to achieving excellent granule embedment for a competent application on the cold-process membrane. With hot-applied membranes, rapid congealing of the flood coat limits the timing for good aggregate embedment to several seconds. But with cold-process, granule-surfaced membranes, this timing tolerance for aggregate application is measured in minutes rather than seconds. A correctly installed hot-applied, aggregate-surfaced membrane should provide better waterproofing quality and durability than a correctly applied, granule-surfaced, cold-process membrane. But it is much easier to get good application with a cold-process membrane.

Besides these general factors favoring cold-process membranes, there are other advantages, depending on the availability of hot asphalt or the size of the roof. In outlying areas, the cost and inconvenience of transporting the fuel and kettle may outweigh the normal economy of hot-applied roofing. And even in a central city or suburb, hoisting a

Fig. 9-4 Aggregate surfacing (left) is sprayed into a top coat of 2 to 2½ gal/square at rate of 50 lb/square, No. 11 mineral granules, the same material used for asphalt roof shingles. Clay-type asphalt emulsion surfacing is either brushed (right) or sprayed.

kettle to a multistory building's roof to cover a relatively small area usually proves uneconomical.

COLD-PROCESS DISADVANTAGES

Cold-process membranes have one failing that makes them generally less satisfactory than conventional hot-mopped bituminous membranes for new construction. Because of the persistent softness of the asphalt cutback adhesive, cold-process membranes are more vulnerable to damage from rooftop traffic than hot-mopped membranes. This limitation makes these cold-process membranes generally more suitable for reroofing than for new construction, with its heavier traffic loads from other trades. To protect the top coat from construction damage, its application is normally delayed on new projects until the other trades have completed rooftop work.

COLD-PROCESS PERFORMANCE CRITERIA

Several years ago, with a combination of field experience and laboratory-testing procedures, a major roofing material manufacturer introducing a new cold-process roof membrane into the market evaluated it in terms of the National Bureau of Standards' (NBS) "Preliminary Performance Criteria for Builtup Roof Systems."[1] Judged by the results of this research, cold-process membranes are comparable to hot-applied bituminous membranes in tested attributes, which included the following:

- Tensile strength
- Coefficient of thermal expansion
- Load/strain modulus
- Thermal shock factor (TSF; computed from the above three attributes)
- Punching shear
- Impact resistance
- Wind resistance
- Fire resistance

[1] R. G. Mathey and W. C. Cullen, "Preliminary Performance Criteria for Bituminous Membrane Roofing," *Build. Sci. Ser.* 55, NBS, November 1974.

Tensile strength in a cold-process membrane differs from that of a hot-applied built-up membrane in the way this strength develops. A hot-applied membrane develops its tensile strength almost immediately, as soon as the mopping asphalt cools. But a cold-process membrane's tensile strength develops slowly over several years, with the slow decrease in the volatile content of the interply adhesive asphalt cutback. A three-ply, cold-process membrane attains the NBS recommended minimum 200 lb/in. tensile strength at a volatile content of about 6 percent (see Fig. 9-5). As the aging membrane continues to lose these volatiles, tensile strength rises still further.

The manufacturer's researchers found that the solvent release in 4 months' accelerated laboratory aging (at 133°F) corresponded to about 5 years' field exposure in the Midwest. Thus a well-applied, three-ply cold-process membrane with coated felts should attain its 200 lb/in. tensile strength within 5 years or so after application. A glass-fiber cold-process membrane would probably attain this strength within 1 year.

Thermal shock factor decreases with age in a cold-process membrane, just as it does in a hot-applied membrane. At 5 percent volatiles, TSF equals 300°F, three times the NBS-suggested minimum of 100°F. This aging drop in TSF occurs despite increasing tensile strength, which appears in the numerator of the fractional TSF formula. Evaporation of the adhesive solvent increases both the load-strain modulus

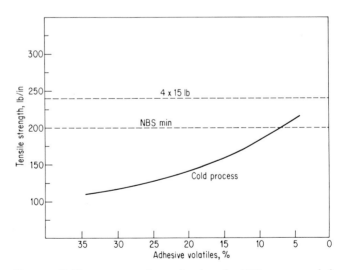

Fig. 9-5 Cold-process membrane develops the NBS-recommended 200 lb/in. tensile strength when volatiles in asphalt cutback adhesive reach 6 percent. (GAF Corporation.)

and the coefficient of thermal expansion, which appear in the denominator of the TSF formula. Fortunately, the loss in volatiles, 4.5 to 5 percent after 5 years, tapers off from a rapid to a slow rate as the membrane ages (see Fig. 9-6).

Note also the effect of temperature on loss of volatiles (see Fig. 9-7). At constant 133°F laboratory temperature, the volatile content drops to 5 percent within 3 months, corresponding to 4 years or less field-service time. Because it would retard the cutback adhesive's solvent loss, a heat-reflective coating is apparently more important for prolonging the life of a cold-process membrane than a hot-applied membrane.

The cold-process membrane meets the NBS-suggested performance criteria for punching shear and impact resistance, but it could not be tested for several attributes, notably tensile fatigue and flexural fatigue.

It passed the standard industry tests for wind and fire resistance. In the Factory Mutual (FM) field-uplift test, the tested cold-process membrane recorded more than 75-psf resistance, against an FM-required 60 psf.

Field investigations of 20 membranes, from 1 to 5 years old, chiefly in the Midwest, confirmed the laboratory indications of good service. These investigations indicated few problems, especially with blistering and splitting, two major problems with hot-applied membranes. Buckling in the organic felt membranes was observed over poured lightweight insulating concrete fills. The organic coated base felts' moisture increased from less than 1 to 3 percent. Expansion of the wetted felt

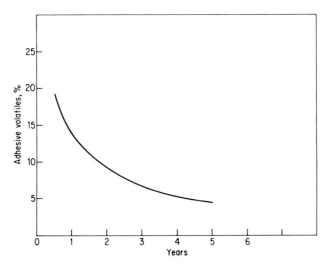

Fig. 9-6   Time required for volatiles to reach 6 percent is about 5 years of service life. (GAF Corporation.)

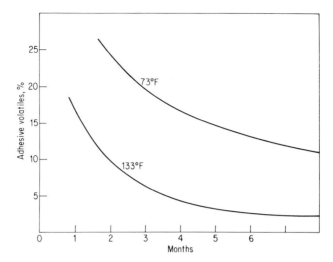

Fig. 9-7   Effect of temperature on loss of volatiles in three-ply, cold-process membrane. (GAF Corporation.)

accounted for the buckling, according to the manufacturing company's researchers.

As one advantage in maintaining dimensional stability, glass-fiber membranes promise to eliminate this buckling problem.

ALERTS

Design

1. Do not specify the chemical-type emulsion for surfacing cold-process bitumen; specify clay-type emulsion where an emulsified bituminous surfacing is used.

2. Limit cold-process roofs to minimum ½-in. slope.

Field

1. Do not apply cold-process emulsion surfacing at a temperature below 32°F or when there is a possibility of rain before the emulsion sets.

2. Require prior cutting and flattening of coated felts used in cold-process membrane.

3. On new construction projects, delay applying the top coat and granule surfacing until other trades have completed rooftop work.

TEN

Protected Membrane Roofs

The protected membrane roof (PMR), also known as the insulated roof membrane assembly (IRMA), inverted roof assembly (IRA), or simply the upside-down (USD) roof, reverses the positions of membrane and insulation in the conventional built-up roof assembly. Instead of its conventional exposed position on top of the insulation, a PMR membrane is sandwiched between the insulation above and the deck below (see Fig. 10-1).

Historically, the prototypical PMR is the sod roof, a centuries-old roof system in northern countries. Birch bark provides a shingled drainage surface off a timber-framed roof structure, and an earth-sod cover is insulation over the shingles.

A still later version of PMR is represented by a United States manufacturer's now-discontinued specification for a "garden" roof, installed directly on a concrete deck, comprising a five-ply, aggregate-surfaced, coal-tar-pitch membrane topped with loose gravel aggregate, earth fill, and sod. Before the flood coat was applied, this roof was test-flooded with 2 in. of water to ensure its watertight integrity before it was covered with earth fill.

Over the past decade, the PMR has commanded steadily increasing international attention as a possible solution for problems encountered with the conventional built-up roof system. By 1980, there were more than 5000 PMRs in service in the United States and thousands more in Canada, Europe, Asia, and the Middle East, performing in the torrid climate of Saudi Arabia and in the Arctic climate of Alaska.[1]

Since Canada has been the world's pioneer in PMR systems, the Canadians' favorable experience is especially significant because of its extent and Canada's rigorous climate. According to a survey by the Division of Building Research, National Research Council of Canada,

[1]K. A. Epstein and L. E. Putnam, "Performance Criteria for the Protected Membrane Roof System," *Proc. Symp. Roofing Technol.*, NBS-NRCA, September 1977.

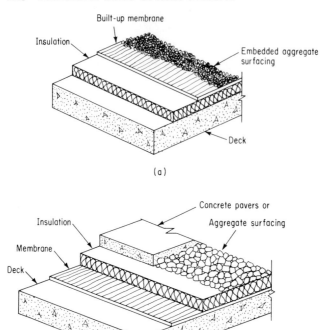

Fig. 10-1 (*a*) Conventional built-up roof system sandwiches insulation between deck and membrane. (*b*) Protected membrane roof (PMR) places membrane directly on deck (or on leveling board for steel decks), insulation above membrane, with concrete pavers or aggregate on top of insulation.

which began promoting the PMR concept back in 1965, PMRs are recommended by the overwhelming majority of architects who have specified and roofers who have built them. They also rate PMRs as superior to conventional built-up roof systems for roof gardens and plazas.[1]

Much United States experience with PMRs has been similarly favorable. Since 1969, when it switched to PMRs, the University of Alaska has built 10 such roofs for its Fairbanks and Anchorage campuses. The experience proved so successful that university planners now specify PMRs exclusively for all new construction. They estimate an additional 25 percent first cost for their version of PMR. But a 70 to 80 percent estimated reduction in operating and maintenance cost, plus longer

[1]M. C. Baker, "Protected Membrane Roofs in Canada—Results of a Survey," *Nat. Res. Counc. Can., Div. Build. Res.*, October 1973.

anticipated service lives, more than pays back the added first cost, making PMR a highly profitable long term investment.[1]

A major city's Board of Education has had similarly favorable experience with PMRs. Over the past few years the Board has installed 30 or more PMRs—some reroofing projects, others new. According to the Board's chief specification writer, these roofs have generally outperformed the conventional built-up roof systems they have replaced. Countless problems with conventional built-up roof systems prompted the switch to PMRs.

The PMR is not the panacea long sought by harassed architects and building owners. However, it is a new technique that already has established its claim to a steadily growing share of the built-up roof market, with unquestionable strengths as well as drawbacks. If not *the* wave of the future, it is nonetheless *a* wave of the future.

WHY THE PROTECTED MEMBRANE ROOF?

The conventional roof-system arrangement is inverted to protect the membrane from the many hazards of the conventional roof's exposed membrane location, notably:

- Accelerated oxidation and evaporation of volatile oils, leading to premature embrittlement and cracking of the bitumen
- Low-temperature contraction splitting
- Roof traffic and hailstone impact
- Warping or delamination from ice contraction
- Blistering and ridging

The PMR concept also avoids the vapor-entrapment problem of conventional roof assemblies, with insulation sandwiched between vapor retarder and membrane. It combines the vapor retarder and membrane into one unit.

The life-shortening hazards to the membrane are drastically reduced in a PMR. Whereas a conventionally exposed membrane may experience a range of 100°F daily and 200°F annual temperature change, a properly designed PMR should normally experience less than 10°F daily and 30°F annual temperature change. Insulation shields the membrane from photo-oxidative reactions that exponentially acceler-

---

[1] Haldor W. C. Aamot and David Schaefer, "Protected Membrane Roofs in Cold Regions," CRREL Rept. 76-2, U.S. Army Corps of Engineers, Cold Regions Research and Engineering Laboratory, Hanover, N.H., March 1976, p. 8.

ate the chemical deterioration of exposed areas of bitumen, especially in the higher temperature ranges.

Illustrating the benefits of this solar shielding, bitumen specimens taken from a conventional 10-year-old asphalt built-up membrane reportedly suffered a roughly 200°F rise in softening point, from 190°F (Type III asphalt) to over 400°F.* In contrast, the Type III asphalt from several PMRs of equal age gained less than 30°F in softening point.[1]

A severe increase in softening point indicates drastic deterioration of a bitumen's waterproofing quality. The affected bitumen becomes more brittle and permeable. Cracks admit water, and the felts wick it deeper into the membrane. Protection from this combined physical and chemical degradation of the waterproofing bitumen constitutes a major benefit of the PMR concept, along with protection from physical abuse.

Another major benefit of the PMR concept is its virtual elimination of thermal stress in the membrane. Subfreezing temperatures, especially subzero (0°F) ones, produce the greatest tensile splitting hazard because of the cold, brittle bitumen's vastly increased coefficient of thermal expansion-contraction at these low temperature ranges. (In the −30 to 0°F temperature range, the coefficient of expansion-contraction for various built-up membranes ranges from about 5 to 10 times as much as the same coefficient in the +30 to +73°F temperature range.)[2]

A PMR membrane apparently remains above freezing temperature even in the coldest weather. Laboratory research and field experience both corroborate this phenomenon. First consider the problem theoretically. As the example in Fig. 10-2 indicates, under the extreme condition of a 55°F setback for interior temperature and just 2 in. of extruded polystyrene, membrane surface temperature is 44°F when outside temperature is −30°F. And although common sense suggests that open joints between insulation boards might act as thermal bridges, with substantially reduced membrane temperatures at those points, laboratory tests and field observations indicate otherwise. In laboratory tests designed to test the effects of open (¼-in.) insulation joints, with simulated outside temperature at −30°F and inside temperature at 80°F, temperatures measured at the membrane directly under the open joints were only 1 to 3°F colder under dry conditions and 3 to 7°F colder under wet conditions than membrane temperatures under the insulation.[3] Insulation in these tests was a maximum 1¼ in. thick. Thicker

---

*Epstein and Putnam, op. cit., p. 51.

[1]Loc. cit.

[2]R. G. Mathey and W. C. Cullen, *Preliminary Performance Criteria for Bituminous Membrane Roofing*, Build. Sci. Ser. 55, NBS, November 1974, p. 6.

[3]M. C. Baker and C. P. Hedlin, "Protected Membrane Roofs," *Can. Build. Dig.*, no. 150, June 1972, p. 150-4.

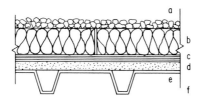

Element	R
a Outside air	0.25
b 2" polystyrene	8.50
c BUR membrane	0.33
d ½" gyp. board	0.45
e Steel deck	0.00
f Inside air	0.61
R_t =	10.14

Fig. 10-2 Under extreme thermal steady-state condition, with outside temperature T_o at $-30°F$ and indoor (setback) temperature T_i at 55 °F, the top of the membrane shown would drop only to 44°F.

$$T_{membrane} = T_i - \frac{\Sigma R_x}{R_t}(T_i - T_o) = 55 - \frac{0.61 + 0.45 + 0.33}{10.14}(55 + 30)$$

$$= 55 - 11 = 44°F$$

Laboratory tests indicate that this temperature might drop as much as 3°F lower at a wide joint. In any event, the membrane would remain well above freezing temperature, so long as it remained dry.

insulation, with narrower joints, should reduce even the slight temperature differences measured in the laboratory test.

Two layers of insulation, with staggered joints, should eliminate the slight thermal bridges created by joints in single-layered insulation. Two-layered insulation also prevents flotation of a loose individual bottom-layer board by interlocking it with surrounding boards.

Even in Fairbanks, Alaska, where the 99 percent design temperature is $-53°F$, properly designed PMRs are free of ice buildup from freezing water. In such a severe climate it is necessary to cover the drains to insulate them (see Fig. 10-3).

The PMR arrangement offers another possible advantage in doubling the membrane's function. To its basic function as the roof system's waterproofing component, the PMR concept adds that of vapor retarder, properly located on the warm side of the roof assembly. In its conventional position above the insulation, the membrane retards the escape of migrating water vapor moving up from the warm interior toward the cold exterior. An exposed membrane thus tends to trap moisture (entering in liquid form through faults in the membrane, or as upwardly migrating vapor) within the system. Especially when there is a vapor retarder under the insulation, the conventionally located membrane helps trap moisture and prevent its escape once it invades the insulation.

The PMR concept has another practical advantage, resulting from the insulation's vulnerable but easily accessible location. If the insulation deteriorates from exposure to sunlight or physical abuse, it can be

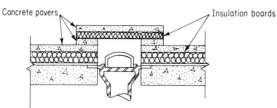

Fig. 10-3   In severely cold climates, covering the drains with pavers protects them from ice clogging.

readily removed and replaced with no disruption in the building's operations because a properly applied protected membrane should live a long service life. This is much easier and more economical than a tearoff-reroofing job, or even a membrane replacement or reroofing over an old roof. Reroofing with a conventional built-up system entails much greater inconvenience and expense than simple replacement of deteriorated insulation on a PMR, especially for a so-called "loose" PMR (defined later).

Still another advantage depends on the easy dismantling of a PMR. On a PMR surfaced with loose aggregate or pavers, you can add insulation by simply removing the surfacing and placing another layer. (Additional surfacing ballast may be necessary to balance the additional buoyancy of the insulation.) It is an easy, convenient way to add insulating capacity as energy costs rise, provided of course that the roof's deck and structural framing can safely carry the added dead load.

Conventional built-up systems have one notable advantage over PMRs in reroofing. With a conventional built-up roof system you can slope a flat or irregular roof surface that ponds water with tapered board insulation or sloped insulating concrete fill. However, with a PMR you are generally stuck with the existing deck slope.

Another noteworthy disadvantage concerns the potential vegetation and fungal growth promoted by the warm, humid environment at a PMR membrane surface. Vegetation may grow through paver joints or loose aggregate ballast; such growth may promote rot and even root penetration of the membrane.

HOW PMR CHANGES COMPONENT REQUIREMENTS

As indicated in the previous discussion, rearrangement of the components in a PMR drastically alters their performance criteria:

- The *insulation* becomes the critical component. Besides its primary function of conserving heating and cooling energy, it must protect the membrane from temperature cycling, physical damage, and solar radiation. It must maintain long term insulating efficiency despite perennial cyclical exposure to water, freeze-thaw cycles, and water-vapor pressure.

- The *membrane* function shrinks to the sole purpose of waterproofing (and possibly vapor retarder). Unlike the membrane in a conventional built-up roof system, it no longer has to withstand extensive temperature change, foot traffic, and solar heat.

- The *surfacing* (aggregate or pavers) must provide ballast against wind uplift and flotation of loosely laid PMR insulation. These functions are added to the surfacing's functions in the conventional built-up roof concept—i.e., shielding against ultraviolet radiation, hailstone impact, foot traffic, and small fire sources.

These changing performance requirements have promoted significant changes in materials. The PMR insulation's exposed location eliminates many insulation materials suitable for the protected sandwich-filler location of the conventional built-up roof system. These more rigorous requirements—notably resistance to moisture absorption—have thus far limited PMR insulation to three foamed plastics: extruded polystyrene, molded-bead polystyrene, and polyurethane.

Conversely, the less rigorous demands on the PMR membrane have expanded the range of suitable materials. A PMR is similar to plaza and below-grade waterproofing, and many products formerly limited to waterproofing can now be used on roofs.

However, the protected location of the membrane does affect the choice of felts for a built-up roof membrane. Inorganic felts probably have a clearer advantage over organic felts in a PMR than in conventional roof systems. An organic felt's cellulosic fibers are highly vulnerable to moisture decay, and the PMR membrane's sheltered, shaded location promotes moisture retention more than the exposed location of a conventional built-up membrane, which can reject some absorbed moisture by solar evaporation. Moreover, there is no positive technical reason (only an economic motive) to use organic felts in a PMR. Superior tensile strength, the major technical reason for specifying organic felts, is generally of less benefit in a PMR than in a conventional built-up membrane. Because a PMR does not experience the temperature extremes of a conventional built-up membrane, a PMR membrane should never undergo significant thermal-contraction stress.

A glass-fiber felt with water-resistant binder is generally the best felt type for a PMR built-up membrane because asbestos felts contain organic material (in the permitted 15 percent organic fibers and starch binder).

PMR DESIGN AND CONSTRUCTION

The PMR concept is compatible with any type of membrane: conventional hot-applied bituminous, fluid-applied or synthetic single-ply sheet, loose or adhered. All systems currently (1981) marketed feature insulation considered, in design concept, loosely laid on the membrane.

The major commercially marketed PMR system is applied in a three-stage construction sequence:

1. Application of membrane

2. Application of insulation (within 3 weeks after membrane application)

3. Application of aggregate (within 3 weeks of insulation application) (See Fig. 10-4.)

In stage 1, a minimum three-ply shingled, built-up membrane is installed, with steep asphalt for interply mopping and a 70 lb/square flood coat. Regardless of whether the insulation is applied on the same day or later, a second (40 lb/square) flood coat is required for embed-

Fig. 10-4 One version of protected membrane roof assembly is installed in three stages: (1) A minimum three-ply built-up membrane is applied to the deck, (2) extruded polystyrene insulation boards are embedded in a second top coat (top), and (3) aggregate ballast is applied. (Dow Chemical Co.)

ding the polystyrene insulation boards. Application of the ballast aggregate (10 psf for 2-in.-thick, 15 psf for 3-in.-thick insulation) can either immediately follow insulation application or be delayed a maximum of 3 weeks.

Although not designed for wind-uplift or flotation resistance, the adhesion of the polystyrene board insulation has a threefold purpose:

1. To restrain the boards' position while stone ballast is applied

2. To maintain tight insulation joints and prevent stones from falling between the boards and jamming open the joints

3. To limit the volume of water flowing at the membrane-insulation interface, restricting this flow as much as possible to the joints

The roofer must carefully control asphalt temperatures to avoid damaging the heat-sensitive polystyrene boards during this application. Obviously, with single-ply synthetic sheet or fluid-applied elastomeric membranes, the boards are simply laid loose on the membrane.

To qualify as a Factory Mutual Class I steel-deck roof assembly, this roof system requires installation of a ½-in. gypsum leveling board (Type X core) anchored by approved mechanical fasteners or adhesives, before the membrane is applied.

Fine Points in PMR Design

A detail improvement, widely used in Europe for loose-aggregate-surfaced PMRs, adds another component: a layer of feltlike, plastic- or glass-fiber fabric between the surfacing aggregate and the insulation (see Fig. 10-5). According to a proponent of this practice, consultant Sybe K. Bakker, of McLean, Virginia, the plastic-fiber cloth (a bonded filament, polyester-based fabric) serves as separator and filter. As a separator it prevents the obstruction of drain paths through insulation

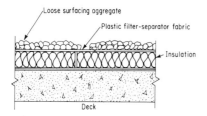

Loose surfacing aggregate
Plastic filter-separator fabric
Insulation
Deck

Fig. 10-5 A plastic fabric placed between the insulation and loose aggregate prevents obstruction of drain paths by filtering water flowing down through the insulation joints. It also "rafts" the insulation system, restraining flotation of individual boards.

joints by aggregate or debris on the roof surface. (Aggregate trapped in insulation joints can even promote insulation board buckling under thermal expansion on hot, sunny days.) Filtration of rainwater prevents buildup of sediment, with clogging of drain paths by fine material that might build up in the insulation joints or on the membrane.

Because a PMR membrane is more like waterproofing than a conventionally exposed roof membrane, Canadian experts M. C. Baker and C. P. Hedlin recommend a nonshingled buildup of alternating layers of felt and bitumen for a PMR membrane.[1] This design approach makes good sense for a ballasted PMR. Defective application of a shingled built-up membrane—e.g., a fishmouth at a felt lap—provides a direct path to the substrate. Membranes laid ply-on-ply are not vulnerable to this diagonal moisture penetration through the entire membrane cross section. Thus, with properly controlled application (a vital proviso) in a securely ballasted PMR—e.g., a system with properly designed and installed pavers providing positive assurance against wind-uplift stresses in the membrane—you could improve the membrane's waterproofing quality by specifying a ply-on-ply felt pattern in the built-up membrane.

As a practical matter, however, this advantage is seldom exploited. Shingling is such a well-established United States roofing practice that it would throw most roofers off stride to require a nonshingled, waterproofing felt pattern. Moreover, the waterproofing pattern poses several other difficulties. It requires phased construction, with its attendant hazards of entrapping water. And especially in cold weather, the application of one felt instead of multiple shingled felts causes faster temperature loss in the hot mopping, thereby increasing the risk of poor interply adhesion. But when it is practicable to apply a nonshingled instead of a shingled membrane, the designer of a PMR should consider this potential design improvement.

Drain Location

Interior drain location is preferable to exterior, scupper-style drains on any built-up roof system, but especially on PMR systems in cold climates. Ice can block perimeter drains. Expanding, freezing water trapped at the perimeter can also displace insulation boards. Interior drains, conducting heat from the building interior, keep drains open and drain water flowing in even the coldest weather.

---

[1] Baker and Hedlin, ibid.

PMR FLASHINGS

Flashings for a PMR are governed by the same design principles as conventional built-up flashings (see Chapter 8). However, there are several differences:

1. Base flashings should extend at least 6 in. above the top of the PMR surfacing. Thus, on a PMR with 3-in.-thick insulation and 2-in. pavers, minimum base-flashing height above the membrane should be 11 in., not the 8-in. minimum generally required above the membrane of a conventional built-up membrane.

2. On a PMR, base flashings should be installed *before* the insulation and surfacing are applied.

PMR flashings are vulnerable to damage from paver edges adjacent to cant strips. Lateral displacement of the paver can puncture the flashing.

PMR INSULATION PERFORMANCE REQUIREMENTS

According to Dow Chemical Company researchers K. A. Epstein and L. E. Putnam, the ideal PMR insulation has (in addition to the obviously required thermal-insulating quality) the following seven properties:

1. Impermeability to water

2. Resistance to freeze-thaw cycling

3. High compressive strength

4. Dimensional stability

5. Resistance to ultraviolet radiation

6. Nonbuoyancy

7. Incombustibility

No commercially available insulation comes close to meeting these ideal requirements, but the PMR concept makes an ideal insulation unnecessary. Protective surfacing can shield the insulation from ultraviolet radiation and provide ballast against flotation and a noncombustible shield against fire brands. Moisture-absorptive resistance, tolerance to freeze-thaw cycling, and compressive strength are indispensable; of these requirements, moisture-absorptive resistance is paramount.

Moisture trapped between membrane and insulation creates an especially serious problem. In cold weather, when the bottom insulation surface is warmer than the top surface, with the vapor-pressure gradient paralleling the temperature gradient, water vapor moves up from the membrane-insulation interface into the insulation. This vapor may condense within the insulation. Alternatively, because the insulation is not impermeable, the water vapor may escape from the top surface, if that surface is properly vented. Concrete pavers laid directly on the insulation's top surface can inhibit water vapor's escape.

PMRs with loose pavers or aggregate surfacing rely totally on the system's self-drying characteristics to limit the insulation's moisture content. To promote self-drying, you can use one or more of the following strategies:

- Slope deck to provide drainage.

- Seal the most vulnerable insulation skin surfaces.

- Design the system to ventilate the insulation.

Laboratory tests confirm the efficacy of all three methods.

Drainage is probably more important for a PMR than for a conventional built-up roof system. Ponding on any roof heightens the risks of leakage. On a PMR, it promotes heat leakage from the interior and heightens the risk of water leakage into the interior. Continual wetting of insulation for long intervals increases its moisture absorption, and wetting of organic felts promotes rot and disintegration.[1]

The ¼-in. slope recommended for conventional built-up membranes (occasionally compromised to ⅛ in. or even 0) should be considered a minimum for a PMR, to ensure positive drainage of the entire roof surface. The Alaska District Engineer, U.S. Army Corps of Engineers, specifies ½-in. slope for PMRs.[2] Federal agencies generally permit ¼-in. slope. In one series of field tests continued over a 5-year period, Canadian researcher C. P. Hedlin demonstrated the drastic reductions in moisture content, by a factor of nearly 10 for urethane insulation, for a 1-in. vs. a ¼-in. slope.[3] The importance of drainage for a PMR is thus a severe handicap on reroofing projects where the existing deck is inadequately sloped.

[1]H. O. Laaly, "Effect of Moisture on the Strength Characteristics of Builtup Roofing Felts," *Symp. Roofing Sys.*, ASTM STP603, 1976, pp. 104–113.

[2]Aamot and Schaefer, op. cit., p. 21.

[3]C. P. Hedlin, "Moisture Content in Protected Membrane Insulations—Effect on Design Features," ASTM STP603, 1976, p. 42.

Besides positive slope, other design features help promote good drainage. Most obvious is proper drain elevation, which is more critical for a PMR than for a conventional built-up membrane. Drains must be depressed below the level of the membrane, well below the elevation of the insulation's top surface. Chamfering the bottom edges of insulation boards promotes good drainage. Chamfering the top corners of insulation boards, in conjunction with side and bottom sealing with asphalt-bonded coated felts and flashing of stone pavers, helps keep the insulation dry.[1]

Ventilation of both top and bottom surfaces of insulation boards reduces moisture content by about one-half. Unsealed insulation boards set on a 1-in.-thick gravel layer on top of the membrane had only half the average moisture content as the same insulation boards placed directly on the membrane. Insulation covered by concrete pavers supported above the top insulation surface on ¾-in. wood strips had only half the average moisture content of insulation covered with loose gravel.[2] Loose-gravel surfacing promoted somewhat better drying than concrete pavers placed directly on the insulation. Raising concrete pavers above the top surface reduces the insulation's average moisture content to roughly one-third the content of the same insulation with pavers placed directly on the insulation. Bottom-surface drainage may, however, increase heat loss because it permits air to flow through.

Extruded polystyrene foam in the higher density ranges (2.3 lb/ft³ or more) exhibits substantially better moisture-absorptive resistance than the other two insulating materials that have been used on PMRs. Independent laboratories and field tests on both sides of the Atlantic substantiate this superiority of extruded polystyrene. Laboratory tests conducted by the Cold Regions Research and Engineering Laboratory of the U.S. Army Corps of Engineers indicate nearly 10 times as much water absorption by beadboard polystyrene compared with extruded polystyrene (2.4 lb/ft³) for 160 days of submergence in water.[3] In field tests conducted by Swedish investigators at the Chalmers University of Technology in Gothenburg, molded, beadboard polystyrene absorbed 5 to 6 percent moisture (by volume), roughly 100 times the 0.05 percent moisture found in extruded polystyrene after both test samples were exposed for 1 year as components in PMRs surfaced with loose aggre-

[1]Hedlin, ibid., p. 46.

[2]Ibid., pp. 7ff.

[3]Chester W. Kaplar, "Moisture and Freeze-Thaw Effects on Rigid Thermal Insulations," Tech. Rept. 249, U.S. Army Corps of Engineers, Cold Regions Research and Engineering Laboratory, Hanover, N.H., April 1974, pp. 7, 9.

gate.[1] Extruded polyurethane is also more water-absorptive than extruded polystyrene.

Resistance to freeze-thaw cycling, the second major property required by a PMR insulation, depends on the insulation's ability to accommodate the expansion of freezing water entrapped in or near its top surface. (As indicated earlier, the bottom surface should never drop as low as freezing temperature.) The property required for freeze-thaw cycling resistance is elasticity—the ability of the closed-cell foam to yield under the expansion of freezing water without brittle cracking, spalling, or flaking of the insulation material. Obviously, the less water absorbed by the insulation, the less hazard exists from the absorbed water's expansion.

Reporting on their research into freeze-thaw cycling, Dow researchers Epstein and Putnam claim that extruded polystyrene performs the best of a number of commonly used insulating materials. Under their test method, ASTM C666-73 (freeze in air, thaw in water, designed for concrete cylinder testing), most tested insulations began to disintegrate between 97 and 458 freeze-thaw cycles, with water absorption of up to 89 percent (by volume).[2] The Dow researchers claim that ASTM Test Method C666-73 is actually less severe than some PMR roof exposures. For example, the test does not allow one side of the insulation to retain water or remain at 70°F. And because both sides of the insulation are at the same temperature, there is no vapor-driving force tending to force vapor from the warm bottom surface toward the cold top surface, the normal, cold-weather situation for a PMR.

Two remaining properties required of a PMR insulation, compressive strength and dimensional stability, disqualify fewer materials than water-absorptive and freeze-thaw cycling resistance. Few currently marketed roof insulating materials lack the required compressive strength because this property is required to just about the same degree for a conventional roof's insulation.

Dimensional stability is less important for a PMR than for a conventional roof's insulation. A high coefficient of thermal expansion-contraction poses no particular problem. The most important aspect of dimensional stability for a PMR insulation is low, long term net contraction or expansion under countless wetting-drying cycles (like the long term shrinkage of organic felts under such cycles). Despite its high

---

[1]Lars-Erik Larsson, Julia Ondrus, and Bengt-Ake Petersson, "The Protected Membrane Roof (PMR)—A Study Combining Field and Laboratory Tests," *Proc. Symp. Roofing Technol.*, NBS-NRCA, September 1977, pp. 86, 88.

[2]Epstein and Putnam, op. cit., pp. 52, 60.

coefficient of thermal expansion-contraction, extruded polystyrene satisfies this requirement.

Thermal Quality of PMR Insulation

The thermal resistance of wet insulation naturally decreases with increasing moisture content. According to the Swedish investigators, the thermal resistance of polystyrene board drops to about 10 percent less than its thermal resistance as part of a conventional built-up roof system exposed to the same weather conditions.[1] (However, if the insulation in a conventional roof gets wet—through membrane leakage or condensed water vapor—then its thermal resistance could drop even lower than a PMR's insulation because evaporation of water in the confined insulation in a conventional built-up roof system is much more difficult than in a PMR.) According to Aamot, the designer should reduce the insulation's published R value by 10 to 20 percent for PMR design.[2] Reduce published R values by 10 percent in dry climates, by 15 percent in moderate climates, by 20 percent in wet climates.

Because of the many complex variables affecting heat transfer through the roof, the so-called thermal efficiency of a PMR may exceed 100 percent.[3] This means that the actual heat gains and losses may be *less* than those calculated from theoretical heat-flow, steady-state, heat-transfer equations, based on the U value and the inside and outside air temperatures. In other words, the roof's thermal performance in retarding heat flow and cutting heating- and cooling-energy bills is actually better than the conventional steady-state heat-flow formulas indicate. (Judged by this criterion, a conventional roof's thermal performance might also exceed 100 percent, depending especially on its surface color.)

In addition to conducted heat flow, the single factor considered in conventional heat-flow calculations for roof design, actual heat flow depends on many other factors: evaporation, condensation, water flow over the membrane, solar radiation intensity and duration, snow cover, and wind.

In cold weather, the forces *increasing* a roof's thermal efficiency are solar radiation (which raises roof-surface temperature and reduces the temperature gradient through the roof) and snow cover (which in sufficient depth can add an insulating blanket to the roof's components). Among the forces *reducing* a roof's thermal efficiency are rainwater

---

[1]Larsson, Ondrus, and Petersson, op. cit., p. 88.

[2]Aamot and Schaefer, op. cit., p. 25.

[3]Ibid., p. 11.

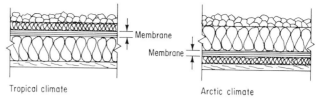

Tropical climate   Arctic climate

Fig. 10-6 Ideal location for membrane in a PMR is *above* the major insulation in a perpetually warm tropical climate, *below* the major insulation in a perpetually cold climate. Thin insulation above the membrane shields a tropical PRM against solar radiation and extreme high temperatures. Thin insulation below an Arctic PMR reduces heat losses from convection and evaporative cooling following a rain.

evaporation (which transfers heat energy from the roof surface into the evaporating water) and cold rainwater or melt water flowing across the roof to the drains. Although detrimental during heating season, these evaporative/convective heat losses are of course helpful during the cooling season because they reduce heat gain.

According to Aamot, thermal efficiency greater than 100 percent was observed about 85 percent of the time in a heat-metered roof in Hanover. On cold, bright days, with intense, undiffused solar radiation, thermal efficiency sometimes exceeded 300 percent.[1] Thermal efficiencies less than 100 percent are observed during rainfall, when the flowing water carries off roof-surface energy, and for several days after, as the evaporating residual moisture dries the insulation and cools the roof surface.

Refinements in Locating PMR Insulation

From the tested performance of PMRs, Aamot worked out a general thermal-design theory that indicates the benefits derived from putting some insulating material under the protected membrane. The thermal performance of such a PMR is higher in a cold climate because internal cooling is reduced. Gypsum board, required as a substrate for fire protection on steel decks, serves this purpose. However, in a predominantly warm, humid climate, most of the insulation should be under the membrane, with a lesser amount above the insulation (see Fig. 10-6). In a warm, humid climate, the temperature and pressure gradients are reversed, so the membrane, doubling as vapor retarder, should be located on the warm side of the dew point, near the top surface of the insulation.

[1]Loc. cit.

In a typically extreme United States continental location (e.g., Kansas City, Missouri), with its cold winters and hot, humid summers, the winter condition should probably govern because the extreme temperature difference between inside and outside is normally greater in winter than in summer.

Placing additional insulation on top of a conventional roof—in effect, converting it into a PMR—can benefit the roof system in two ways: (1) by the obvious increase in thermal resistance and (2) by raising the membrane's cold-weather temperature, thereby reducing the threat of moisture condensation at the underside of the membrane. Note that any such addition requires a structural review to ensure that the additional dead load of the PMR insulation and ballast is not excessive.

SURFACING PROTECTED MEMBRANE ROOFS

Loosely laid aggregate or pavers resist flotation forces plus wind uplift and scour. For design techniques to avert these hazards, see Chapter 14, "Wind Uplift." As that chapter indicates, it may be necessary to specify pavers instead of loose gravel for vulnerable corner and other peripheral areas subjected to high wind velocities and uplift pressures. As an additional advantage, pavers ensure uniform weight distribution, whereas loose aggregate can vary.

In addition to its greater wind-uplift protection, pavers may be economically justified to serve three other purposes:

- As a foot-traffic surface for terraces, plazas, or for a roof requiring frequent maintenance trips for access to mechanical equipment and so forth

- Easier access to repair the roof (because removing pavers is more convenient than removing loose aggregate)

- Better appearance

When specifying pavers as a PMR surfacing, the designer should also consider soffit treatment—corrugations, channeling, or simply support on ¾-in. wood strips or insulation pads—to facilitate evaporation from the top surface of the insulation.

Ballast weight against flotation is about 5 psf/in. thickness of insulation. Because water weighs 5.2 psf/in. and the lightest insulations weigh only 0.2 psf/in., 5 psf/in. is a minimum figure to prevent flotation. Solid concrete pavers weigh about 12 psf/in. thickness in air (7 psf/in. thickness submerged).

The principle of "rafting", i.e., using a porous mat over the insulation boards or foamed polystyrene boards with a thin, integral concrete topping and tongue-and-groove edges, permits slightly reduced ballast weight because the boards tend to float in large units rather than as individual boards. The rafting principle springs from the reduced probability of a large area being subjected to full flotation pressure. Successful rafting prevents displacement of individual boards.

LIFE-CYCLE COSTING

Protected membrane roofs generally, although not always, cost more than conventional built-up roof systems, as substantiated by the previously cited Canadian survey of architects and roofers. Increased cost for a PMR may result from one or more of the following factors:

- Additional structural cost to support ballast dead load

- Additional cost for greater amount of aggregate, minimum 10 psf loose aggregate vs. 3 to 4 psf for conventional built-up system aggregate surfacing

- Additional cost for leveling board on steel deck

Benefits, with possible cost savings, that may accrue from a PMR are:

- Avoidance of wind-uplift stresses on the structure (as a result of the ballast)

- Greater protection, functional enhancement (as a roof terrace), and improved roof appearance of pavers

- Need for only one membrane, which doubles as a membrane and vapor retarder

- Reduced installation work and fewer weather constraints on application of loose-laid PMR system

Some converts to PMR roofs find increased first cost economically justified on a life-cycle (long term) cost basis. From a survey of conventional built-up maintenance costs in Alaska, Aamot estimates annual maintenance cost at 5 percent of initial cost, vs. 1 percent for a PMR. At interest rate = 12 percent for a 20-year life cycle, a 4 percent annual maintenance saving with 10 percent cost escalation is equivalent to a 66 percent increase in initial cost [computed on a Present-Worth (PW) basis].[1] In other words, a PMR is more economical than a conventional built-up roof system if its initial cost is less than 1.66 times the conventional roof system's initial cost.

Although the dramatic savings estimated for an Alaskan location may not hold for more moderate climates, the owner concerned with long term economy should investigate the PMR concept.

ALERTS

1. Provide slope for positive drainage, ¼-in. minimum, ½ in. if practicable.

2. Locate roof drains within heated part of building. (Exterior drains periodically freeze shut in cold climates.)

3. Depress drains below membrane elevation.

4. Check roof structural load, especially when reroofing existing buildings, for both strength and deflection under ballast load.

5. Flood test completed membrane before installing insulation and surfacing. (If impracticable to flood test *before* insulation goes down, flood-test afterward.)

6. Reduce theoretical dry factor of insulation by 10 to 20 percent (to allow for moisture-caused reduction in thermal resistance and interior cooling from rainwater flow).

7. Specify two insulation layers, with staggered joints, for best thermal performance.

8. Design for ventilation of insulation covered with loose aggregate or concrete pavers.

9. Favor glass-fiber felts over asbestos or organic felts for a built-up bituminous PMR membrane.

10. Consider a nonshingled, separate-layer waterproofing pattern for felts in a built-up bituminous PMR membrane (but only if good interply mopping adhesion is ensured).

---

[1] On a 20-year life cycle, the PW of the 4 percent annual maintenance-cost saving (expressed as a percentage of initial cost C) is computed as follows:

$$PW = 0.04C \frac{a(a^n - 1)}{a - 1}$$

where $a = (1 + m)/(1 + i) = (1 + 0.10)(1 + 0.12) = 0.982$
m = maintenance-cost escalation = 10%
i = interest rate = 12%
n = number of years = 20
$PW = 0.04C \times 16.62$
$= 0.66C$

11. Do not specify an adhered membrane on top of a poured deck subject to shrinkage or thermal cracking.

12. Design loose aggregate size to resist wind uplift and scour. (Specify heavy. pavers in high wind areas or along roof perimeter.)

13. Consider a filter-separator fabric between loose aggregate surfacing and top surface of insulation.

14. Note flashings' increased vulnerability to damage, especially at cants adjacent to pavers.

ELEVEN

Synthetic Single-Ply Membranes

In the late 1960s, during the writing of this manual's first edition, the new synthetic single-ply membranes comprised a minor fraction of 1 percent of the low-slope roofing market. By 1980, these membranes had multiplied their market share to 5 to 10 percent, according to one estimate.[1] Moreover, the single-ply rate of increase was accelerating. Clearly, these new synthetics appear to be the wave of the future, and their more zealous boosters forecast a tidal wave. According to one manufacturer's representative, the synthetics should take 50 percent of the low-slope roofing market by 1988, and some projections are even more optimistic.

Behind this second-generation revival of the single-ply synthetics (following their less successful first-generation appearance in the late 1960s) is a combination of economic and technological factors:

- Skyrocketing petroleum prices have made the formerly more expensive synthetic materials cost-competitive with conventional bituminous built-up systems. (Asphalt prices, $125/ton in 1979, reached $200/ton in 1980, and $250/ton by mid-1981.)

- Continued dissatisfaction with built-up bituminous roof performance provided a ready market of unhappy building owners and architects eager to try something different.

- Successful performance of the new synthetics in Europe, where they account for a much greater share of the industrial-commercial-institutional roofing market, compiled a track record testifying to the new materials' potential durability.

- Lessons learned in the United States (and Europe) from failures in materials and quality of work with the first-generation

---

[1]"The Roofing Contractor's Single-Ply Starter Kit," *RSI Magazine*, November 1980, p. 34.

synthetics gave the second-generation synthetics a more solid base for development of improved materials.

Among the improvements promoting better performance by the second-generation synthetics are improved chemical formulations, improved manufacturing processes, thickened sheets, improved reinforcing, and better control of fieldwork. The following anecdote illustrates the importance of control over fieldwork. One material withdrawn from the market in the early 1970s, polyvinyl fluoride sheets, required a 5-minute wait after the contact adhesive was applied before adjacent sheets could be field-adhered. Premature joining of the seam prevented escape of the adhesive solvent, whose subsequent evaporation could blister the membrane. In one notorious blistering instance, in response to the manufacturer's representative's question, "Did you follow the installation directions?", the roofing foreman's reply was, "What directions?"

Improved fieldwork conditions constitute another reason for the trend to the new synthetics. As one of their basic appeals, the synthetic sheet or fluid-applied membranes offer an escape from messy bitumens, with their heavy equipment, air-polluting kettles and tankers, "hot" buggies, and felt-laying machines. Recruits for this drudging fieldwork are becoming ever scarcer. Attesting to this fact are roofing contractors' complaints about personnel shortages, despite roofers' high pay. Their crews are steadily aging, say these contractors, with few replacements from the young.[1] In contrast with the heavy work associated with built-up roof systems, the new synthetics offer light work, with shears, cutters, rollers, squeegees, and sprays replacing the mops, kettles, felt-laying machines, aggregate spreaders, and other cumbersome equipment used on hot-applied bituminous roofs (see Fig. 11-1).

Tightening pollution controls and Occupational Safety and Health Administration (OSHA) regulations governing use of the traditional hot-applied bitumens, with their polluting fumes and fire hazards, constitute still another force behind the trend toward single-ply synthetics.

PROS AND CONS

Synthetic membranes have the following potential advantages:

- Ultralightweight (for unballasted, adhered membranes)

- Adaptability to irregular roof surfaces

---

[1] "Strong Market Ahead for Reroofing, Insulation," *RSI Magazine*, February 1978, pp. 57ff.

Fig. 11-1 Rolling is one of three ways (along with brushing and spraying) to apply fluid chloroprene-chlorosulfonated polythylene coatings. (E. I. du Pont de Nemours & Co.)

- Isotropic (or nearly isotropic) physical behavior (as opposed to the anisotropic weakness of organic and asbestos felt membranes in the transverse direction)
- Superior architectural quality (notably color)
- Generally superior heat reflectivity (for cooling-energy conservation)
- Better elongation (up to 800 percent at 70°F)
- Superior performance at subzero temperatures (flexibility down to −75°F for some materials)
- Easier application (plus greater potential for shop prefabrication of large sheet membranes, flashing, hoods, and so forth)
- Fewer weather restrictions on application (permissible at subfreezing temperatures for some loose-laid sheets)
- Less hazard from moisture entrapment during installation (for loose-laid, permeable synthetic sheets)
- Easier repair of punctures, splits, tears
- Easier flashing application at corners and irregular surfaces, where stiffer built-up materials are difficult to form
- Possibly greater reliance on factory-manufactured material quality

- Easier detection of leak source (in smooth-surfaced, adhered type)

- Easier removal (for temporary roof converted to floor and so on)

Synthetic-material manufacturers generally claim another advantage for their material: its alleged superior performance under ponding. Ponding, according to this argument, presents no problem for synthetic materials that have performed well for years as tank, canal, and pondliners.

Rejecting this argument in their study of synthetic roof membranes, National Bureau of Standards (NBS) researchers R. G. Mathey and W. J. Rossiter recommend ¼-in. slope for synthetic as well as built-up membranes.[1] A roof's performance requirements are more rigorous than a pondliner's. A roof membrane, especially one that is exposed, is subjected to thermal and structural stresses that a pondliner may never experience. Ponded water could increase these stresses, and it certainly deposits dirt as sediment, affecting architectural appearance and reducing a smooth-surfaced roof's heat reflectivity. And when a leak occurs in roof-ponded water, it increases the water-damage hazard to the building and its contents. It may be economically necessary to tolerate ponding, but it should be recognized as an inevitably added risk.

The single-ply synthetic manufacturers do, however, score an impressive point in their claims that ponding will not affect their systems' performance. They generally omit the ponded-water exclusion clause that appears in most built-up membrane bonds or guarantees.

Here are the synthetic membranes' probable disadvantages:

- Greater importance of good workmanship

- More limited range of suitable substrates (especially for fluid-applied systems)

- Less puncture resistance (and consequently greater vulnerability to traffic damage)

- Lack of performance and design criteria comparable to those available for built-up systems

- Poor reporting of technical data

Note that some disadvantages of synthetic single-ply membranes in other contexts offer advantages. Against "easier application" listed

[1]W. J. Rossiter and R. G. Mathey, "Elastomeric Roofing: A Survey," NBS Tech. Note 972, July 1978, p. 14.

among the new systems' assets, you must weigh the "greater importance of good workmanship" listed among the liabilities. Good workmanship is vital for all roofing application, but especially so for the new synthetics. A multi-ply built-up membrane has some safety margin against one poorly applied felt layer; a single-ply membrane has none. Failure at a seam in a single-ply membrane means an almost certain leak.

The protected membrane roof (PMR) concept mitigates one liability of single-ply synthetics. With the insulation shielding the membrane from roof traffic, the single-ply synthetics' lower puncture resistance is less significant.

The last two items in the liability list—lack of performance criteria and poor reporting of technical data by manufacturers—are two aspects of a larger problem. With some notable exceptions, single-ply synthetics remain quasi-experimental, subject to unknown problems like those forcing removal of several first-generation materials from the market. Architects harassed with malpractice lawsuits have justifiably lagged in specifying these new materials; our legal system offers few incentives, but plenty of deterrents to architectural or engineering innovation.

Research Needs

Illustrating the poor reporting of data is the manufacturers' widespread practice of reporting sheet tensile strength in psi. This term is essentially meaningless in roofing technology. Researchers report built-up membranes' tensile strengths in lb/in. units, the relevant data for calculating thermal shock factor (TSF) or any other tensile-strength property of roof membranes. With strength reported in psi, you must multiply the strength by the membrane thickness to get the meaningful lb/in. strength unit. A 48-mil sheet with 1400-psi tensile strength is stronger than a 30-mil sheet of 2000-psi tensile strength (67 lb/in. vs. 60 lb/in.). A superficial comparison of the two materials might obscure that fact.

More important than poor reporting of technical data is the simple lack of such data. "Little quantitative data for evaluating the performance of elastomeric roofing have been published. In general, the performance of elastomeric roofing systems has been determined by in-use experience and not laboratory research."[1]

As an example of such missing data, Rossiter and Mathey cite the current widespread practice of measuring mechanical properties at room temperature. This practice may be totally inadequate for ultimate elongation. At subfreezing temperature, when it will usually require

[1]Ibid., p. 4.

elasticity, a synthetic membrane's ultimate elongation may drop significantly. Moreover, a weathered or aged synthetic membrane may also lose a substantial part of its room-temperature elongation. The combined negative effects of subfreezing temperature and TSF with aging may drastically reduce the membrane's ultimate elongation and result in poor performance in cold climates. Some manufacturers do report low-temperature data. Roof designers should request it.

Built-up membranes suffer drastic reduction in breaking strain and in thermal shock factor with aging. As discussed in Chapter 7, these membranes' deterioration is documented by research reports. We need comparable data for the new synthetic membranes, for rational comparisons of these radically different materials. (See Table 11-1 for the NBS researchers' suggested performance criteria.) It will be a difficult job because as Rossiter and Mathey point out, there are no criteria for determining what are the required tensile strengths and elongation for synthetic membranes. Reported values for elongation range from 100 to 800 percent. But what is the required value? Obviously it must be higher for an adhered system than for a loose-laid system. But how much higher? There are no answers yet.

Thus the new synthetics suffer from the inevitable drawbacks of any new material when it enters the marketplace. Moreover, the variety that gives roof designers such wide choices compounds designers' problems in selecting materials. Not only is there a much greater diversity in available roof systems and variations on basic roofing systems, but the properties of the same basic material may vary among different manufacturers because of different chemical formulations and manufacturing techniques.[1]

Rossiter and Mathey make another important point: "The ultimate test of weather resistance is in-service performance, but this is not practical in the case of new materials. Thus, laboratory tests are used in attempts to predict performance. The relationships between laboratory tests and weather resistance are difficult, if not impossible to establish."[2]

Further compounding the problem of evaluating synthetic membranes is the extremely complex task of comparing laboratory test data. Attempting to compare different manufacturers' products, you enter a technological Tower of Babel. One manufacturer reports the results of tensile strength of his nylon-reinforced EPDM (ethylene propylene diene monomer) sheet in psi, per ASTM Test Method D412; another reports his nonreinforced EPDM strength in lb/in. (the correct way for roofing membranes), per ASTM Test Method D751. One manufacturer

[1]Ibid., p. 10.
[2]Ibid., p. 31.

Table 11-1 Suggested Performance Criteria for Elastomeric Membranes

Performance characteristic	Suggested preliminary performance criterion	Test method	Comment
Adhesion between layers	Not established	Not established	Multilayered membranes should resist delamination, blistering, or peeling between layers.
Chemical resistance	Should be resistant	ASTM D543	Elastomeric membranes should be resistant to solvents and chemicals to which they may be exposed during both application and service life.
Color stability and reflectivity	Percent allowable change not established	Not established	Color stability is considered an essential performance characteristic when the color of the membrane is chosen to reflect solar radiation. In these cases the reflectivity of the membrane should not be altered by color change or dirt retention.
Creep resistance	Not established	Not established	This level of performance may depend upon the design of the roofing system.
Fire resistance	Class A, B, or C	UL 790	UL 790 has been accepted by the roofing industry as a test procedure for indicating the fire resistance of roofing membranes.
Fungus resistance	Should be resistant	ASTM G21	
Impact resistance	Resistant to a 1½-in. (38-mm) hailstone falling at 112 ft/s (34 m/s)	Hail test	This level of performance has been recommended for bituminous roofing membranes and may be applied to elastomeric membranes. Lower limits may be established for areas not subjected to severe hailstorms. Resistance depends on the substrate.

Table 11-1 Suggested Performance Criteria for Elastomeric Membranes (Continued)

Performance characteristic	Suggested preliminary performance criterion	Test method	Comment
Resistance to foot traffic, wind-blown particles, and similar forces:			
○ abrasion	Not established	Not established	
○ hardness	Not established	ASTM D2240	
○ punching shear	Not less than 250 lbf/in.^{2} (1.7 MPa)	NBS test method	The test described in Building Science Series 55 is conducted at 73°F (23°C) using a ¾-in. (19-mm)-diameter probe. This level of performance has been recommended for bituminous membranes and may be provisionally applied to elastomeric membranes. Resistance depends on the substrate.
Strength of seams (sheet systems)	Not less than 6 lbf/in. (1 kN/m)	ASTM D1876	This level of performance is selected from Martin, who suggested that seams in sheet membranes should be considered suspect for use in exposed applications, if the strengths of the seams are less than 6 lbf/in. (1 kN/m) as measured by the T-peel test.
Stress resistance:			The level of performance for stress resistance may depend upon design of the roof system. It may be necessary to establish levels of performance for stress resistance, which is a function of both tensile strength and extensibility.
○ tensile strength	Not established	ASTM D412	
○ ultimate elongation	Not established	ASTM D412	
○ tensile fatigue	Not established	Not established	
Tear resistance	Not established	Not established	ASTM methods of test D624, D751, D1004, and D1922 have been listed as means for measuring the tear resistance of elastomeric membranes.
Thermal expansion	Not established	ASTM D696	This level of performance may depend upon the design of the roofing system.

Water resistance:			
o absorption	Not established	ASTM D570	This level of performance depends upon the design of the roofing system, which may require either a nonpermeable or permeable membrane.
o vapor transmission	Not established	ASTM E96	
Wind resistance	Class (30), (60), or (90)	UL test procedure	This level of performance is believed adequate for most areas in the United States. Other tests, such as those recommended by Factory Mutual, may be suitable.
Weather resistance	Should be resistant for a minimum of 15 years	Experience and laboratory tests in conjunction with engineering judgment	It is not unreasonable to expect a minimum service life of 15 years because elastomeric membranes have performed satisfactorily for longer periods. Maintenance is necessary during the service life. The ultimate test of weather resistance is in-service performance; laboratory tests may be conducted to predict weather resistance, but the relationships between the laboratory tests and weathering are difficult if not impossible to establish. The decrease in performance characteristics after weathering should be limited.
o Accelerated aging	Not established	ASTM D2565	These tests may provide an indication of the weather resistance of elastomeric membranes. Relationships between test results and performance need to be determined. The level of performance for brittleness may depend on expected winter temperatures.
o Brittleness	Not established	ASTM D2137	
o Dimensional stability	Not established	ASTM D1204	
o Heat aging	Not established	ASTM D573 or D865	
o Ozone resistance	Not established	ASTM D1149	
o Pollutants	Not established	Not established	
o Volatile loss	1% maximum	ASTM D1203	Loss of plasticizer may lead to embrittlement.

Source: W. J. Rossiter and R. G. Mathey, "Elastomeric Roofing: A Survey," NBS Tech. Note 972, July 1978.

measures sheet tear resistance per ASTM D751, another via ASTM D624; still another uses a German test, Deutsche Industrie Normal (DIN) test method. Another manufacturer omits tear resistance altogether and reports other properties—tensile strength, ultimate elongation, and so forth—without even identifying the test methods.

Development of standards enabling the designer to make more rational product selections will obviously take years. Meanwhile, as test criteria are advanced and roof track records compiled, the designer's chief guide will be the manufacturers' recommendations, the products' previous track records (incomplete though they may be), and the manufacturers' records of honoring their guarantees or warranties.

CURRENT STANDARDS DEVELOPMENT

Largely because of the single-ply manufacturers' general policy of pursuing their own narrow self-interests, praising their own products' alleged superiority, denigrating their competitors' products, competing more on the basis of consumer sales techniques than on demonstrated scientific performance, development of standards enabling the designer to make more rational product selection is a lagging process. ASTM Committee D8 has had a task force on single-ply synthetics since 1972, but nearly a decade later (in early 1981), it was still struggling to produce tentative recommendations for performance criteria. One task force, on EPDM and polychloroprene, had made fair progress in its "Tentative Standard for Reinforced and Non-Reinforced Vulcanized Rubber Sheeting for Roofing." But another task force group, on plasticized PVC sheet membranes, had achieved next to nothing. Virtually every performance criterion for such properties as "adhesion between plies," "fire resistance," "impact resistance," "tear resistance," and so on was noted as "not established." In contrast with the ASTM's futility, the Canadian General Standards had established values for these properties.

Accelerated weathering tests constitute a major part of this effort to establish performance criteria for single-ply membranes. In early 1981, three major PVC membrane manufacturers, with products that had endured long service exposure, were attempting to develop a recommendation on accelerated weathering tests.

Meanwhile, the National Roofing Contractors Association (NRCA) has been independently pursuing its own accelerated weathering test. This NRCA effort focuses on an accelerated weathering test developed by the Norwegian Building Research Institute and reported by Einar M. Paulson in the International Symposium on Roofs and Roofing, Brighton, England, 1974. Designed to test any roof system—conventional built-up bituminous as well as single-ply synthetics—the testing apparatus features an upper chamber equipped to simulate destructive weather phenomena: ultraviolet and infrared radiation, water spray, and pulsating wind pressure. The test allegedly compresses these natural effects into a 15/1 ratio; that is, 1 year's testing simulates 15 years' natural weathering. It can be adjusted to simulate the temperature cycling, rainfall, wind-uplift pressures, and solar radiation for a full spectrum of climates.

CHEMICAL NATURE OF SYNTHETIC MEMBRANES

Although often classed generically as "elastomerics," the new synthetic, single-ply membrane materials actually cover a broader range. An elastomer is a synthetic polymer with rubberlike properties: i.e., an elastic material that rapidly regains its original dimensions upon release of deforming stress. Synthetic rubber materials—neoprene, butyl, EPDM—satisfy this definition. But other materials used for single-ply membranes, e.g., plasticized polyvinyl chloride (PVC) or rubberized or modified asphalt, do not. These materials exhibit some permanent elongation after tensile stressing. Still other materials only partially satisfy the definition of an elastomer. Thus arises the general category synthetic to distinguish these new single-ply membrane materials from the bituminous materials. Despite their subsequent processing, bitumens are basically natural materials whose molecular structure is only slightly modified by manufacturing processes. Synthetics, however, are totally artificial.

Chemically, the new synthetics are polymers. Their high tensile strength depends upon long chain molecules built up from monomers (the basic molecular units). Polymerization increases monomeric molecular weights from the 30 to 150 range to so-called "macromolecules," 100 to 10,000 times as large. Tensile strength depends on the degree of polymerization; the longer the molecular chain, the stronger the polymer.

A major chemical distinction differentiates thermosetting from thermoplastic polymers. Synthetic rubber materials (the elastomers) are thermosetting. They harden permanently when heated, like an egg. Thermoplastics soften when heated (like butter) and harden when cooled. With thermoplastic materials, thermal cycling can repeat indefinitely; viscosity changes accompany temperature changes.

This contrasting behavior stems from a basic difference in molecular structure. Thermosetting resins start as tiny threads. Heat promotes chemical reactions that cross-link these tiny molecular threads, creating a permanently rigid matrix (see Fig. 11-2). This molecularly cross-linked structure makes a thermosetting material more resistant to heat, solvents, general chemical attack, and creep (i.e., plastic elongation).

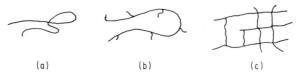

(a) (b) (c)

Fig. 11-2 Schematic representation of chainlike, polymeric molecules shows the polymerization process: from (a) linear to (b) branched to (c) cross-linked molecules that make the material more heat- and creep-resistant and generally stronger.

Thermoplastics comprise long, threadlike molecules, so intertwined at room temperature that they are hard to pull apart. However, when heated they slide past one another, like liquid molecules.

Never content with relative simplicity, the modern chemist has further complicated matters by creating intermediate polymers—neither exclusively thermosetting nor thermoplastic. Some thermoplastic materials are cross-linked for special uses, and some cross-linked resins are available in thermoplastic form. Examples used in roofing are chlorinated polyethylene (CPE), chlorosulfonated polyethylene (Hypalon), polyisobutylene (PIB), and polyvinyl fluoride.

As one tentative general conclusion about thermosetting vs. thermoplastic materials, you might consider thermosetting materials as potentially superior to thermoplastics on a direct material-vs-material basis. But thermoplastic materials offer superior, easier field joint-sealing processes. Thermosetting, synthetic rubbers require use of contact adhesives, which require a wait before the field-seaming process can be completed. Thermoplastic sheets can be joined by solvent or heat-fused welds, performed with no delay, with the basic material itself in partial solution. Solvent-welded seams actually fuse the sheet's base material.

As a practical consequence, you might favor a thermosetting material—EPDM, for example—on a roof with few openings and other penetrating elements because this roof requires fewer field seams. (The membrane could be prefabricated into large sheets; see later section on EPDM sheets.) A thermoplastic material—plasticized PVC, for example—might prove better for a roof with numerous penetrations and consequently numerous field seams.

SYNTHETIC MEMBRANE ROOFING SYSTEMS

Just as they comprise a much broader range of materials than conventional built-up bituminous materials, so do the synthetics offer a broader range of roofing systems (see Table 11-2). In addition to classification by application technique—sheet- vs. fluid-applied membranes—synthetic membranes fall into any one of three basic design categories:

- Fully adhered
- Partially adhered
- Loose-laid ballasted

Fully adhered and *partially adhered* systems follow the basic concept of conventional built-up roof systems. Loose-laid, ballasted systems make a clean break with conventional built-up roof systems.

Fully adhered systems are favored for many irregular, nonplanar roof surfaces: domes, hyperbolic paraboloid shells, or where light weight is important. They require hard, solid substrates—cast-in-place concrete, precast concrete, plywood, or rigid steel decks—for minimal deflection. Fluid-applied membranes are restricted to a more limited range of substrates than prefabricated sheet membranes.

Fully adhered systems are more expensive than loose-laid, ballasted systems. Proper bonding of the contact adhesive anchoring the membrane to the substrate requires meticulous preparation of a clean, dry, crack-free substrate. It also imposes greater weather restrictions, comparable to the restrictions on built-up membrane application. Fully adhered membranes have the advantage of negligible weight, measured in oz/ft^2. Lacking a protective ballast cover, however, they may require some kind of surface protective coating.

Partially adhered systems also limit roof dead load. As its primary function, a partially adhered membrane allows the membrane to float free over a crack or joint in the substrate and to distribute stress in the membrane between bonded areas. In the most popular version, mechanical fasteners driven in a grid pattern of 20-in. maximum spacing clamp circular disks of PVC sheet between metal washers above and a felt slip sheet below. The PVC membrane sheet is bonded by an adhesive, brush-applied to the PVC disks.

Washington's Dulles International Airport Terminal Building, completed in 1962, is a successful example of the partially adhered membrane concept. Bond-breaker tapes applied over the joints between cable-supported precast concrete deck planks prevent adhesive contact and consequent stress concentration of the neoprene-sheet, Hypalon-coated membrane. Between joints, the neoprene sheet is adhered to the precast concrete plank.

Loose, ballasted systems are the synthetic manufacturers' favorite when the roof structure can take the 10 to 15-psf loose aggregate or 25-psf concrete paver ballast loading. In Europe, where structural concrete

Table 11-2 Single-Ply Synthetic Roof Membranes*

Type of membrane	Membrane material	Acceptable substrates	Thickness	
			mil†	mm
Liquid applied‡	Acrylic	Concrete, plywood, spray-in-place urethane foam, remedial roofing	20	0.5
	Butyl	Concrete, plywood, insulation board, remedial roofing	15–30	0.4–0.8
	Chlorosulphonated polyethylene	Spray-in-place foam, weatherproof coating for elastomeric membranes	20–45	0.5–1.1
	Neoprene/chlorosulphonated polyethylene	Concrete, plywood, spray-in-place urethane foam	20	0.5
	Polyvinyl chloride (PVC) and vinyl	Concrete, plywood, spray-in-place urethane foam	15–30	0.4–0.8
	Rubberized asphalt	Concrete	150–180	3.8–4.6
	Silicone	Spray-in-place urethane foam	20	0.5
	Urethane§	Concrete, plywood, spray-in-place urethane foam, remedial roofing	20–60	0.5–1.5

Preformed sheets¶			
Chlorosulphonated polyethylene/ asbestos backed	Plywood decks on industrialized or modular construction	35	0.9
EPDM (ethylene propylene diene terpolymer)	Concrete, plywood, insulation board, remedial roofing	45	1.1
Neoprene	Concrete, plywood, insulation board	63	1.6
Polyvinyl chloride (PVC)	Concrete, spray-in-place urethane foam, remedial roofing	48	1.2
Composite preformed sheets			
Nylon-reinforced PVC backed with neoprene or butyl	Concrete, plywood, insulation board, remedial roofing	30	0.8
Nonwoven glass reinforced PVC	Concrete, plywood, insulation board	47	1.2

*The information in this table has been assembled from the literature of various manufacturers. It is presented for purposes of comparison, to give the reader an overview of the materials available and the substrates to which they may be applied. Individual products should be applied to substrates and in thicknesses recommended by the manufacturers.

†1 mil = 0.001 in.

‡Liquid membranes are commonly applied by a number of techniques, including brush, roller, squeegee, trowel, and conventional and airless spraying.

§A number of widely varying membrane systems are classified under the title urethane, including asphalt and coal tar-modified urethanes.

¶Sheet membranes are applied bonded to or loose-laid on the substrate.

Source: W. J. Rossiter and R. G. Mathey, "Elastomeric Roofing: A Survey," NBS Tech. Note 972, July 1978.

is a more popular deck material, the loose-laid concept dominates. But in the United States, the popularity of long-span steel framing with flexible, lightweight steel-deck systems designed for minimal structural cost severely limits the market for loose-laid ballasted systems.

As its name implies, a loose-laid membrane lies loose on its substrate, purposely slack to prevent tensile stress. Insulation in a loose-laid roof system is similarly loose-laid: on the deck in a conventional, sandwich-style roof system, on top of the membrane in a PMR.

However, a loose-laid membrane is anchored at the perimeter and openings. Such anchorage serves several purposes:

- Prevents stress concentration at flashings, which could either split or pull loose from their backings and become vulnerable to puncture. (Loose-laid systems receive maximum membrane tensile stress at perimeters and other terminations, according to field strain tests.)

- Restrains contraction in materials subject to long term shrinkage.

- Prevents membrane billowing from wind impulses in areas— e.g., building corners—where ballast can be displaced by wind-uplift pressures.

- Prevents membrane wrinkling from obstructing drainage.

The loose-laid concept either alleviates or eliminates three major problems—splitting, blistering, and ridging—that afflict adhered roof systems. It reduces the splitting risk by isolating the membrane from substrate movement that results from structural deflection, temperature change, or drying shrinkage, all of which can produce stress concentrations at critical spots. And a loose-laid membrane cannot blister or ridge because there are no confined voids between membrane and substrate for air–water-vapor-pressure buildup and a single-ply membrane cannot have interply voids. Thus there are fewer worries about entrapped moisture in the insulation. This moisture, when evaporated by solar heat, can either vent downward to the building interior through a permeable deck or diffuse upward through a relatively permeable membrane material. Relatively high vapor permeability is thus an advantage for a loose-laid membrane.

As another major advantage, the loose-laid concept permits fast, economical installation, eliminating the costly, time-consuming mopping or adhesive-application operation required for built-up membranes or adhered synthetic membranes. Not only is the application process simplified, it also becomes less dependent on good weather;

there is no need to stop work for a rain shower (because moisture can generally be tolerated). And work can go on even at subfreezing temperatures, depending only on the ability to join the sheet seams.

Under the correct conditions—i.e., where the roof structure can take the additional load—the loose-laid, ballasted synthetic membrane promises to be the most problem-free type of flat roof.

Sheet-Applied Membranes

Unlike fluid-applied materials, which are chemically thermosetting synthetic rubbers, synthetic sheet membranes belong to all three previously discussed chemical categories: thermosetting, thermoplastic, and intermediate polymers.

Major synthetic, single-ply sheet materials include:

- EPDM

- Neoprene

- Butyl rubber

- Plasticized PVC

- CPE

EPDM, a material originally developed for automobile window gaskets and (in 1980) the most popular of the synthetic rubber materials for roofing membranes, has a unique combination of properties. In some formulations it remains flexible from −75 to 300°F. It suffers relatively slight loss in tensile strength and elongation after heat aging. It also has superior ozone resistance (an index of general weathering resistance), but it lacks resistance to petroleum oils or gasoline.

Synthesized from ethylene, propylene, and a small proportion of diene monomer, EPDM comes in nylon-reinforced or unreinforced sheets, normally black, but also in heat-reflective white. Chemically, it resembles butyl rubber, which it has largely replaced in exposed roofing membranes because of its superior weathering resistance. As another advantage over butyl rubber, EPDM's high permeance permits escape of air/water vapor entrapped below the membrane. (Butyl rubber has extremely low permeance, a good property for waterproofing and intertubes but a possible liability for roof membranes.)

In addition to sharing butyl rubber's vulnerability to petroleum oil and gasoline, EPDM is difficult to splice to itself. Joints are sealed with a contact adhesive that requires a 5- to 15-min wait between the application of the adhesive and the pressure joint-sealing process (performed with a roller). Precise length of required waiting time depends on ambient temperature and humidity.

Following proper joint-sealing procedure is critical. Tests conducted by Engineering Research Consultants (ERC) of Madison, Wisconsin, as part of a research program into single-ply synthetics for the Midwest Roofing Contractors Association (MRCA), demonstrate this importance. In one test sample, ERC followed the manufacturer's instructions, using white gasoline to remove talc dust (placed on EPDM rolls, as on conventional, asphalt-coated felt rolls, as a releasing or "nonstick" agent). Another test-sample joint was made *without* removing the talc, contrary to the manufacturer's instructions. At −20°F, the improperly made joint tested at less than half the strength of the properly made joint. (The properly made joint also recorded ultimate elongation greater than 100 percent.)

Prefabrication into sheets as large as 45 × 150 ft (67 squares) drastically reduces the number of these sensitive field joints. Precut in the shop to 1- to 3-in. oversize tolerance, sheets are rolled onto 6-in.-diameter reinforced cardboard cylinders before being shipped to the jobsite and hoisted to the deck.

Flashing used with EPDM sheet membranes is uncured neoprene sheet, a plastic material lacking "memory" and thus readily conforming to corners and other flashed surfaces. Uncured neoprene lacks the cross-linked polymeric chains that characterize elastomers, giving them the normally desirable rubberlike quality of returning to their original shape upon removal of a distorting stress. Later in its service life, the originally uncured neoprene flashing material does cure. Within a year or so after installation, it should attain the same essential properties as factory-cured EPDM or neoprene.

Neoprene, available in sheet as well as fluid form, comes in a black, weathering-grade sheet and a nonweathering light-colored sheet vulnerable to ultraviolet radiation. Like its fluid-applied counterpart, properly formulated sheet neoprene has good resistance to petroleum oils, solvents, and heat. More expensive than EPDM, it has generally slight advantages over EPDM, except on roofs requiring (1) superior chemical resistance or (2) superior fire resistance.

A partially adhered, single-ply membrane made of sheet neoprene surfaced with liquid Hypalon has performed successfully at the Dulles International Airport terminal building in Chantilly, Virginia, since late 1961. The neoprene sheet is applied directly to the surface of precast concrete deck units, which incorporate foamed polystyrene insulation under the deck slabs and between their ribs. To accommodate membrane stresses resulting from deflections in the long-span inverted concrete arches, the architect specified a 6-in.-wide neoprene adhesive bond-breaker strip on the concrete surface at the overlapped edges of the sheets. He also omitted adhesive from the central portion of the sheet. Two coats of liquid Hypalon (one aluminum, one gray) followed the sheet application.

Plasticized polyvinyl chloride (PVC), one of the most popular of the new synthetic sheet membranes, comes in an unusually wide variety of chemical formulations, in both reinforced and unreinforced sheets. PVC sheet is a thermoplastic polymer synthesized from vinyl chloride monomer. It resists acids, alkalis, and many chemicals. Most PVC sheets are vulnerable to asphalt and coal tar pitch, though some commercially available PVC sheets are compatible with bitumens.

As its major advantage, PVC sheet offers comparative ease of solvent-welding or, better yet in the opinion of some roofing experts, heat-welding of watertight joints in the field. This property springs from its solubility in PVC solvent-cements. Unlike synthetic rubber materials, which require the workers to wait while the contact adhesive's solvent evaporates before joining the seam, a PVC seam can be made immediately, either with the solvent-welding technique or 500 to 600°F hot-air, heat-fusion method. Properly done, the field seam—a true fusion of base and jointing material, analogous to a structural steel weld— makes a watertight joint (see Fig. 11-3).

A corollary advantage accompanies PVC's solvent-welding ability. PVC-coated flashing metal enables you to solvent-weld the membrane directly to perimeter gravel-stop fascia and other flashed building components.

As their major liability, PVC sheet membranes may become embrittled and shrink from loss of plasticizer. Without a plasticizer, PVC is rigid and brittle. Addition of a plasticizer makes it softer, more flexible, more extensible, and tougher. Loss of plasticizer increases thermally

Fig. 11-3 Hot compressed air at 500°F fuses PVC sheet field seam.

induced contractive stress in a stretched PVC sheet. Low-cost, monomeric plasticizers are mere additives. Because they do not chemically enter the PVC polymeric chain, they can migrate. Requirements of these impermanent monomeric plasticizers are (1) compatibility with the base material, (2) high boiling point (to prevent evaporation), and (3) physical stability within the PVC. Some loss of plasticizer inevitably accompanies aging, and exposure to sunlight accelerates the loss rate.[1] Plasticizer loss follows an exponential curve, starting fast and slowing as the material ages.

Unlike the less expensive monomeric plasticizers, high-cost polymeric plasticizers are more or less permanent. They unite chemically with the PVC, becoming part of the PVC copolymer chain.

PVC sheet manufacturers generally blend a little of each type of plasticizer—low-cost monomer and high-cost polymer. Precise formulation changes with in-service experience with the membranes.

Selecting the precise combination of plasticizers is an extremely difficult problem. Plasticizer loss can occur not only from evaporation but from contact with other plastics, asphalt, or rubber. The problem of blending plasticizers for low-temperature flexibility shows the kind of required compromises. Brittleness temperature of an 80-durometer phthalate compound can be as high as 32°F (using dibutyl phthalate), whereas the best phthalate compound—di(n-octyl, n-decyl)phthalate (NODP) has a brittleness temperature of −30°F.* Other plasticizers can reduce brittleness still further, to even −90°F. But these benefits may require unacceptable volatility and oil extraction, thus requiring a compromise—normally some combination of phthalate with a diester.[2]

Plasticizer loss varies significantly for different manufacturers' PVC sheet formulations—to a factor of 2 for two different unreinforced, 48-mil sheets, according to high-temperature exposure tests conducted by researcher R. M. Dupuis.[3] As one curious result, ballast-covered, loose-laid PVC sheet membranes apparently recorded a higher loss of plasticizer (in plasticizer-extraction tests) than exposed sheets. Membrane temperatures of ballast-covered membranes were, significantly, field-measured at 130°F+.

Plasticizer loss, however, is not the only factor causing PVC membranes to contract. Unreinforced PVC sheet has an inordinately high coefficient of thermal contraction-expansion, ranging up to three times

---

[1]Ibid., p. 8.

*"Additives for Plastics—Plasticizers," *Plast. Eng.*, January 1976, p. 17.

[2]Ibid.

[3]R. M. Dupuis, "An Update on Single-Ply Roof Systems," *RSI Magazine*, November 1980, p. 46.

that of an asphalt organic-felt membrane in the 0°F range. Membrane shrinkage has reportedly caused PVC membranes to pull flashings out of peripheral walls. But these membranes were obviously installed incorrectly by roofers unskilled in the proper technique of allowing for shrinkage. Folding the membrane along its perimeter can allow slack for future shrinkage without rupture or stress on perimeter flashing.

Tensile reinforcement can virtually eliminate shrinkage from loss of plasticizer. With woven glass-fiber mat or polyester scrim reinforcement placed directly in the center of the sheet's cross section, PVC sheets shrink only in local areas, where plasticizers have escaped. Reinforcement provides dimensional stability because it prevents longitudinal orientation of the polymer molecules, which in cooling could "freeze" and create internal strains (so-called "memory effect"). It also drastically reduces the PVC's coefficient of thermal contraction-expansion by a factor of nearly 10, according to one manufacturer's published figures.

Unreinforced PVC sheet is vulnerable to solar radiation, which can relax internal stresses and contract the sheet permanently. Aging, evidenced primarily by loss of plasticizer, further shrinks unreinforced sheet. According to Canadian researcher H. O. Laaly, some unreinforced calendared PVC sheets shrink 5 percent or more when heat relaxes the tensile memory stresses locked into the sheet during the calendaring process. The corresponding figure for nonwoven fiberglass-reinforced PVC is 0.01 percent.

In summary, a PVC sheet's resistance to shrinkage depends on six factors:

- Precise plasticizer formulation
- Manufacturing process
- Sheet reinforcement
- Sheet thickness (generally, the thicker the better)
- Protection from ultraviolet radiation
- Weight of gravel ballast

Fluid-Applied Membranes

Fluid-applied synthetic membranes are a special class of fully adhered membrane systems with special assets and limitations. Chemically, these fluid-applied systems are synthetic rubbers. Some, notably neoprene, neoprene/Hypalon, or butyl, require a solvent. Others, silicone rubber, polysulfide, and polyurethane rubber, require an added cata-

Fig. 11-4   The fluid-applied acrylic membrane coating for the sprayed-in-place polyurethane roof system is sprayed in two stages: first, a gray base coat designed for contrast, then the white top coat.

lyst. They come in either single- or two-component materials, applied by brush, roller, trowel, squeegee, or spray (see Fig. 11-4).

Single-component, fluid-applied membranes solidify by either (1) evaporation of the solvent (or water) or (2) chemical curing. Two-component membranes solidify by chemical curing, which starts as soon as the two components are mixed. Curing rates of single-component materials depend on temperature; the higher the temperature, the faster the solidification. Curing rates for two-component materials, however, depend less on atmospheric conditions (temperature, humidity) and more on chemical formulation than single-component curing rates.[1]

Fluid membranes are classed as either *low* or *high* solids. A low-solids fluid contains a high proportion of solvent, required to deposit the solids left as membrane material after the solvents evaporate. Fluid-applied polyurethane exemplifies the high-solids fluid. Depending on chemical curing rather than evaporation to deposit a solid material membrane, polyurethane contains 100 percent solids. None of its two-component fluid material boils off during the curing process.

Membrane thickness obviously depends on the amount of fluid applied and on the solids content. Uniformity depends on the (1) quality of the work and (2) substrate surface roughness (the smoother the better).

This dependence of the waterproofing film's uniformity on the substrate's surface roughness illustrates the narrow limits to which fluid-

[1]Rossiter and Mathey, op. cit., p. 5.

applied membranes are confined. Satisfactory substrates for fluid application are limited to concrete, plywood, and sprayed-in-place polyurethane foam. Insulation boards are unsatisfactory because they soak up the fluid as it is applied, creating an irregular film and resulting in premature embrittlement of the solidified membrane.

Within their narrow range of applicability, fluid-applied membranes have these notable advantages:

- Conformance to irregular roof surfaces with good adhesion and little waste
- Continuous, seamless waterproofing (including base flashing at penetrations and edges for some systems)
- Good adhesion
- Ease of handling, transporting, and application

At the heart of the fluid-applied synthetics' many disadvantages is their overwhelming dependence on high-quality fieldwork. Heading the list of these disadvantages is elaborate substrate preparation. In a concrete deck, cracks must be located, marked, and either sealed or taped (see Fig. 11-5). One manufacturer requires that cracks and seams greater than ¹⁄₆₄ in. be sealed, an indication of the fastidious care required for substrate preparation. A viscous mastic material seals cracks between ¹⁄₆₄ and ⅜ in. wide, and because of mastic shrinkage, satisfactory joint filling may require two mastic applications. Cracks or joints over ⅜ in. require taping. Cast-in-place concrete decks must be steel-troweled and then cured for at least 4 weeks to reduce repairs of

Fig. 11-5 Sealing cracks in concrete shell roof (previously primed for fluid-applied elastomeric, fluid-applied membrane) is essential to prevent membrane splitting. (E. I. du Pont de Nemours & Co.)

post-application shrinkage cracks. And to prevent moisture entrapment in the fluid-applied membrane, allow at least 2 days between the latest rainfall and membrane application.

A plywood deck requires similarly elaborate preparation. To keep such a deck dry, prime it immediately after installation. Either tape the joints, or alternatively, install a glass mat over the entire plywood deck to reinforce the membrane.

As another critical aspect for preparation of a fluid-applied membrane's substrate, it must be clean. Good adhesion of a fluid-applied membrane demands a substrate surface free of oil, wax, grease, dirt, dust, and other loose particles such as rust and spalled concrete fragments.

Training a roofing crew to provide meticulous cleaning is a tremendous challenge. Cleanliness offends typical construction workers' practical religion; they are accustomed to getting things done, not getting things clean. Moreover, the vagaries of weather—e.g., gusty winds blowing dust and debris onto and around the deck—can frustrate the efforts of even a well-trained, responsible crew attempting to do a good job on substrate preparation.

Without this elaborate substrate preparation, a fluid-applied membrane starts off with heavy odds against it. You need a smooth membrane surface paralleling a smooth, even substrate surface to ensure uniform thickness. A rough-surfaced substrate promotes variable film thickness and even pinholes. Variable film thickness increases the probability of membrane puncture from foot traffic. For some rough substrates, a primer, carefully chosen for its compatibility with both membrane and substrate, can smooth out a rough substrate and improve adhesion.

Beyond the requirement for a meticulously cleaned substrate, fluid-applied systems carry other liabilities:

- Hazards of post-application substrate cracking (because the membrane is fully adhered and thus vulnerable to substrate stress concentrations)

- Relatively short anticipated service life, often requiring recoating within 5 years

- Most severe weather restrictions of any roofing system, generally limited by prospect of temperature below 40°F; imminent rain (which can ruin an application); wind (which can blow dirt, dust, and debris onto the substrate); and minimum relative humidity (for moisture-curing compounds)

- Difficulties in controlling application for uniform film thickness (stretching the abilities of most roofing mechanics)

- Limitations on fluid's shelf life (for storage) or pot life (for field application)

With these many disadvantages, fluid-applied membranes are seldom a roof designer's first choice. They are generally a second choice—in practical effect, a last resort forced on the designer as the least undesirable alternative: the solution to a problem that cannot be practicably solved by the inherently preferred sheet-applied membrane.

The most popular fluid-applied membrane combines two synthetic rubbers: neoprene (chemically, polychloroprene) and Hypalon (chlorosulfonated polyethylene). Neoprene/Hypalon exploits their complementary qualities. Two neoprene coatings provide waterproofing body; two surface coatings of Hypalon provide superior resistance to ozone, ultraviolet radiation, heat, abrasion, and color stability.

There are two basic neoprene/Hypalon membranes. A general-purpose membrane, suitable for dry, hard substrate, consists of a primer, two coats of neoprene, and two coats of Hypalon. A heavy-duty version, for use where added tensile strength may be required, adds a glass-fiber reinforced mat embedded in the first neoprene coating. A variation merely substitutes a continuous roving for the glass-fiber mat.

Where substrate cracking poses no real problem—e.g., on a thin-concrete shell roof under compressive stress—an ultrathin neoprene/Hypalon membrane may prove satisfactory. One such ultrathin fluid membrane comprises two coatings of Hypalon on a neoprene primer.

Fluid-applied polyurethane (not to be confused with sprayed-on, foamed polyurethane insulation) comes in a single- or two-component system. The single-component membrane cures in the presence of moisture. Because of the wide variety of chemical reactants suitable for its production, urethane can be synthesized with widely varying properties, including resistance to petroleum oils, solvents, oxidation, ozone, and weather. Bitumen-modified urethane systems, containing either asphalt or coal tar pitch, are available in both weathering and nonweathering grades.

For maximum ultraviolet resistance, polyurethane coatings should be made with aliphatic isocyanate. In its reaction with polyol, aromatic isocyanate produces a polyurethane more vulnerable to ultraviolet radiation.

Fluid-applied silicone rubber is the best general membrane coating for sprayed polyurethane foam insulation. But largely because of its generally unsatisfactory performance over other substrates (chiefly its high dirt retention), silicone is generally limited to that one substrate.[1]

---

[1]Rossiter and Mathey, op. cit., p. 16.

The semiorganic silicones have molecular chains comprising alternate silicon and oxygen atoms modified with various organic groups linked with the silicon atoms. Silicones retain flexibility over a far wider temperature range than any United States roof will ever experience—from −85 to 300°F. They offer outstanding general weather resistance and durability and ozone resistance. The Si—C chemical bonds in silicones approximate the strength of C—C chemical bonds, partly accounting for silicone's great stability.[1]

However, silicone elastomers are vulnerable to chemical attack from hydrocarbon solvents, acids, strong alkalis, and steam, weaknesses that disqualify them in some industrial atmospheres. Moreover, they attract and retain atmospheric dust, which mars their appearance and reduces their heat reflectivity.

Most silicone membranes comprise a primer and two coats: a 10- to 15-mil undercoat and a 6- to 12-mil surface coat, brushed, rolled, or sprayed.

Fluid-applied silicones are a notable exception to the temperature limits imposed on most fluid applications. They can be applied from 0°F to 100°F.

Modified-Asphalt Membranes

Modified asphalt, long used for waterproofing concrete structures, has become a popular single-ply synthetic roof membrane, first in Europe, now in the United States. Field joints are heat-fused with a propane torch at temperatures reaching nearly 3000°F.

As its chief function, the asphalt modification process preserves its flexibility, elasticity, and ductility down into the subzero range—in some instances as low as −15°F. Chemically, there is a broad range of treatments. In one European application, a rubberized asphalt core sheet consists of regenerated rubber from old automobile tires, plus attactic polypropylene (APP), the most popular of several synthetics used to plasticize asphalt.

Despite its improved properties, modified asphalt is still vulnerable to solar radiation. The membrane is protected via one of the following three methods:

- Mineral aggregate surfacing

- Insulation (in a protected membrane roof)

- Aluminum-foil surfacing

---

[1]*The Encyclopedia of Chemistry*, 2d ed., G. L. Clark (ed.), Van Nostrand Reinhold Company, New York, 1967, p. 968.

In its most popular version in the United States, the single-ply modified-asphalt membrane comes in 43-in.-wide rolls, 33 ft long (see Fig. 11-6). This membrane comes in two versions:

- A "standard," 0.16-in.-thick, three-layer laminate with a flexible 3.5-mil polyethylene-based core sheet sandwiched between layers of modified asphalt. (Thin, 1-mil polyethylene facing sheets act as a releasing agent, preventing the concentric layers of the roll from sticking.)

- A 0.12-in.-thick aluminum-foil-faced sheet, differing from the standard sheet chiefly in that aluminum foil substituted for the polyethylene facing on the exposed surface.

In both standard and aluminum-foil-faced membranes, the modified asphalt is a catalytically blown asphalt of high softening point (well over 200°F) but relatively high flexibility at low temperature. It has a minor amount of powdered mineral stabilizer (less than a conventional coated felt). The two asphalt strata are waterproofing elements. The central sandwich-filler plastic core (a modified polyethylene sheet designed to improve flexibility and elongation at low temperature) enhances the membrane's waterproofing integrity. It also supports and reinforces the two asphalt strata, maintaining uniform thickness and preventing excessive lateral flow during the heat-fused jointing process to be described later.

The two membrane types are designed for different uses:

- Standard membrane is for low slope (under 1½-in.), loose-laid systems with aggregate surfacing, anchored only at the roof periphery and around roof penetrations

Fig. 11-6 Modified-asphalt laminate membrane, with embossed, aluminum-foil surfacing, comes in 43-in.-wide rolls, 33 ft long. (Koppers Co., Inc.)

• The aluminum-faced (smooth-surfaced) membrane is a fully adhered (with cold adhesive) system, for slopes of ½ in. or greater

For the standard sheet, there are two different recommended surfacings: (1) loose aggregate (either crushed gravel in a 1-in. minimum glass-fiber batting, designed to protect the membrane from sharp edges; or smooth, rounded gravel without protective batts, limited to 1-in. slope) and (2) bituminous, graded aggregate topping embedded in a bituminous asphalt emulsion.

The aluminum-faced sheets (for smooth-surfaced roofs of ½-in. or greater slope) are protected from aging embrittlement by the aluminum facing, which retards evaporation of the asphalt's volatile oils. An embossed pattern permits expansion-contraction without breaking the bond between the two materials because of different thermal coefficients. In addition to protection against ultraviolet radiation, the aluminum foil also acts as a highly efficient heat-reflective surface. Field-applied coatings are also available.

A high-heat propane torch melts the plasticized asphalt for both membrane seams and substrate adhesion in loose systems, where the membrane is adhered at the roof periphery and at roof penetrations. For substrate bonding, the worker heats the underside of the membrane sheet along a 4-in. strip until it is plastic and quickly forces the sheet against the substrate surface. Slight runout of bitumen from under the edge indicates a properly fused joint.

For interply fusion, the worker simultaneously applies the flame to both surfaces until they are plastic and presses them together until they form a fused seam of 4-in. minimum width. Again, a slight runout of bitumen indicates proper fusion.

On the aluminum-faced sheet, the propane torch fuses the bottom surface to an exposed 4-in. asphalt selvage edge (see Fig. 11-7). The top aluminum-foil facing is bent over to cover the mated edges.

Fig. 11-7 Propane torch topseals joints between adjacent sheets of modified asphalt laminate membrane. (Koppers Co., Inc.)

"Top sealing" completes seams on the standard sheets, through heating of a 2- to 3-in.-wide section at the edge of the top sheet and troweling of this plastic asphalt into the previously fused joint.

Flashing for both standard and aluminum-faced membranes is the aluminum-faced sheets, which are adhered to flashed surfaces with the membrane adhesive. (This foil-faced flashing is also used with conventional built-up bituminous membranes.)

ALERTS

(Use in conjunction with manufacturer's recommendations.)

Design

New Roofs

1. Require minimum ¼-in. slope, unless manufacturer agrees to guarantee a ponded roof.

2. Review material properties, limitations, advantages and disadvantages, and past performance of all systems under consideration.

3. Consider wind and fire resistance during this preliminary survey.

4. Check substrate for suitability: For sheet membranes, satisfactory substrates are structural concrete (cast-in-place and precast), plywood, and rigid insulation board; for fluid-applied membranes: structural concrete (cast-in-place and precast), sprayed-in-place polyurethane foam, and properly prepared plywood.

5. For cast-in-place concrete decks, check concrete curing agents to ensure compatibility with synthetic membranes. (Some curing agents can cause poor adhesion or even premature deterioration of the membrane.)

6. Limit plywood to smooth-surfaced exterior grade, with edge support limiting differential deflection of adjacent panels to ⅛ in. at panel edges and ¹⁄₁₆ in. at ends.

7. Consider lightweight insulation concrete fills, gypsum deck, and semirigid insulation boards as suitable substrates only for mechanically fastened or loose-laid sheet membranes. They are not suitable for totally adhered membranes (which eliminates them from consideration for fluid-applied membranes).

8. Favor sheet- over fluid-applied synthetic membranes when substrate joint movement is anticipated.

9. Check fluid-applied membrane flashing material for two grades: higher viscosity grade for vertical surfaces, lower viscosity grade for horizontal surfaces.

Remedial Roofing

1. Before specifying a synthetic, single-ply system as a remedial membrane over an existing bituminous membrane substrate, check for the synthetic material's compatibility with asphalt or coal tar pitch.

2. Require the following preparation of an existing built-up roof system as a substrate for a new synthetic membrane:
   (a). Remove *all* aggregate if new membrane is applied directly on top of the existing membrane. (If a protective insulation board or venting base sheet is applied to the existing membrane as substrate for the new membrane, remove only *loose* aggregate.)
   (b). Repair defects in existing membrane (blisters, ridges, splits, fishmouths, and so on).
   (c). Apply new materials (protective insulation or membrane) only after existing built-up substrate is clean, dry, smooth, and free of loose particles. (Consider application of a primer or a single-ply base sheet to improve existing substrate.)
   (d). For loose-laid synthetic membranes, require smooth, firm substrate. (If protective insulation board is applied over existing roofs, it can be laid loose.)

3. Check roof's structural framing and deck for additional gravel or paver ballast load for loose-laid, ballasted roof systems.

Field

Sheet-Applied Membranes

1. Cover substrate joints with masking tape or elastomeric strips. Prepare joints (and cracks) on same day as membrane application, to prevent moisture entrapment between joint preparation materials and membrane.

2. Joint sealing:
   (a). Let contact adhesive solvent evaporate before sealing joint.
   (b). Seal solvent-welded joints before solvent evaporates and solvent cement dries.

3. Maximize sheet size (to minimize number of vulnerable, field-spliced seams).

4. Prohibit stressing or stretching of sheet during application. (If sheet cannot easily conform to substrate contour, cut and seam. Use semicured sheet materials when available.)

5. Do not reposition sheet after contact adhesion is made. (Attempts to smooth wrinkles or fishmouths can damage the sheet.)

6. Cut fishmouths, readhere, and patch.

7. During ballast application, take care to prevent puncturing or tearing of membrane. Prohibit loaded wheelbarrows from rolling over previously graveled membrane. (Use plywood load distributers.)

Fluid-Applied Membranes

1. Prepare joints and cracks on same day that membrane is applied, to prevent moisture entrapment between joint preparation and membrane.

2. Prepare plywood substrate for fluid-applied membrane as follows:
   (a). Apply tape or narrow strip of synthetic material.
   (b). Cover joint with fluid material.
   (c). Add reinforcing fabric to fluid.
   (d). Coat top of fabric with second fluid application.
   (e). Apply high-performance sealant in beveled plywood joint. Cover with liquid membrane.

3. Check shelf life of fluid-applied material. Store at required temperature. Aqueous emulsions must not freeze. Check two-component fluid systems for specified pot life of mixture.

4. Apply fluid membranes uniformly, at recommended thickness. Take care to prevent overspray.

5. For two-stage fluid-applied membrane, apply second coat at 90° with first coat. Use different color, if practical, to ensure detection of incomplete coverage.

6. Do not apply fluid membrane if there is a fair chance of rain.

7. Check manufacturer's weather requirements (some limit outdoor temperature to 40°F minimum).

8. Check minimum recommended relative humidity for moisture-curing materials.

9. Prohibit thinning or heating of fluid to reduce viscosity, except on manufacturer's instructions.

10. Require application of fluid-applied membrane on sprayed polyurethane foam on same day as insulation foam application.

TWELVE

Premature Membrane Failures

Well-constructed, well-maintained bituminous membranes may far outlive the conventional 20-year life expectancy. When properly designed and maintained, protected from degradation by exposure to solar radiation and moisture, bitumens and felts age gracefully. Despite some embrittlement and hardening with age, bitumens can perform their waterproofing function for half a century or more.

Premature membrane failures generally indicate some sort of physical failure within the built-up membrane or a malfunction within the roof system rather than simple accelerated chemical degradation. There are three major modes of built-up membrane failure:

- Blistering

- Splitting

- Ridging

Less widespread, but still potentially destructive, are

- Slippage

- Delamination

- Alligatoring

- Surface erosion

BLISTERS

Blisters are by far the most common ailment afflicting built-up membranes. Blistering is roughly twice as common as splitting, the second

279

Fig. 12-1   Blisters occur with especial frequency in Florida and other southern states, where the semitropical climate provides the heat and humidity conducive to blister formation and growth. This roof was a poorly designed and constructed project in south Florida.

most common problem, according to a survey of roofing contractors by the National Roofing Contractors Association (NRCA).[1]

Blisters can range in size from barely detectable spongy spots to bloated humps 6 in. high and 50 ft^2 in area (see Fig. 12-1). Most built-up membranes can be expected to have at least some minor blistering. If the blisters remain small and few in number, the roof can readily live out its full service life, although it may require some minor repair.

The prognosis for a roof with numerous large blisters is shortened membrane service life. Blisters drastically increase the membrane's vulnerability to physical and chemical degradation. A large blister is easily punctured by foot traffic, dropped tools, and wheeled equipment loads. A blister's sloping sides make gravity a relentless agent of deterioration. Erosion of aggregate surfacing exposes the bitumen flood coat to ever-renewed photochemical oxidation, which accelerates to a rate 200 times faster in sunlight than in the dark.[2] Deterioration of the exposed bitumen, which develops cracks caused by accelerated embrittlement, exposes the membrane to direct invasion of moisture. When the combined chemical degradation and erosion of the bitumen expose bare felt, the membrane's deterioration enters an accelerated stage,

[1]"The Shape of Roof Construction," *The Roofing Spec.*, November 1979, p. 40.

[2]K. G. Martin, "Evaluation of the Durability of Roofing Bitumen," *J. Appl. Chem.* (Australia), vol. 14, 1964, p. 427.

with felts open to direct attack by water and atmospheric ultraviolet radiation.

Blistering Mechanics

All blisters originate with the formation of a void, or unadhered area, in the mopping bitumen, either between the felt plies or between the substrate and the membrane (see Fig. 12-2). Regardless of other disagreements in their theories of blistering mechanics, roofing experts generally agree on this point: A void in either the interply mopping or at the insulation-membrane interface is essential to the development of a blister in a built-up membrane. And this void generally, *but not necessarily*, dates from the application of the hot bitumen. That is why continuous, void-free interply moppings are so important: Mopping voids create the potential for a crop of future blisters.

Voids can result from a host of causes. Architect Justin Henshell, AIA, provides the following list:

- Moisture in or on the felts (either top or bottom ply being mopped)
- Use of felt rolls crushed into oval shape by storage in horizontal instead of vertical position
- Uncoated felt surfaces resulting from mop skips or clogged holes in bitumen-dispensing machine
- Failure to broom out entrapped air
- Distorted insulation boards (warped or misaligned)

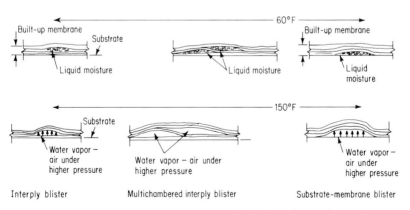

Fig. 12-2   Blister configurations range from single-chambered interply or substrate-membrane blisters to multiple, interply blisters.

- Moppings of improper viscosity (usually too viscous, but sometimes too fluid)

- Trapping of foreign material: gravel, matchbooks, broken insulation, paper wrappings, and so on

- Tenting or ridging caused by expanding insulation or expanding blisters in plies below

- Unfilled voids in insulation substrate surface—e.g., broken corners of insulation boards

- Improperly set, upturned metal flanges of curbs and so forth

- Unfilled edges (toes) of cants, tapered edge strips, or blocking

- Coated felts' side or end laps remaining uncoated when mop skims by

- Fishmouth or wrinkle (particularly in coated felts)

Most blisters probably grow from the evaporation of liquid moisture and the expansion of the resulting water vapor contained within the void. Because of liquid moisture's tremendous expansion when it evaporates, its presence within a void is the readiest agent promoting the spectacular growth of blisters often observed on built-up membranes. In late spring or summer, under midday and early afternoon sun, membrane temperature can easily rise from 70 to 150°F, with a tremendous increase in pressure (see Fig. 12-3).

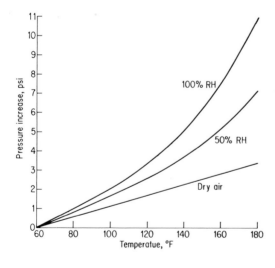

Fig. 12-3   Vapor-pressure graph shows exponential increase in water-vapor pressure vs. linear increase in dry air pressure with constant volume and rising temperature. *Note:* Graph ignores effect of increased pressure (less than one standard atmosphere at 180°F) on the humidity ratio W = lb water vapor per lb dry air, which would slightly reduce above values.

Under standard atmospheric pressure, this 80°F temperature rise expands water about 1500 times its original liquid volume. However, if it were confined within a constant volume, the water-vapor pressure would rise by more than 4 psi, about 600 psf. In an actual blister, the result lies between these two physical extremes: that is, the membrane resists expansion of the evaporated moisture, and so the blister's volume is far less than the volume associated with unrestrained expansion at atmospheric pressure. Nonetheless, the blister volume greatly expands, relieving the pressure that would occur within an unyielding, constant-volume void space.

Although liquid moisture accelerates blister growth most rapidly, it is not indispensable to blister formation. Trapped air–water-vapor mixture can also promote blister growth. Here is how it happens. When the hot-mopped bitumen cools and hardens, it seals in the trapped air–water-vapor mixture at a temperature probably exceeding 100°F, depending on ambient as well as bitumen temperature. Cooling of the hot-mopped bitumen creates negative pressure within the void, and the still-flexible felt plies tend to press together tightly, appearing as if they are bonded. Unless additional air and/or moisture enter the void, subsequent heating of the membrane cannot produce positive pressure within the void until the temperature of the trapped air–water-vapor mixture exceeds the temperature at which it was originally entrapped.

But the negative pressure naturally occurring within a mopping void promotes entry of additional air and water vapor. This entry can occur via two mechanisms: (1) infiltration through tiny membrane cracks and (2) simple gaseous diffusion.

In either case, the void behaves like a bellows. It expands and contracts vertically, but it cannot expand horizontally until the total internal force is great enough to break the bond between felt and bitumen at the void perimeter.

Basic Blister Theory

Obviously, the smaller the void, the less probability it will grow into a blister. Mathematical equations demonstrate this fact (see Fig. 12-4). You can calculate the force F (lb) tending to break the peripheral bond around a circular void and enlarge the blister from the following formula:

$$F = p\,\frac{\pi d^2}{4} \qquad (12\text{-}1)$$

where F = force tending to enlarge blister (lb)
p = total internal (air-vapor) pressure (psi)
d = void diameter (in.)

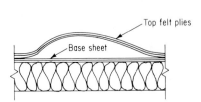

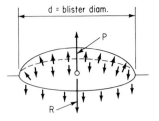

Fig. 12-4 *Blister growth mechanics.* For the interply circular blister shown, blister growth requires a cyclic prying action, in which the total interply pressure P exceeds the prying resistance at the blister's perimeter. Because the internal pressure is uniform in all directions, its total vertical component P must exceed R, the peripheral resistance, for the blister to expand its base area. Then

$$P > R$$

$$\frac{p\pi d^2}{4} > s_p \pi d$$

$$p > \frac{4s_p}{d}$$

where p = internal gas pressure (psi)
$\quad s_p$ = prying resistance (lb/in.)
$\quad d$ = blister diameter (in.)

For a circular blister, the pressure required to expand its area (and volume) varies inversely with the diameter. This same rule holds for an elongated oval (e.g., an ellipse), so long as the blister retains its basic shape (i.e., the same length/width ratio).

The force resisting this expansion, R (lb) equals the product of the mopped bitumen's peripheral bond strength by the void circumference:

$$R = s_p \pi d \qquad (12\text{-}2)$$

where R = resisting force (lb)
$\quad s_p$ = peripheral bond strength at void perimeter (lb/in.)

Equating these expressions, $F = R$, yields the following:

$$p = \frac{4s_p}{d} \qquad (12\text{-}3)$$

Equation (12-3) explains why a small void is less likely to grow into a blister than a larger void. Internal void pressure required to expand the blister varies inversely with void diameter. Thus, for example, it takes five times as much pressure to expand a 1-in.-diameter void as it does to expand a 5-in.-diameter void. Note also that the precise blister shape—regardless of whether it is a circle or, like most blisters, an elongated oval—does not affect the basic theory. So long as the blister retains its basic shape (i.e., its length-width ratio remains constant), the pressure required to expand it decreases in inverse proportion to its width or length.

Experimental Proof of the Mathematical Theory

E. H. Rissmiller, a senior research associate at Jim Walter Research Corporation, conducted a series of tests to verify this mathematical approach. His apparatus was simulated blisters of varying diameters, constructed as follows:

> Circular patches of No. 15 asphalt-saturated felt and varying diameter were laid over the surface of an insulation board and covered with two additional felt plies mopped solid in steep asphalt. The resulting void was connected to a system of aspirator bottles through a copper tube cemented into the insulation. Internal void pressure was regulated by adjusting water-level heights in the two bottles. Membrane temperature was maintained at 160°F by a bank of infrared lights [see Fig. 12-5].

Gradually increased pressure within each differently sized blister was recorded just when the blister began to expand. Substituting observed values of pressure and diameter for each of the various-sized blisters in Eq. (12-3), Rissmiller solved for the corresponding peripheral bond strengths (tabulated in Fig. 12-5). Within the limits of probable

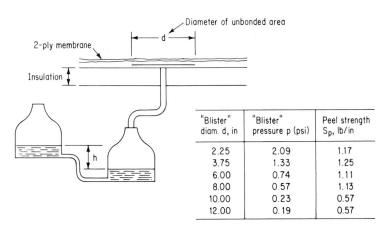

"Blister" diam. d, in	"Blister" pressure p (psi)	Peel strength S_p, lb/in
2.25	2.09	1.17
3.75	1.33	1.25
6.00	0.74	1.11
8.00	0.57	1.13
10.00	0.23	0.57
12.00	0.19	0.57

Fig. 12-5 Laboratory-test setup simulates blister formation. Unbonded plastic patch, of diameter d, represents wet, unbonded interface between insulation and membrane. Water vapor under pressure head h exerts blister pressure. Tabulated values of blister vapor pressure p are recorded for varying sized "blisters" of diameter d. Prying strength s_p is calculated at equilibrium condition, just before blister started to grow.

error for this rather crude experiment, peripheral bond strength at 160°F appears to be a constant 1.0 lb/in. (for steep asphalt). (The poor agreement of calculated peripheral bond strengths for the larger blisters probably results from the difficulty of reading the very low pressures required to initiate growth and the relatively large volume of gas necessary to maintain internal blister pressure.)

The value of 1.0 lb/in. for peripheral bond strength of steep asphalt at 160°F was later verified in a separate experiment, which yielded values ranging from 0.99 to 1.2 lb/in.

The relatively large quantity of air available in this experiment produced rapid blister growth, which usually continued until the blister vented at the edge of the test specimen. On an actual roof, the increasing volume that constitutes the blister's growth simultaneously reduces the pressure exerted by the trapped gases. Thus, in the field, a blister can continue to grow only by (1) further heating, which raises the blister's internal pressure, and/or (2) entry of additional air or water vapor into the blister void.

Blister growth encounters several other limitations indicated by the formulas. In a small blister, membrane weight is insignificant compared with internal pressure, which may rise to 2 psi or so, 50 times the weight of an aggregate-surfaced membrane. Moreover, the buckled membrane can act as a dome or arch for spans measured in inches. But reduced pressure and longer dome spans accompanying blister growth make membrane weight an increasingly important factor limiting further growth. In fact, for an aggregate-surfaced membrane weighing 6 psf, the theoretical limit for a circular blister is 8 ft because that is the point at which the membrane weight exceeds the theoretical pressure required to overcome peripheral bond resistance.[1]

As still another practical limitation on blister growth, a blister's increasing area exponentially increases the probability that a small membrane puncture will vent the blister and thus relieve internal pressure. (When you apply foot pressure to a spongy blister, you can sometimes hear the expelled gases hiss as they escape through tiny fissures.) The theoretical approach thus helps establish the void theory of blister origin.

Practical Implications of Blistering Theory

The theory of blister development yields useful, practical conclusions about void size and the role of liquid moisture in originating a blister.

---

[1]For a membrane weighing 6 psf ($=$ 0.042 psi), the blister diameter required to balance membrane weight $= d = 4s_p/p = 4 \times 1/0.042 = 96$ in. (8 ft). Thus for growth beyond 8-ft diameter, an aggregate-surfaced membrane's weight becomes the theoretically limiting factor rather than a 1 lb/in. peripheral bond resistance.

The most basic point is: *It takes at least a dime-sized (roughly ¾-in.-diameter) void to originate a blister.*

A ¾-in.-diameter blister requires about 5 psi to enlarge it, which is roughly the pressure gain within a saturated (100% RH) constant-volume space for a temperature rise from 60 to 140°F. Because any enlargement of a blister relieves pressure (by increasing the volume), it is highly unlikely that a blister could originate with less than a ¾-in.-diameter void. It is inconceivable that an isolated ⅛-in.-diameter void could grow into a blister because it would require 32-psi gage pressure (i.e., pressure above atmospheric pressure) to expand the blister. Such a pressure would almost certainly require a temperature associated with fire.

A second basic point is: *A blister can originate in a smaller void containing liquid moisture than in a void containing only an atmospheric air–water–vapor mixture.*

This inference follows from Eq. (12-3), $p = 4s_p/d$. Because the internal pressure required to expand the void varies inversely with void diameter, it logically follows that the higher pressure resulting from a vapor-saturated condition produced by liquid moisture can expand a smaller void than the lower pressure resulting from an unsaturated air-vapor mixture. (A void containing even a saturated air-vapor mixture, without liquid moisture, would become unsaturated, with consequently less pressure rise than a saturated mixture, when its temperature rose.) Thus to avoid blistering, you need better, more nearly continuous mopping for wet felts than for dry felts. But, by a sort of technological Catch 22, you are far less likely to get uniform, void-free mopping of wet felts; it is in fact next to impossible because the sudden boiling and foaming of the felt-entrapped moisture create interply mopping voids.

The hazard of wet felts as blister-initiating agents is highlighted in a provocative research paper by T. A. Schwartz and C. G. Cash of Simpson, Gumpertz & Heger, Cambridge, Massachusetts, consultants.[1] This paper convincingly refutes some roofers' assurances that hot mopping bitumen evaporates *all* moisture contained in the felts. Hot bitumen removes, at best, 50 to 60 percent of this moisture from felts containing excessive moisture. And some of this liberated moisture may remain trapped within the interply adhesive layer.[2] This moisture entrapment can occur if the next felt ply is applied before all the bubbles in the hot asphalt can burst and release their water vapor to the atmosphere. Moreover, the laboratory tests yielding these moisture-

---

[1]T. A. Schwartz and Carl G. Cash, "Equilibrium Moisture Content of Roofing and Roof Insulation Materials and the Effect of Moisture on the Tensile Strength of Roofing Felts." *Proc. Symp. Roofing Technol.*, NBS-NRCA, September 1977, pp. 238–243.

[2]Ibid., p. 242.

Table 12-1   Equilibrium Moisture Content (EMC), % Felt Dry Weight

Felt or fabric type	ASTM designation	Conditioning environment	
		45% RH	90% RH
Organic felt	D226-77, D227-78, D2626-73, D3158-78	7.2	15.1
Asbestos felt	D250-77, D3378-80, D3672-78, Type I	1.7	2.7
Glass-fiber felt	D2178-76, D3672-78, Type II	1.0	2.3
Woven burlap fabric	D1327-78	8.0	16.5
Cotton fabric	D173-80	5.4	9.4
Woven glass fabric	D1668-80	0.3	0.5

Source: Carl G. Cash reporting round-robin testing from seven laboratories to ASTM Committee D8.04 in December 1980. Note that quantity of absorbed moisture depends only on felt fiber content and type, not on quantity of bituminous saturant or coating. An asphalt coating retards the rate of moisture gain or loss in a given RH environment but has no effect on ultimate quantity of absorbed moisture.

release figures generally favored moisture liberation: Felts were warm and asphalt temperature high (375 to 425°F). Because field conditions are almost always worse than these laboratory conditions, the figure achieved by the Schwartz and Cash experiments will normally exceed the moisture release you can reasonably anticipate in the field.[1]

The Schwartz and Cash paper also highlights the hazards of phased construction, in which the base ply is applied in one operation, followed by the top plies several days (or weeks) later.[2] Storing felts in a 90% RH environment results in excessive moisture content (see Table 12-1). Felts left exposed even overnight are often exposed to 100% RH when condensation forms at cool night and early-morning temperatures. Glaze-coating the felts with a light asphalt mopping (about 10 lb/square) can help protect the felts from water absorptance. Early-morning condensation, prevalent in humid climates, can however promote mopping voids if it does not evaporate before the shingled felts are applied.

Moisture entrapped in insulation can also produce blisters, at the interface between the insulation and the membrane's base ply. Under the normal upward vapor-pressure gradient, wet insulation can supply moisture—liquid or vapor—to the membrane above. If this moisture

[1]Loc. cit.

[2]Loc. cit.

enters a void at the insulation-membrane interface, it provides the agent of future blister growth.

The presence of a vapor retarder under the insulation also helps promote blisters originating in mopping voids at the insulation-membrane interface. Insulation materials generally contain enough moisture to saturate the air within any roof-system voids. A 2-in.-thick fiberglass with 2 percent moisture (by weight) contains more than enough moisture to keep the entrapped air saturated above 160°F. A vapor retarder, or a solidly mopped concrete deck, impedes underside venting, thus promoting pressure buildup and blister growth (see Fig. 12-6). But a self-drying deck vents water vapor to the interior when the summer sun creates a downward pressure gradient through the roof system.

Although liquid moisture is generally considered the chief cause of blisters, additional air may sometimes be required to make a blister grow. Consider, for example, a blister of 0.05-ft³ volume, containing five or six drops of moisture (about 1 grain per drop) at 160°F. Addition of more water cannot increase the blister's size. (Because the air is already saturated, excess water must remain in liquid state, occupying negligible volume.) And because the temperature is not likely to exceed 160°F, further growth requires the entry of additional air to increase internal pressure.

Why Blisters Grow

What keeps a blister growing is a daily cyclic pumping action, with the daily volume of air inhaled into the blister chamber exceeding the daily volume of exhaled air. To see how this cyclic pumping mechanism works, consider the daily pressure changes within a membrane mopping void. This mopping void originally sealed as the bitumen cooled between mopping temperature (say, 350°F) and ambient temperature (say, 70°F). Let us assume that the sealing temperature is 120°F, slightly below the 135°F softening point of Type I asphalt. (The precise assumed temperature is unimportant to this argument.) Now, if this sealed mopping void—a potential blister chamber—was at atmospheric pressure at 120°F, it will be at negative pressure whenever membrane temperature drops below 120°F and at positive pressure when membrane temperature exceeds 120°F. As shown in Fig. 12-6, even on a hot, sunny day, membrane temperature exceeds 120°F only for 6 h or so (roughly 12 noon to 6 p.m.) and falls below 120°F during the remaining 18 h. Because of these changing pressure differentials, air and water vapor tend to diffuse *out* of the sealed blister chamber during a short part of the day (6 h at most) and *into* the blister chamber during most of the day. On cloudy and rainy days, suction into the blister will probably continue throughout the day because membrane temperature

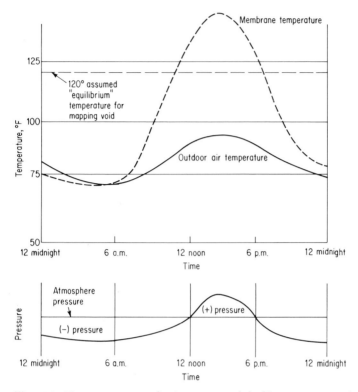

Fig. 12-6 How a vapor retarder (or nonvented deck) promotes growth of blisters at the insulation-membrane interface. (1) During the 12 to 16 h with negative air pressure within the roof sandwich, air can enter the blister void via diffusion and infiltration through tiny cracks in the membrane. These cracks form because the membrane is only partially elastic—i.e., it does not contract as much for an 80°F temperature drop as it expands for an 80°F temperature rise. Moreover, below 80°F the bitumen has a higher coefficient of thermal expansion-contraction than the reinforcing felts. Thus it contracts more than the felts at low temperature, and this additional contraction would open minute cracks. If the insulation contains a normal 2 percent or so moisture (by weight) there is enough moisture to keep the trapped air–water-vapor pressure saturated at temperatures above 160°F. Thus the major gain during the 16-h negative pressure period is air, not water vapor. Daily additions of dry air can maintain blister growth through cyclic pumping action until the natural limiting factors stop it. (2) During spring or summer daytime hours, solar radiation heats the membrane to 150°F or so, raising pressure within the void and within the insulation. A vapor retarder raises this internal pressure because it obstructs underside venting of the roof. This internal positive pressure is maintained during the 8- to 12-h period when the membrane is warmer than the outside air because the heated, plastic bitumen expands and flows into the microscopic cracks, sealing them against escape of the internal air–water-vapor mixture.

290

probably will never reach the 120°F assumed equilibrium sealing temperature.

Excess daily air intake into the blister chamber expands the blister's volume. On hot, sunny days, when the membrane is heated to its peak 130 to 155°F temperature, the cumulative, increasing pressure breaks the peripheral bond at the blister boundary until the expanding volume relieves the pressure and restores a new equilibrium of forces within the blister. Thus, by a continual, rachetlike pumping action, a blister expands by tiny daily increments, possibly ⅛- to ¼-in. diameter a day during the hottest weather, with each day's prying action expanding it a little farther.

The conditions promoting blister growth are really worse than those postulated for the simplified model. When membrane temperature rises to 150°F or so, the softened, semifluid bitumen seals microscopic membrane cracks and reduces membrane permeability. By reducing the membrane's exhalation, this process increases internal blister pressure. When the membrane cools to 65 or 70°F, contraction of the stiffened bitumen reopens the microscopic cracks, thus increasing membrane permeability and consequent air intake. So the blister tends to suck in more air than it expels under daytime pressure.[1]

Still another mechanism promoting blister growth is the coalescing of adjacent mopping voids, which often occur in clusters, into large blister chambers. As individual voids expand, they stress the mopping bitumen that adheres the felts in patches of membrane separating these numerous, smaller voids. (These resisting lines of asphalt make the familiar embossed alligatoring pattern often observed within the blister, discussed later as "legs," or "stalactites" and "stalagmites.") In extreme cases, linked chains of tunnel blisters occur along lines extending 30 ft or more.

Aging membrane blisters resist contraction, thus tending to stiffen into permanently blistered shapes, for several reasons. The stronger, reinforcing felts have a much lower coefficient of thermal contraction-

---

[1]This inference, that the longer "inhalation" time implies a greater volume of "inhaled" than "exhaled" air from the blister chamber, assumes roughly comparable negative and positive pressures. Note by the previous argument that internal blister pressure decreases as the blister grows. (Pressure varies inversely with blister diameter.) Because internal blister pressure decreases with increasing blister size, further growth by cyclic pumping action, resulting from a daily excess of inhaled over exhaled air, grows easier with increasing size once the blister's internal pressure starts it growing from a small originating void. We thus have a rational evolutionary explanation of blister growth: It requires high internal pressure to start blister growth. Then, once the blister starts growing and high internal pressure is no longer necessary, reduced internal pressure promotes greater inhalation of air into the blister, perpetuating the cyclic pumping action that keeps the blister growing.

expansion than the bitumen. The accelerated aging produced by exposure hardens and embrittles the bitumen, which cracks and loses its original contractive strength. And wetting-drying cycles, made more probable by the blister, also tend to stiffen the felts. As noted earlier, the blistering process ultimately exposes the membrane to direct invasion of moisture, which permanently swells organic or asbestos felts. (Organic felts submerged in water expand up to 1.5 percent—⅜-in. in a 36-in.-wide roll.) Finger wrinkling, additional evidence of water soaking, is sometimes observable on exposed, asbestos felt, smooth-surfaced roofs (see Fig. 18-8).

Do Blisters Reseal?

Resealing of blisters, propounded by some roofing experts, is highly dubious. As it ages, and especially in its first year, asphalt progressively loses the fluidity required for resealing. Even more destructive to the resealing hypothesis is the coincidence of the blister's highest internal pressure, which tends to break the peripheral bond at the blister boundary and enlarge the blister, with the highest membrane temperature, which increases bitumen fluidity. Resealing would require the opposite condition: negative pressure tending to force the blistered felt plies together. Instead, there is positive pressure tending to pry the felt plies apart. As pointed out in the earlier discussion on internal cyclic pumping action, negative pressure promoting resealing occurs when membrane temperature falls below 120°F or so. For optimum mopping asphalt viscosity, application temperature should range no lower than 390°F for ASTM Type III asphalt, 345°F for Type I asphalt, according to NBS researchers.[1] Good adhesion can still be achieved at temperatures as low as 275°F, according to Celotex researchers. But that is more than 100°F hotter than the highest surface temperature normally experienced by even a black-surfaced roof. Moreover, resealing must occur without brooming or unrolling of the felt, which adds pressure to the heat, tending to bond the felts during application.

Resealing thus appears to be a totally unsubstantiated hypothesis, lacking corroborative theoretical and empirical evidence.

Other Factors Promoting Blister Growth

The other notable factors tending to promote blister formation and growth are

- Use of coated instead of saturated felts as shingled plies

---

[1]W. J. Rossiter and R. G. Mathey, "The Viscosities of Roofing Asphalts at Application Temperatures," *Build. Sci. Ser.* 92, NBS, December 1976, p. 14.

- Use of steep (Type III) asphalt for interply moppings (instead of dead-level, Type I asphalt)

- Cold-weather application

Coated felts create a substantially greater risk of interply mopping voids and consequent blistering than uncoated felts, for the following reasons:

1. The thicker, stiffer, coated felt is more difficult to broom solidly into uniform contact with its underlying hot-mopped surface, thus making it more likely to contain voids within the interply mopping. Continuous voids often form at base sheet laps where the next felt, especially if it is coated, can bridge over the vertically discontinuous joint (see Fig. 12-7). Long blisters have been observed in a pattern of 36-in. spacing where bridging voids have occurred at every base sheet lap. Strings of small voids, possibly coalescing into longer voids, can occur along these side laps, even along end laps.

2. Coated felts usually lack the perforations that allow heated air/water vapor entrapped during the application process to escape. (At least one manufacturer has apparently solved this problem by providing ¼-in.-diameter holes on a 2-in. grid in coated felts.)

3. Because of their higher heat conductivity and heat-storage capacity, coated felts promote faster cooling of hot-mopped asphalt than saturated felts during the heat exchange accompanying the felt-application process.

4. The mineral surface dusting (usually talc or mica) on coated felts may inhibit good adhesion.

Widespread blistering and splitting in the now virtually extinct two-ply coated-felt membranes resulted in their removal from the market.

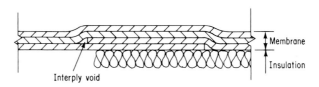

Interply void   Membrane   Insulation

Fig. 12-7 Coated felts can promote the formation of long membrane blisters at base sheet laps where a coated felt, stiffer than a lighter, more flexible saturated felt, bridges over the edge of the base sheet, forming a continuous interply void.

Another factor affecting blister formation concerns the type of asphalt—steep (ASTM Type III) or dead-level (ASTM Type I)—used for interply mopping bitumen. This complex question involves a still unresolved roofing industry controversy. However, several general points can be asserted about the dead-level vs. steep asphalt controversy:

- *Properly applied* steep asphalt generally provides better resistance to blister growth than dead-level asphalt or coal tar pitch.

- Steep asphalt requires more rigorous field-application control than dead-level asphalt to ensure solid, continuous interply adhesive.

Properly applied steep asphalt provides greater resistance to blister growth because it is less temperature-susceptible than dead-level asphalt. The longer maintained, hotter blowing temperatures required to produce steep asphalt make it a stronger, more rubberlike material than dead-level asphalt. Of greatest importance in resisting blister growth, increased elasticity makes steep asphalt stronger and less fluid at high temperatures than dead-level asphalt.

Proper application of steep asphalt, however, requires substantially tighter application controls—both in temperature control of the mopping asphalt and in the timing of the felt-laying operation—than proper application of dead-level asphalt. This is especially true in cold or windy weather. In the NBS Building Science Series 92 report on roofing asphalt viscosities, recommended application temperature for steep asphalt is nearly 50°F hotter than that for dead-level asphalt: 390 to 450°F for ASTM Type III asphalt vs. 345 to 405°F for ASTM Type I, (see Fig. 12-8).* Because of the high temperatures required to attain proper viscosity (50 to 150 cSt) for mopping steep asphalt, the tolerance for overheating Type III asphalt without reaching the asphalt's flash point, or vapor-ignition temperature, is substantially less than the cor-

*Rossiter and Mathey, ibid.

Fig. 12-8 (*Opposite*) ASTM D312 Type III asphalt requires higher application temperature than Type I asphalt to attain correct mopping viscosity (50 to 150 cSt), as shown in graph (*a*). Correct mopping viscosity restricts mopping thickness to desired 0.03 to 0.04-in. range, with mopping weight 15 to 20 lb/square, as shown in graph (*b*). Excessive viscosity, resulting from cooled or inadequately heated asphalt, thickens the mopping film, impairs uniformity, and increases the probability of mopping voids. (From W. J. Rossiter and R. G. Mathey, "The Viscosities of Roofing Asphalts at Application Temperatures," *Build Sci. Ser.* 92, NBS, December 1976, pp. 12, 13.

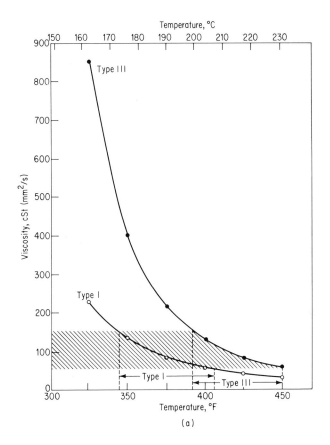

(a)

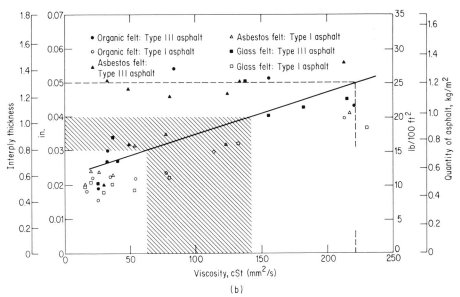

(b)

responding tolerance for overheating Type I asphalt. Steep asphalts are also more viscosity-sensitive than dead-level asphalt to variations from proper application temperature.[1]

Regardless of whether it is too viscous or too fluid, thus producing overweight or underweight interply adhesive layers, mopping asphalt of improper viscosity (i.e., outside the NBS-recommended 50- to 150-cSt range) can produce (along with other deficiencies) interply voids and consequent blistering.[2]

The difficulty of maintaining proper temperature control of any asphalt, but especially steep asphalt, was demonstrated by field studies carried out by Engineering Research Consultants, Inc., of Madison, Wisconsin. At one of the four test sites, in Waukesha, Wisconsin, where a 10-mph wind dropped the 22°F ambient temperature to an effective wind-chill temperature of 5°F, the ground-level kettle temperature of the asphalt was 560°F, far above the 475°F or so maximum temperature permitted by most major roofing manufacturers.[3] Yet the asphalt's "peak-contact" temperature (i.e., the highest temperature attained during application) was 350°F, 40 to 100°F below the NBS-recommended range for steep asphalt's application temperature.[4] In fact, not one of the roofers on any of the cold-weather test projects reached the minimum 390°F application temperature recommended by the NBS researchers for steep (Type III) asphalt.[5] And on one of these cold-weather projects the ambient temperature was a relatively mild 43°F, with light winds. These roofers did, however, generally reach the 345°F minimum temperature found suitable for Type I asphalt. These field tests demonstrate the difficulty of attaining uniform, void-free interply moppings with steep asphalt applied in cold weather.

The Engineering Research Consultants' study also revealed an important, but generally overlooked, advantage of shingled felt application as a means of promoting continuous bitumen moppings, both at the substrate and between plies. Measuring the mopping-asphalt cooling rate for a coated base sheet applied in a single operation and then the mopping-asphalt cooling rate for three shingled saturated felts, the ERC researchers found much faster cooling in the single mopping for the base sheet than in the three moppings for the shingled plies. Math-

[1]Ibid., p. 8.

[2]Ibid., p. 1.

[3]R. M. Dupuis, J. W. Lee, and J. E. Johnson, "Field Measurement of Asphalt Temperatures During Cold Weather Construction of BUR Systems," *Proc. Symp. Roofing Technol.*, NBS-NRCA, September 1977, pp. 278, 279.

[4]Rossiter and Mathey, loc. cit.

[5]Dupuis, Lee, and Johnson, op. cit., p. 5.

ematically, in their temperature-decay formula, the decay (i.e., temperature loss) constant for the individually applied coated sheets was more than twice the value for the three shingled, saturated felts, indicating the base sheet's faster cooling.

This disparity obviously results from the lesser amount of hot bitumen (only one-third as much for bonding the base sheet as for bonding the three shingled plies). A mopping of 20 lb/square naturally cools much faster than three moppings weighing 60 lb/square. Here then is another reason to shingle *all* membrane felts for simultaneous application and to avoid phased construction. The single shingling operation concentrates the heat of several hot bitumen moppings. The slower cooling rate of the hot-mopped asphalt enhances the chances of achieving good interply adhesion, thereby reducing the chances of voids and consequent blistering (see Fig. 12-9).

Blistering over Urethane Insulation Board

Blistering in membranes placed directly on top of urethane insulation board had become a widespread problem in the late 1970s. As a preventive remedy, NRCA issued its Technical Bulletin No. 4, recommending use of a thin layer—say, ½ in.—of fiberboard, perlite board, or fiberglass roof insulation interposed between the hot-mopped membrane and the urethane. This recommendation remains essentially valid, although it was superseded by Technical Bulletin #7, which recommends either a porous insulation board or a venting base sheet interposed between the membrane and urethane board.

Moisture, the ubiquitous roofing troublemaker, apparently explains the phenomenon of blistering over urethane boards. The escape of Freon gas trapped in the urethane foam cells was blamed at one time, but a coordinated test program sponsored by the NRCA and MRCA indicated the fallacy of that theory and the guilt of the old enemy, moisture. Laboratory tests, backed by field tests, demonstrated the familiar frothing, bubbling, and formation of tiny foaming craters resulting from the sudden boiling of liquid water trapped in the urethane board's felt facers. Laboratory application of hot asphalt to the "control" insulations—fiberboard, perlite board, and fiberglass insulation board—produced no foaming in the asphalt film. Application of hot asphalt to the felt-faced urethane boards, however, produced either immediate or slow foaming.

The theory of Freon diffusion into the asphalt film was rebutted by gas chromatography, which indicated insignificant quantities of Freon 11 (the urethane foaming agent) among the bubble-forming gases. This discovery is totally consistent with theoretical considerations. To diffuse from surface layers into the asphalt film, the Freon 11 would have

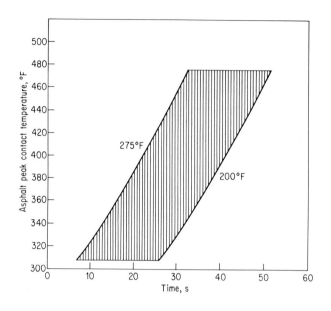

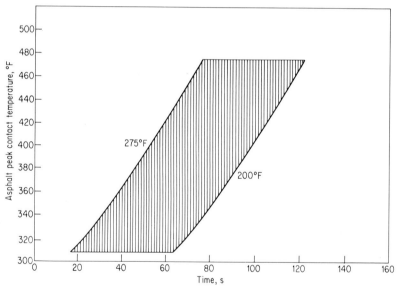

Fig. 12-9 These curves show the advantage of felt shingling in reducing the cooling rate of mopping asphalt. The mopping asphalt for a singly applied coated felt base sheet cools much faster (top) than three shingled saturated felts (bottom). (*Proc. Symp. Roofing Technol.*, NBS-NRCA, September 1977, p. 280.)

to diffuse against a strong temperature-pressure gradient accompanying the hot asphalt application. Surface moisture requires no such anomalous behavior. Its high expansion ratio when flashed to steam by contact with the hot asphalt explains the foaming craters.

The urethane blistering phenomenon cannot be attributed solely to its felt facers. Although fiberboard and perlite boards do not have felt facers, fiberglass board does. Urethane, however, tends to promote hot-mopped asphalt foaming because of the following properties:

- Low thermal conductivity

- Low heat capacity

- Low permeability (in contrast with the porous insulations' high permeability)

Because of its low thermal conductivity and low heat capacity, the urethane board raises the transient peak temperature accompanying the hot mopping of the asphalt. Higher temperature at the insulation-asphalt interface raises the pressure of heated moisture. The urethane board's low permeability aggravates the situation by obstructing the downward flow of the suddenly created water vapor, which then escapes via the only open paths: lateral or upward into the hot-mopped asphalt film, where it forms bubbles or channel voids as the cooling asphalt congeals.

Another, less common type of membrane blistering over urethane board insulation occurs via a totally different mechanism: from urethane's long term expansion. The long term growth of urethane boards can produce blisters in an unusual way that conflicts with the conventional theory of blister formation. According to this conventional theory, held by some roofing experts, mopping voids are the only source of blister-forming cavities. Some unusual problems with membrane blisters over urethane board insulation indicate, however, that the blister-originating voids are not necessarily mopping voids. These voids can develop later, a consequence of differential movement between the urethane board substrate and the membrane. This differential movement is caused by expansion and warping of the urethane insulation boards (see Fig. 12-10).

The first urethane boards, introduced around 1970, were made solely of urethane, with no other composite material. Some of these urethane materials exhibited tremendous water absorptivity, exceeding 2000 percent (by weight) in some instances. The expanding boards curl at their ends, producing a void at the insulation-membrane interface (or at the interface between the base sheet and the top felt plies above), where

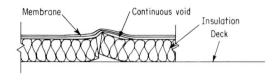

Fig. 12-10 Curled polyurethane insulation, from moisture expansion and warping at joint can create a blister void. Some early polyurethane boards, introduced about 10 years ago, exhibited tremendous water absorptivity.

the forced ridging of the membrane breaks the bond with the urethane substrate (or with the base sheet).

Clues to Blister Origins

A blister's interior, cut open in the field or laboratory, usually contains telltale clues pointing to its origin, notably:

- Tiny foaming craters in interply moppings
- Smooth, shiny bitumen surfaces
- "Legs" (sometimes called "stalactites" or "stalagmites")
- Bare, uncoated felt spots
- Foreign substances—e.g., aggregate—entrapped between plies

Foaming craters (⅛ in. or so in diameter) provide circumstantial evidence that liquid moisture was present during the felt-laying operation. To boil water at standard atmospheric pressure requires a minimum temperature of 212°F, 150°F or more below the temperature of properly heated bitumen and well above the maximum 190°F or so that the hottest, black-surfaced, sun-baked roof will experience (short of fire).

A *shiny bitumen surface* may (but not necessarily) indicate a lack of interply mopping cohesion dating from the original application (see Figs. 12-11 and 12-12). These shiny surfaces often contain foaming craters, which indicate the sudden boiling of liquid moisture. From these two clues, you can infer that the sudden expansion of boiling liquid moisture, augmented by the expansion of heated, saturated air, produced the shiny-surfaced blister cavity.

Shiny bitumen does not, in itself, necessarily prove lack of original adhesion, *unless* there is a dry top felt (thus indicating that the bitumen cooled before the felt was laid and thus failed to adhere). The interply bitumen may develop a shiny surface merely from flow induced by solar heating.

So-called legs, or stalactites and stalagmites, usually form in an alligatored pattern when the adjacent felts are pulled apart. These legs reveal the heated bitumen's resistance to internal blister pressure prying adjacent felts apart (or prying a bottom felt off its substrate). Rising membrane temperature *increases* the blister's internal pressure and simultaneously *reduces* the bitumen's tensile strength. As the temperature approaches the bitumen's softening point, the increasingly fluid material flows like taffy or chewing gum, with little or no increase in stress. Legs are thus lines of contact between the blister's separated plies inside the blister's periphery, where prying action has succeeded in breaking contact between plies. The legs break only when the blis-

Fig. 12-11 This exposed blister chamber exhibits two telltale clues to the blister's origin: (1) countless foaming craters in the shiny bottom asphalt surface indicate the presence of liquid moisture on the base felt or on the felt above when the hot asphalt was mopped; (2) bare stretches of uncoated base felt (¾ in. wide, at least 6 in. long) indicate the origin of mopping voids, probably resulting from stopped holes in the asphalt dispenser. Laboratory test samples also revealed overweight interply moppings, a probable consequence of the application of overcooled asphalt.

Fig. 12-12 This blister occurred on the same project as the blister cut open and photographed in Fig. 12-11. It shows how blisters can promote ponding even on a well-sloped roof: in this instance, 1 in./ft slope.

ter's vertical growth exceeds the bitumen's ductility. (Ductility diminishes with falling temperatures that may not significantly contract the blister.)

A *bare, uncoated felt spot* often marks the original blister void. A radial, starlike uncoated felt pattern emanating from the original void indicates the sudden boiling of liquid moisture in the bare spot when it was hit by 300 to 400°F asphalt (see Fig. 12-13).

Foreign substances—aggregate, insulation fragments, broken matchbooks—are sometimes found in the original void. These are the most direct circumstantial evidence for a blister's origin.

Other hypotheses—notably incompatibility between mopping bitumen and felt coating or saturant—are sometimes offered as explanations of blister formation. By a fairly general expert consensus, the incompatibility theory is highly dubious.

In perhaps the most unscientific hypothesis yet presented as an explanation, extensive blistering has even been attributed to the formation of excess carbon dioxide gas by the manufacturer's use of crushed seashells for the stabilizing particles in the coated felts used in the built-up membrane. In one such case, the reported discovery of three times ambient atmospheric carbon dioxide inside some membrane blisters purportedly accounted for the blisters.

In the case built to validate this theory, the proponent chemists repealed Dalton's law of partial pressures: i.e., each gas in a mixture

Fig. 12-13   The bare, uncoated, half-dollar-sized spot in the felt (see arrow) in the exposed mopped asphalt film is surrounded by shiny asphalt, with legs, marking the void origin of a blister. The shiny asphalt of a larger blister chamber is discernible at the middle–bottom of this 8 × 18 in. laboratory test sample. Cut from a 5-ft-diameter membrane blister, the test sample was frozen at the NBS laboratory to −40°F with carbon dioxide in an environmental chamber and then delaminated for visual inspection. Absence of foaming craters in the interply moppings indicated that the felts were not excessively wet during application. At one smooth, shiny, interply surface there was evidently no interply adhesion. The apparent cause for this blister was excessively cooled mopping asphalt, probably below 280°F, when the felts were applied.

exerts its share of total pressure in direct proportion to its *volume*. Normal carbon dioxide atmospheric content is about 0.03 percent by volume. The excess carbon dioxide inside the blisters thus accounted for $(3 - 1) \times 0.03 = 0.06$ percent of total internal pressure. How 0.06 percent, or even 100 times that amount, could be positively identified as the incremental pressure that caused the blisters remained an unexplained mystery.

SPLITTING

Membrane splitting, the second most frequent roofing problem according to the previously cited NRCA survey, is obviously the most serious of the three major membrane failure modes. Splitting occurs most frequently in cold climates, more often on large roof areas, with widely spaced contraction-expansion joints, than on small area roofs. Like ridging (discussed later in this chapter), splitting usually occurs above the continuous longitudinal joints of board insulation.

Membrane splits almost always occur parallel to the longitudinal (machine) direction of the felts, for two reasons:

1. For all types of built-up membranes, the thermal coefficient is higher in the transverse direction (three times as high for asphalt-impregnated organic felt membranes in the $+30$ to $-30°F$ range—see Table 12-2).

2. Built-up membranes are weaker in the transverse than in the longitudinal direction. (Asphalt-impregnated organic felts are twice as strong in the longitudinal direction, whereas the new glass-fiber felts are almost isotropic—see Table 12-3).

Splitting has many causes, acting singly or in combination:

- Thermal contraction

- Insulation movement

- Water absorption in felts

- Drying shrinkage of wet felts

- Shrinkage cracking of poured decks

- Deck deflection

- Stress concentrations

Table 12-2 Coefficients of Thermal Expansion-Contraction for Four-Ply Built-Up Roof Membranes

Type of membrane	Membrane designation	Coefficient of thermal expansion, $°F^{-1}$ temperature range*					
		73–30°F		30–0°F		0––30°F	
		L†	T‡	L	T	L	T
Organic felt and coal tar	A	−3.3	4.4	22.3	36.0	19.3	29.5
Organic felt and asphalt	B	−3.4	−6.6	2.7	12.6	13.9	37.4
Asbestos felt and asphalt	C	2.3	9.2	4.8	18.1	19.5	37.5
Glass felt (Type I) and asphalt	D	−7.1	−5.3	8.9	10.1	35.1	46.4
Glass felt (new product) and asphalt	E	−7.3	3.9	−4.2	10.7	29.0	39.0

*Values in table to be multiplied by 10^{-6}.

†L denotes longitudinal or machine direction of felts.

‡T denotes transverse or cross-machine direction of felts.

Source: R. G. Mathey and W. C. Cullen, "Preliminary Performance Criteria for Bituminous Membrane Roofing," *Build. Sci. Ser.* 55, NBS, November 1974, p. 6.

Table 12-3 Tensile Strength of Four-Ply Built-Up Roof Membranes

Type of membrane	Tensile strength, lb/in.							
	73°F		30°F		0°F		−30°F	
	L*	T†	L	T	L	T	L	T
Organic felt and coal tar	126	62	395	217	468	265	410	237
Organic felt and asphalt	141	60	396	186	506	267	592	283
Asbestos felt and asphalt	120	36	301	123	448	182	479	165
Glass felt (Type I) and asphalt	86	70	190	161	175	144	184	123
Glass felt (new product) and asphalt	202	159	510	408	448	365	372	301

*L denotes longitudinal or machine direction of felts.

†T denotes transverse or cross-machine direction of felts.

Source: R. G. Mathey and W. C. Cullen, "Preliminary Performance Criteria for Bituminous Membrane Roofing," *Build. Sci. Ser.* 55, NBS, November 1974, p. 7.

Of these factors, insulation movement is by far the major cause of membrane splitting. It can split a membrane unaided by any other contributing factor. Insulation movement produces one type of stress concentration (at insulation-board joints), but other types of stress concentration can also produce membrane splitting unaided by contributing factors. Notable among these are shrinkage cracking of poured decks to which the membrane has been hot-mopped and omission of an expansion-contraction joint along a line where the deck changes span direction or changes material. The other factors listed apparently require some other factor in combination to split a healthy membrane.

Their brittle, glasslike nature at low temperatures is what makes built-up bituminous membranes vulnerable to splitting. Just how vulnerable can be demonstrated by a comparison of various materials' breaking strains. At $-30°F$, the breaking strain (transverse direction) of any built-up membrane drops to a range of ½ to 1 percent, depending on felt type.[1] A ductile material such as hot-rolled steel sheet has a breaking strain of 32 percent, about 30 times the breaking strain of an asphalt-organic felt membrane, 40 times the breaking strain of an asphalt-asbestos membrane, 50 times the breaking strain of an organic coal tar pitch membrane. Some elastomeric membranes have breaking strains ranging around 300 percent. At $-30°F$, a 3-in. length of built-up membrane can tolerate less than ⅟₃₂-in. strain without splitting; a 3-in. steel sheet can tolerate nearly 1-in. strain; and some 3-in. elastomeric sheets can tolerate 9-in. strain without splitting.

Thermal Contraction

Contraction splitting is a problem chiefly in cold climates, a fact supported by theory and corroborated by laboratory data and widespread field experience. In a survey of built-up roof systems in Alaska in 1957, NBS researcher W. C. Cullen reported a 50 percent premature failure rate, with many instances of membrane splitting.

Even in the coldest inhabited climates, thermal contraction by itself probably cannot split a healthy built-up membrane; it takes another factor acting in combination with thermal stress. Among the complementary factors that can combine with thermal stress to split a built-up membrane is stress concentration, usually over an insulation-board joint or at an edge detail where the insulation has moved. Another split-promoting factor is drying shrinkage of previously wetted felts, a weakening hazard for both organic and asbestos felts.

---

[1]P. M. Jones and G. K. Garden, "Properties of Bituminous Membranes," *Can. Build. Dig.*, no. 74, May 1968, p. 74-4.

Several investigators have affirmed the theory that thermal stress alone cannot split a well-constructed membrane.[1] In his study of thermal contraction in built-up membranes, Cullen computed a contraction strain of 0.18 percent for a 60°F temperature drop (+30 to −30°F) for the most vulnerable tested membrane. That is only 40 percent of the membrane's breaking strain (0.45 percent) at −20°F.* Engineering Research Consultants got similar results in experiments with temperature-induced loads of membrane samples held at a constant length and subjected to a 90°F temperature drop (from 70 to −20°F). A four-ply membrane developed 71 lb/in. thermal stress, about one-quarter the tensile strength of an average four-ply asphalt-saturated organic felt membrane. A two-ply coated-felt membrane developed 52 lb/in., about 35 percent of its ultimate tensile strength.[2]

Two-ply built-up bituminous membranes subjected to thermal stress have failed at temperatures around −100°F in tests reported by Clemson University researchers.[3] That is far below the lowest temperature ever recorded anywhere in the United States, including Alaska, so this research confirms the previous conclusion: Temperature-induced stress alone probably cannot split a healthy built-up membrane.

Note however, that even this tentative conclusion requires qualification. The viscoelastic behavior of bituminous built-up membranes makes splitting an extremely complicated phenomenon. Membrane strength increases with rate of stress application, and laboratory rates of stress application generally exceed the natural rate of thermal-stress application. As a consequence, in-service membranes may split at higher temperatures than identical laboratory-tested membranes subjected to faster temperature drops.

In the meantime, however, until this problem is adequately investigated, with laboratory stress rates correlated with service stress rates, the hypothesis that thermal stress alone can split a healthy membrane remains unproven. At this stage of roofing technology, the burden of proof lies with proponents of the hypothesis.

Overweight interply bitumen moppings increase thermal stress by raising the membrane's thermal coefficient. As shown in Fig. 12-15, asphalt has a much higher thermal coefficient than felt at subzero tem-

[1]W. C. Cullen, *Effects of Thermal Shrinkage of Builtup Roofing*, NBS Monog. 89, March 1965. J. W. Lee, R. M. Dupuis, and J. E. Johnson, "Experimental Determination of Temperature-Induced Loads in BUR Systems," *Proc. Symp. Roofing Technol.*, NBS-NRCA, September 1977, p. 43.

*Cullen, op. cit., p. 5.

[2]Lee, Dupuis, and Johnson, loc. cit.

[3]Joel P. Porcher and Herbert W. Busching, "A Study of Thermal Splitting of Roofing Membranes," Department of Civil Engineering, Clemson University, July 1979, p. 21.

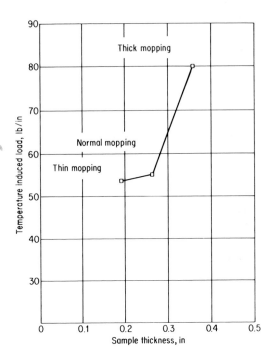

Fig. 12-14 Thick asphalt mopping raises thermal stress in two-ply membranes. (From J. W. Lee, R. M. Dupuis, and J. E. Johnson, "Experimental Determination of Temperature-Induced Loads in BUR Systems," *Proc. Symp. Roofing Technol.*, NBS-NRCA, September 1977, p. 47.)

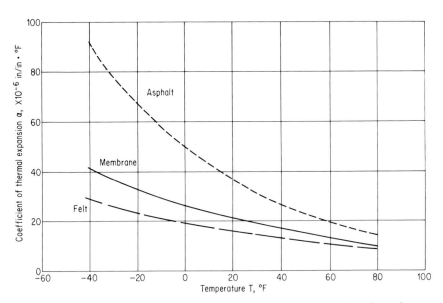

Fig. 12-15 Coefficient of thermal expansion-contraction for felt, asphalt, and membrane. (From J. W. Lee, R. M. Dupuis, and J. E. Johnson, "Experimental Determination of Temperature-Induced Loads in BUR Systems," *Proc. Symp. on Roofing Technol.*, NBS-NRCA, September 1977, p. 46.)

308

peratures. As a composite of these two constituent materials, the membrane has an intermediate coefficient. A thick mopping (roughly double weight) increased the induced thermal load by about 50 percent in two-ply test samples (see Fig. 12-14). A slightly underweight mopping had no significant effect on thermal stress.

The same phenomenon, a proportionately greater quantity of asphalt in the membrane, explains the disproportionately high thermal stress previously reported for a two-ply coated-felt membrane vs. a four-ply (base plus three shingled saturated felts).

Thermal shock factor (TSF) is an index of a membrane's ability to withstand thermal stress. This widely misunderstood concept is worth discussing in some detail. The term "thermal shock," by itself, merely refers to sudden membrane cooling—e.g., when a rain shower follows hot sunshine and membrane temperature suddenly drops from 150 to 70°F. But TSF refers to something entirely different. It is in fact a calculated property (°F) representing the temperature drop required to split a membrane totally restrained from contracting and initially under slight (essentially 0) tensile stress.

TSF is computed from the following formula (see Fig. 12-16 for derivation):

$$TSF = \frac{P}{M\alpha}$$

where P = membrane tensile strength (lb/in.) at 0°F
M = load/strain modulus (lb/in.) at 0°F
α = coefficient of thermal contraction-expansion, 0 to -30°F

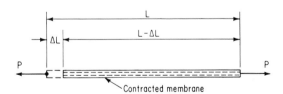

Fig. 12-16 Thermal shock factor (TSF) derivation. If it were unrestrained, the membrane segment of length L would contract by ΔL under the contraction stress accompanying a temperature drop ΔT. P is the tensile strength (lb/in) required to restrain the membrane from contracting. By analogy with the well-known formula for the modulus of elasticity E = unit stress/unit strain, $M = P/(\Delta L/L)$ and $\Delta L/L = P/M$. By definition, $\alpha = \Delta L/L\Delta T$. Then $\Delta T = \Delta L/L\alpha$ and $\Delta T = P/M\alpha$. Since TSF is the ΔT required to break the membrane, TSF $= P/M\alpha$.

From this formula we can infer desirable membrane characteristics for good thermal-shock resistance as follows:

- *High* tensile strength
- *Low* load/strain modulus
- *Low* coefficient of thermal contraction-expansion

Here is a typical computation of TSF, for an organic-felt, coal-tar-pitch membrane, transverse direction. (Data from tables in "Preliminary Performance Criteria for Bituminous Membrane Roofing.")

$$TSF = \frac{265}{7.4 \times 10^4 \times 29.5 \times 10^{-6}}$$
$$= 121°F$$

Mathey and Cullen tentatively set 100°F as the minimum generally recommended TSF. In mild climates, a lower value may suffice.[1]

A TSF = 100°F means that a tested specimen of built-up roof membrane held by tensile forces at constant length should be able to resist a 100°F temperature drop in the subfreezing temperature range. Obviously, few membranes need to resist such a severe thermal stress (from +30 to −70°F). Even Fairbanks, Alaska, the nation's coldest city, has only 22 h at or below −51°F in a normal winter, and no other city in the continental United States approaches such intense cold. And even if it were subjected to a 100°F temperature drop, a membrane in service would not be held fixed in length because its contracted substrate would relieve the thermal stress.

What the TSF concept does is offer a standard for comparing different membranes solely for their resistance to thermal stress. It is essentially an index of the membrane's reserve strength for resisting tensile stresses from other sources—substrate movement, stress concentrations of one sort or another, and so on. The higher a membrane's TSF, the greater its reserve strength to resist these other forces. As evident from Table 12-4, most commercially marketed built-up membranes easily satisfy this criterion when new.

Membranes do, however, steadily decline in TSF as they age, a consequence of age-hardening and consequent rise in load/strain modulus (or modulus of elongation), which varies inversely with TSF. In fact, a group of 21-year-old coal-tar-pitch membranes investigated by NBS

---

[1]R. G. Mathey and W. C. Cullen, "Preliminary Performance Criteria for Bituminous Membrane Roofing," *Build. Sci. Ser.* 55, NBS, November 1974, pp. 6, 7, 10.

Table 12-4 Thermal Shock Factors (°F) for Four-Ply Built-Up Membranes

Type of membrane	Thermal shock factor, °F	
	L*	T†
Organic felt and coal tar	360	120
Organic felt and asphalt	640	200
Asbestos felt and asphalt	290	90
Glass felt (Type I) and asphalt	170	120
Glass felt (new product) and asphalt	550	420

*L denotes longitudinal or machine direction of felts.

†T denotes transverse or cross-machine direction of felts.

Source: R. G. Mathey and W. C. Cullen, "Preliminary Performance Criteria for Bituminous Membrane Roofing." Build. Sci. Ser. 55, NBS, November 1974, p. 7.

researchers Mathey and Rossiter had TSF factors ranging between 32 and 40 percent of the suggested 100°F minimum.[1]

All three properties involved in the TSF react adversely to aging. "Marked" decrease in breaking loads of weather-exposed membranes is reported by Cullen and Boone.[2] They also report increased values of load/strain modulus and coefficient of thermal contraction and expansion, both of which lower the TSF. Comparisons of weathered field samples and laboratory-prepared samples yielded considerably lower TSF values for the weathered samples.

Besides the complicating effects of aging, the difficulties of predicting individual membrane behavior further complicate the problem of using TSF as a precise design tool. A built-up membrane is a field-fabricated element whose two basic components—felts and bitumen—are themselves subject to far greater variation in basic properties than such isotropic materials as structural steel or even structural concrete. Felts are generally anisotropic. Bitumens vary in viscosity, penetration, and other significant properties. These properties change drastically under the normal range of temperatures encountered on roof surfaces. Add to the foregoing complexity the inevitable variations in field-application techniques—e.g., the impossibility of maintaining interply bitumen mopping thickness at close to the dimensional tolerances easily achieved with other building materials. All these factors combine to make precise prediction of membrane strength and other critical properties a totally impracticable goal. Nonetheless, despite the limitations, TSF can still serve its purpose as a comparative index of membrane resistance to thermal-contraction stress at low temperature.

Another aspect of the TSF concept is the time-temperature effect, which explains the occurrence of membrane splitting in extremely cold weather. Stressed at fast loading rates, a built-up membrane behaves elastically. But at the slow rates generally characteristic of service conditions, plastic or viscous flow relieves tensile stress. If strain is maintained over a long period, tensile stress diminishes until the membrane completely relaxes. This membrane-stress relaxation depends primarily on temperature (the higher the temperature, the greater the stress relaxation), and to lesser degree on the loading rate (the slower the loading rate, the greater the relaxation).[3]

---

[1]R. G. Mathey and W. J. Rossiter, "Properties of 21-Year-Old Coal-Tar-Pitch Membranes: Comparison with the NBS Preliminary Performance Criteria," *Proc. Symp. Roofing Technol.*, NBS-NRCA, September 1977, p. 31.

[2]W. C. Cullen and T. H. Boone, "Thermal Shock Resistance for Builtup Membranes," *Build. Sci. Ser.* 9, NBS, August 1967, p. 11.

[3]R. G. Turrene, "Shrinkage of Bituminous Roofing Membranes," *Can. Build. Dig.* no. 181, April 1979, p. 181-2.

On severely cold nights, when radiative cooling can drop membrane temperature 10°F below ambient temperature, splitting can be promoted by a combination of mutually reinforcing factors:

- Rising coefficient of thermal contraction, as temperature drops

- Relatively rapid temperature change, as a solar-heated membrane cools down to −20°F or lower nighttime temperature

- More nearly elastic behavior, as the bitumen becomes a glasslike solid at low temperature

The extraordinary incidence of membrane splitting discovered after the record-setting cold of the winter of 1976–1977 in the midwestern, northeast, and mid-Atlantic states is empirical corroboration of the theoretical factors just outlined.

Insulation Movement

One of the most correctly belabored rules of built-up roof construction is the vital importance of anchoring the insulation to the deck. An unstable insulation substrate can cause membrane splitting unaided by any other factor. A major function of any sandwiched insulation is to restrain membrane movement, especially thermal movement. When insulation is not firmly anchored to the deck, it cannot perform this function, and the result is often a badly split built-up membrane.

Loose, unanchored insulation is responsible for the vast majority of membrane splits, according to broad expert consensus. Johns-Manville researchers investigating 24 split membranes in the north central United States found 75 percent with totally unanchored insulation in the area of the splits.[1] Fig. 12-17 shows the mechanics of such splitting.

Loose insulation promotes a progressive contraction that ultimately splits the membrane (see Fig. 12-18). Under cold-weather temperature cycling, especially when the temperature drops below 0°F, the contracting membrane drags loose insulation toward the securely anchored sections. During warmer daytime temperatures, the membrane cannot expand to its previous length because it has negligible compressive strength. Subjected to rising temperature, the membrane can only buckle, usually over insulation joints, where the insulation-membrane bond is weakest. Cumulative, contraction-buckling cycles, with loose

[1]Johns-Manville, *Roof Topics*, vol. 1, no. 3, August 1978.

Fig. 12-17 (*Opposite*) These diagrams depict the vital importance of firmly anchored insulation, showing how unanchored insulation can cause membrane splitting from concentrated thermal-contractive stress at an insulation-board joint. Assume a four-ply asphalt organic membrane (with strain properties reported by P. M. Jones, "Load-Strain Properties and Splitting of Roof Membranes," *Engineering Properties of Roofing Systems*, ASTM STP409, 1967, p. 83). Average calculated temperature drop in insulation = 30°F as outside air temperature (and membrane temperature) drops 60°F, from + 30 to −30°F. Assume the following properties for membrane and foamglass insulation: Insulation coefficient of thermal expansion α_i = 5 × 10^{-6}; membrane coefficient of thermal expansion α_m = 25 × 10^{-6}; insulation modulus of elasticity E_i = 150,000 psi × 1.5 = 225,000 lb/in.; membrane modulus of elasticity E_m = 36,000 lb/in.; insulation modulus of elasticity in shear $G_i = \dfrac{1.5 \times E_i}{2(1 + 0.3)}$ = 58,000 psi. If the original joint opening = 0.125 in., then membrane elongation across this joint can be computed from the deformations shown in the diagrams as

$$\text{Membrane elongation} = \frac{\Delta - \Delta_s}{\Delta_i}$$

Ignoring the warping tendency of the loose insulation board (see left-hand diagram), we can compute the intermediate deformation Δ between Δ_m (unrestrained contraction of the membrane from a temperature drop of 60°F) and Δ_i (unrestrained contraction of the foamglass insulation from a temperature drop of 30°F). The asphalt film at the membrane-insulation interface transmits the prestressing forces, and S_m, membrane tensile stress = S_i, insulation compressive stress.

Further assuming that the membrane develops its ultimate tensile strength at the joint (280 lb/in. for a four-ply asphalt-organic membrane), we can compute Δ_s (shear strain in the foamglass insulation) = 0.006 (see page 315). Then

$$\text{Membrane elongation} = \frac{0.016 - 0.006}{0.125} = \frac{0.010}{0.125} = 0.08 = 8\%$$

The computed strain is thus about eight times the ultimate strain of a four-ply asphalt-organic felt built-up membrane at −30°F, nearly 10 times the ultimate strain of a 4-ply asphalt asbestos membrane. Note also that the 42 lb/in. prestress induced at the membrane-insulation interface is about 15 percent of the membrane's ultimate strength, thus reducing its reserve strength.

Though the above analysis ignores stress relief in the joint vicinity, it nonetheless illustrates how stresses and strains concentrate over the insulation joint. Warping of the loose insulation board (or boards) obviously aggravates the situation, adding flexural stress plus an indirect tensile component to the basic contractive stress.

insulation moving across the deck and some boards restrained by curbs and other obstructions, inevitably open some insulation joints, with consequent membrane tensile stress.

To reduce this explanation to the bottom line, I offer this common-sense explanation of why an opening insulation-board joint concentrates stress and strain in a built-up membrane. At the cross section where the membrane is bonded to the insulation, insulation contraction

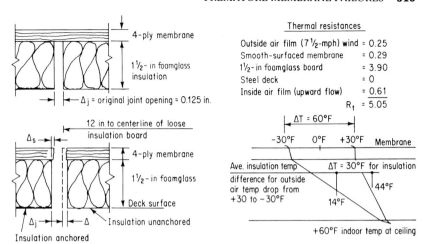

4-ply membrane

1½-in foamglass insulation

Δ_j = original joint opening = 0.125 in.

12 in to centerline of loose insulation board

Δ_s

4-ply membrane

1½-in foamglass

Deck surface

Δ_j Δ Insulation unanchored

Insulation anchored

Thermal resistances

Outside air film (7½-mph) wind = 0.25
Smooth-surfaced membrane = 0.29
1½-in foamglass board = 3.90
Steel deck = 0
Inside air film (upward flow) = 0.61
R_t = 5.05

$\Delta T = 60°F$

-30°F   0°F   +30°F   Membrane

Ave. insulation temp $\Delta T = 30°F$ for insulation
difference for outside
air temp drop from
+30 to -30°F   14°F   44°F

+60°F indoor temp at ceiling

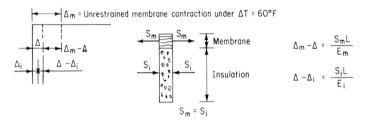

Δ_m = Unrestrained membrane contraction under $\Delta T = 60°F$

Δ $\Delta_m - \Delta$

Δ_i $\Delta - \Delta_i$

S_m S_m Membrane

S_i S_i Insulation

$S_m = S_i$

$\Delta_m - \Delta = \dfrac{S_m L}{E_m}$

$\Delta - \Delta_i = \dfrac{S_i L}{E_i}$

$$\Delta = \alpha_m L \Delta T_m - \frac{S_m L}{E_m}$$

$$\Delta = \frac{S_m L}{E_i} + \alpha_1 L \Delta T_i$$

$$S_m(\frac{1}{E_i} + \frac{1}{E_m}) = \alpha_m \Delta T_m - \alpha_i \Delta T_i$$

$$S_m = \frac{(25 \times 10^{-6} \times 60) - (5 \times 10^{-6} \times 30)}{(4.4 \times 10^{-6}) + (27.7 \times 10^{-6})} = 42 \text{ lb/in.}$$

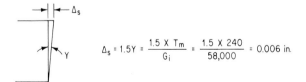

Δ_s

Y

$$\Delta_s = 1.5Y = \frac{1.5 \times T_m}{G_i} = \frac{1.5 \times 240}{58,000} = 0.006 \text{ in.}$$

reduces membrane stress and strain. But wherever the insulation-board joint widens, insulation contraction *increases* membrane stress and strain.

Loose insulation occurs frequently around openings, where the workers must field-cut insulation boards. During the extra time

Fig. 12-18 1⅛-in gap, shown at the folded-back insulation boards, indicates the widened joint formed between the insulation boards under a long membrane split that occurred during subzero weather in the northeast during the severe winter of 1977–1978. Large areas of steel deck had no adhesive whatsoever; the insulation boards had been simply laid loose on the deck. At one test cut through this roof, matching of random adhesive marks on the deck below and the insulation board above showed lateral movement of 2½ in. from the board's original location. What permitted this large movement was a large unadhered area, coupled with wide joints between the loose boards when they were originally laid.

required to perform this task, the workers may let the adhesive congeal, with consequent loss of adhesion.

Countless failures from poorly adhered, or unadhered, insulation have occurred from the use of extruded polystyrene boards as a sandwiched insulation in conventional built-up roof systems. Ubiquitous problems with this material finally forced its withdrawal from the United States market as a below-membrane insulation. (It works successfully as an above-membrane insulation in protected membrane roofs; see Chapter 10). In extreme instances, this insulation pulled away as much as 10 in. from its original contact with a peripheral nailer, with the flashing pulled out from masonry wall copings. Unbonded boards were actually precompressed at interior portions of the roof, and the membrane split and ridged, as is typical with this type of membrane failure.

Behind the widespread failures of extruded polystyrene board insulation are two corresponding causes:

1. The extreme difficulty of bonding polystyrene in hot-mopped asphalt to a steel- (or other) deck surface

2. Polystyrene's extreme thermal coefficient of expansion-contraction

Carl G. Cash, an associate with Simpson, Gumpertz & Heger, has propounded a theory of thermal warp of insulation boards as an explanation for membrane splitting over unadhered insulation boards.[1] This

---

[1] Carl G. Cash, "Thermal Warp—A Hypothesis for Builtup Roofing Splitting Failures," *Symp. Roofing Sys.*, ASTM STP603, 1976.

theory accounts for not only the contraction but the concave warping of loose insulation boards, which tend to curl away from the deck at their ends because of bending stresses induced by the temperature gradient through the insulation cross section at low outside temperatures. Membrane-splitting stress then becomes a combination of tensile resistance to thermal contraction, plus a tensile (extreme fiber) flexural stress resisting the insulation board's warping tendency.

Cash's paper reaffirms the paramount importance of good anchorage. He goes on, however, to recommend certain insulation qualities implied by thermal-warp theory. Apart from the obvious quality of low thermal conductivity, an "ideal" insulation would offer two structural qualities:

- Low modulus of elasticity
- Low thermal coefficient of expansion-contraction

Of these two properties, the more important in preventing membrane splitting is low modulus of elasticity. But as Cash warns, this thermal-warp model is highly simplified, focusing solely on the desirable properties of an insulation poorly adhered to the deck; it ignores other important properties of insulation.

Membrane-splitting risk increases exponentially where loose insulation boards are cantilevered over steel-deck flutes, violating the rule to provide minimum 1½-in. bearing on the deck flange (see the Alerts in Chapter 5). When insulation boards are so cantilevered, the contractive splitting hazard along the line of membrane-stress concentration above a continuous insulation-board joint is combined with flexural stress from the warping of the cantilevered board ends and, perhaps even more hazardous, with the relative vertical movement of the cantilevered boards, which may even break from concentrated rooftop traffic loads (see Fig. 12-19).

Discovery of such cantilevered insulation boards constitutes prima facie evidence of negligent field application. This condition can be readily prevented merely through field trimming the insulation boards

Fig. 12-19 Where insulation boards cantilever over a steel-deck flute, the membrane splitting hazard from inadequately anchored insulation multiplies exponentially. In addition to contractive tensile stress, this condition can induce flexural stress from insulation-board warping and even vertical shearing stress from differential movement at the ends of the cantilevered boards. Minimum bearing of insulation boards on steel deck flange is 1½ in.

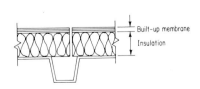

if cumulative dimensional errors reduce bearing on the deck flanges below the minimum 1½-in. requirement.

Water Absorption by Felts

Splitting failures are promoted by water absorption in organic and asbestos felts. Wetting drastically reduces these felts' strength, a fact corroborated by several investigators.[1]

After being saturated in water, asphalt-saturated *and* coated felts, both asbestos and organic, lose two-thirds or more of their strength.[2] (See Table 12-5.) Conditioning at 90% RH substantially reduces ultimate strength (by about one-half for saturated asbestos felts).

Similar strength reductions are reported by Laaly. After just one wetting-drying cycle, asphalt-saturated organic felt loses 20 percent of its ultimate tensile strength. Immersion in water cuts this strength by more than 80 percent.*

To reduce RH to the 40% recommended by Schwartz and Cash requires either raising storage temperature from 25 to 30°F above *wet-bulb* temperature (25°F above 40°F wet bulb, 30°F above 78°F, design condition for such hot, humid locations as Washington, D.C., and St. Louis) or dehumidifying the storage space.

Freeze-thaw cycles carry this strength-destroying process even further, according to Laaly. Although freeze-thaw cycles do not drastically reduce the strength of an already seriously weakened, water-saturated organic-felt membrane, they might be more damaging at intermediate levels of water absorption. Freeze-thaw cycles cut the membrane strength of immersed samples from 19 to 14 percent of dry strength for 1 cycle, to 11 percent for 10 cycles. That is a nearly 90 percent reduction of maximum tensile strength.[3]

Such drastic reductions in felt strength have a nearly proportionate effect on membrane strength because the reinforcing felts contribute about 90 percent of total membrane strength. Thus a membrane weakened by water absorption could split from thermal contraction or other tensile stress easily resisted by a strong, dry-felted membrane. Water-

---

[1]H. O. Laaly, "Effects of Moisture on the Strength Characteristics of Builtup Roofing Felts," *Symp. Roofing Sys.,* ASTM STP603, 1976, pp. 104ff. K. Tator and S. H. Alexander, "The Effects of Moisture on Builtup Roofing Membranes," *Engineering Properties of Roofing Systems,* ASTM STP409, 1967, pp. 187ff. Schwartz and Cash, op. cit., pp. 238, 241.

[2]Schwartz and Cash, ibid., p. 241.

*Laaly, loc. cit.

[3]H. O. Laaly, "Effects of Moisture and Freeze-Thaw Cycles on the Strength of Bituminous Builtup Roofing Membranes," *Proc. Symp. Roofing Technol.,* NBS-NRCA, September 1977, p. 244.

Table 12-5 Ultimate Tensile Strength (lb/in.) of Roofing Felts after Conditioning in Various Constant-Humidity Environments

	Conditioning environment				Percentage reduction in tensile strength		
Felt category	Oven dry*	40% RH	90% RH	Water-saturated	40% RH	Oven dry to 90% RH	Saturated
OA-MD†	36	34	32	12	6	11	67
OA-XMD†	13	‡	‡	4	‡	‡	69
AA-MD	35	31	26	5	11	26	86
AA-XMD	17	7	6	3	59	65	88
ACO-MD	50	41	31	17	18	38	66
ACO-XMD	24	24	20	9	0	17	63
ACA-MD	49	38	31	16	22	37	67
ACA-XMD	23	18	12	5	22	48	78

*"Oven dry" is exposure to 105°C for 24 h.

†MD = machine direction; XMD = cross-machine direction.

‡Not determined.

Source: T. A. Schwartz and Carl G. Cash, "Equilibrium Moisture Content of Roofing and Roof Insulation Materials and the Effect of Moisture on the Tensile Strength of Roofing Felts," Proc. Symp. Roofing Technol., NBS-NRCA, September 1977, p. 241.

weakening of felts provides another of the many arguments for draining the roof and protecting the membrane from penetration by ponded water.

Drying Shrinkage

Drying shrinkage of wetted felts is another way in which water absorption by organic and asbestos felts promotes membrane splitting. As discussed in the "Ridging" section, water saturation can expand an organic felt by more than 1 percent. Drying shrinkage lags behind the drying of the felt, but it ultimately stresses the membrane. A 1 percent shrinkage is about five times the calculated contraction of organic-felt membrane under a 100°F temperature drop, from 70 to −30°F. Note, too, that wetting of asbestos or organic felts further weakens the membrane, making it even more vulnerable to shrinkage stress.

The probability that drying shrinkage is a more important factor than thermal contraction in membrane splitting emerges from a paper by Carl G. Cash.[1] Cold-weather drying shrinkage of organic and asbestos-

[1]C. G. Cash, Low-Temperature Shrinkage of Membranes Due to Moisture Loss, paper delivered at the Second International Symposium on Roofs and Roofing, Brighton, England, September 1981.

felt membranes can exceed the breaking strain, according to Cash's research. (Glass-fiber–felt membranes experience insignificant drying shrinkage.) Cash's study demonstrates the vital importance of keeping organic and asbestos felts dry during and after installation.

Shrinkage Cracking of Poured Decks

Shrinkage cracking in poured-in-place decks that serve as a membrane substrate—structural concrete or gypsum or lightweight insulating concrete fill—can easily split a built-up membrane, with no help from any other factor. Solidly mopping a membrane to a poured, monolithic substrate exposes the membrane to an extremely high splitting risk, which is averted only if the substrate material is considerate enough not to crack after the membrane is applied. At $-30°$F, a good asphalt-saturated organic-felt membrane has less than 1 percent breaking strain. Solid-mopped to a substrate that develops a ⅛-in. crack, a membrane would have to accommodate about 5 percent strain (five times its breaking strain), even if we assume uniform yielding of 1¼ in. of the solid-mopped adhesive on each side of the crack, allowing 2⅜ in. of membrane to absorb ⅛-in. elongation. In the subfreezing range, with the bitumen frozen hard into a glasslike state, membrane elongation over a ⅛-in. crack could approach 20 percent, and the membrane would certainly split. Only if the crack formed gradually, widening from initial shrinkage in warm weather when the bitumen is relatively fluid and to a much smaller dimension than ⅛ in., would the membrane have a reasonable chance of avoiding a split over the crack.

Mechanically fastening the membrane to a poured substrate distributes strains over a larger area and consequently reduces membrane stress.

Insulation, preferably in two layers sandwiched between deck and membrane, is a much more reliable method of relieving the membrane from splitting strains caused by deck cracking.

Deck Deflection

Deck deflection or movement can promote membrane splitting, directly or indirectly. Flexible steel decks are a major indirect contributor, especially where deck span exceeds 6 ft. If deflection breaks the adhesive bond between the deck and insulation, the insulation fails in its function of restraining membrane movement. (See earlier section, "Insulation Movement.") Type III (steep) asphalt is especially vulnerable to horizontal shear failure from steel-deck deflection.

Steel decks account for a disproportionately high share of deck-deflection problems that result in the insulation bond breaking. Omission of side lap fasteners can cause excessive local deflection between adjacent deck units. Excessive bending deflection can result from stockpiling aggregate, felt rolls, and other materials, from dynamic roof-top equipment loads, from faulty structural details supporting drains and drainpiping directly on the steel deck instead of supporting structural framing.

Excessive roof deflection may produce membrane splits directly, when the roof structural framing comprises simple-span joists or other members that can rotate at their bearing ends and produce excessive membrane strains directly above the support points.

Mechanical fastening to steel decks provides far more dependable horizontal shearing resistance than adhesives (hot-mopped or cold), according to a recent research report.[1] Thus, in addition to more effective wind-uplift resistance, mechanical fastening also provides more effective membrane-splitting resistance.

Stress Concentrations

Cracking of poured decks is the most dramatic illustration of the stress-concentration phenomenon, but there are other sources in built-up membranes.

Stress concentration occurs at 270° corners, where contraction stresses intensify on a diagonal line working out from the corner. They also occur along peripheral terminations of the membrane, especially with poorly adhered insulation.

Lines where the structural deck changes span direction, or where deck material changes, also create complex stress concentrations. The architect should always provide an expansion-contraction joint along these lines because they are vulnerable to three kinds of differential movement:

- Vertical displacement (from deflection of the parallel structural member)
- Horizontal opening
- Horizontal shearing movement

Membrane splitting is extremely common over lightweight insulating concrete decks on lightgage steel centering where the steel deck changes direction and the required expansion joint is omitted. Where a 2- or 3-in. lightweight insulating concrete deck is poured continuously over a break in a typical, high-strength lightgage steel deck that changes

[1] J. E. McCorkle, R. M. Dupuis, and R. A. LaCosse, "Membrane-Splitting Resistance Depends on Quality of Entire Roof Assembly," *Proc. 6th Conf. Roofing Technol.*, NBS-NRCA, 1981, p. 34.

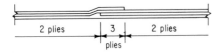

2 plies | 3 plies | 2 plies

Properly mopped and lapped joint

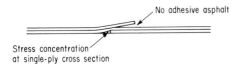

No adhesive asphalt

Stress concentration at single-ply cross section

Fig. 12-20 Stress-concentration line at unbonded lap in two-ply membrane can split the membrane. (Consultant Roger G. Riedel.)

span direction, the tremendous reduction in tensile strength through the composite cross section creates a vertical plane of weakness right through the cross section where the steel is discontinuous. The steel furnishes 50 to 80 times as much tensile strength as the insulating concrete.

Stress concentrations within the built-up membrane itself can also promote splitting. To fulfill their tensile-reinforcing function, felts must be properly lapped, like reinforcing bars in structural reinforced concrete members. For an obvious reason, a two-ply membrane is especially vulnerable to this type of failure (see Fig. 12-20).

Another possible source of split-promoting stress concentration in a built-up membrane is interply mopping voids.[1] Such stress concentrations could become especially important at low temperature as the embrittled bitumen gains strength and coefficient of thermal contraction-expansion but loses ductility. As is well known to stress analysts, a hole in an otherwise uniform cross section subject to uniform stress concentrates stresses at the edge of the hole.[2]

How to Improve Membrane-Splitting Resistance

There are many techniques for reducing the risks of membrane splitting. Here, from a previously cited paper, is a summary of the major split-preventing techniques:

- Specify mechanical fasteners (not hot-mopped or cold-applied adhesives) to anchor insulation to steel decks

[1]Herbert Busching, "Effects of Moisture and Temperature on Roofing Membranes in Thermally Efficient Roofing Systems," *Proc. 5th Conf. Roofing Technol.*, NBS-NRCA, 1979, p. 54.

[2]S. Timoshenko and Gleason H. MacCullough, *Elements of Strength of Materials*, Van Nostrand Rheinhold Company, New York, 1940, pp. 26ff.

- Design a rigid deck

- Specify four- instead of three-ply (or worse yet, two-ply) membrane

- Specify Type I or II, not Type III asphalt, for membrane inter-ply moppings.[1]

More extraordinary split-preventive techniques include

- Phasing the membrane application, with the two top shingled felt plies oriented perpendicular to the lower two shingled plies

- Use of a roof area divider or relief joint to avoid stress con-centrations at reentrant (270°) corners.

RIDGING

Ridges generally appear in parallel lines above continuous, longitudi-nal insulation-board joints. When ridges also appear above the trans-verse insulation-board joints, the pattern is called "picture framing" (see Fig. 12-21). In a less orderly pattern, ridges may resemble long blis-ters. But the mechanics and causes of ridging differ from the causes of blistering. A blister depends upon the volume expansion of a heated

[1]McCorkle, Dupuis, and LaCosse, op. cit.

Fig. 12-21 Ridging normally occurs over the joints of dimensionally unstable insulation boards. The staggered grid pattern is sometimes called "picture framing," a phenomenon generally limited to smooth-surfaced membranes.

air–water–vapor mixture entrapped within a closed space (generally an interply membrane void); a ridge normally depends on the moisture-absorbing expansion of the felts.

Ridging appeared as a major problem 20 to 30 years ago, when insulation sandwiched between deck and membrane became a normal part of the roof system. It has become far less common, since the use of coated base sheets appeared as a solution in the early and mid-1960s.

Ridging Mechanics

Ridges (sometimes called "wrinkles") usually form in, the following way. Hot asphalt applied over the insulation boards leaks into the joints between boards, and because of this bitumen leakage, the base felt is only lightly coated, or even bare, along the insulation-joint line. When water vapor from the building interior flows upward through the open insulation joint, it condenses within the base felt. With cumulative moisture absorption, organic or asbestos felts swell and buckle, forming ridges directly above the insulation joints (see Fig. 12-22).

Ridging failure can result from membrane flexural fatigue. Repeated flexing, caused by cyclic elongation and contraction accompanying wetting-drying of the felts, ultimately cracks the membrane. (Flexural fatigue resistance is 1 of the 20 attributes cited by NBS researchers Cullen and Mathey as part of their recommended performance criteria for built-up bituminous membranes. See Chapter 7, "Elements of the Built-Up Membrane".)

Moisture appears to be virtually the sole cause of ridging expansion of the felts, with little or no contribution from temperature cycling. In an experiment demonstrating ridging mechanics, Professor E. C. Shuman, formerly of Penn State University, found no ridges whatever formed on a test sample made with dry material under daily temperature cycling from −20 to 160°F continued over a 6-month period.[1] Expansion of a thoroughly wetted organic-felt membrane is 1.2 percent, compared with a computed expansion for a 180°F temperature rise of about 0.2 percent.

Shuman's studies also explain the persistence of ridging even after a wetted felt dries. Plotting curves with water absorption (lb/ft³) as abscissa vs. linear change (percent) as ordinate, Shuman found that felt expands rapidly with wetting but contracts very slowly with drying. After a 115-day wetting of a three-ply organic-felt membrane had expanded the felt by 1 percent, it took more than 100 days to contract

---

[1]E. C. Shuman, "Moisture-Thermal Effects Produce Erratic Motions in Builtup Roofing," *Engineering Properties of Roofing Systems*, ASTM STP409, 1967, p. 64.

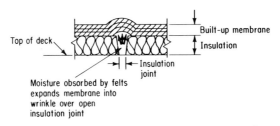

Fig. 12-22 Ridging mechanics.

the strips anywhere near their original lengths, and most of this contraction occurred in the last 3 weeks.[1] (See Fig. 12-23.) Natural drying thus offers little hope of flattening membrane wrinkles.

Physically, the explanation for this slow drying contraction is the presence of *inter*cellular water preventing the removal of *intra*cellular water. Intracellular expansion accounts for the major part of a wet organic-felt membrane's expansion.

Contributing Factors

The widespread practice of unrolling felts parallel to the continuous longitudinal insulation-board joints promotes ridging. Wet felts expand much more in the transverse than in the longitudinal direction. Aligning the felt rolls parallel to the longitudinal axis of board insulation thus maximizes ridging because the greatest insulation substrate movement occurs at these longitudinal joints. Organic and asbestos felts expand and contract much more than glass-fiber felts and thus ridge more readily.

Insulation movement also promotes ridging. Single-layer insulation, as opposed to the more desirable two-layer, vertically staggered insulation-joint pattern, can create instant ridging if the boards are already warped at the time they are installed.

Preventive Measures

Ridging-abating measures include

- Use of a coated felt as membrane base sheet

- Orienting the felts *perpendicular* to the longitudinal insulation-board joints

[1]Ibid., p. 55.

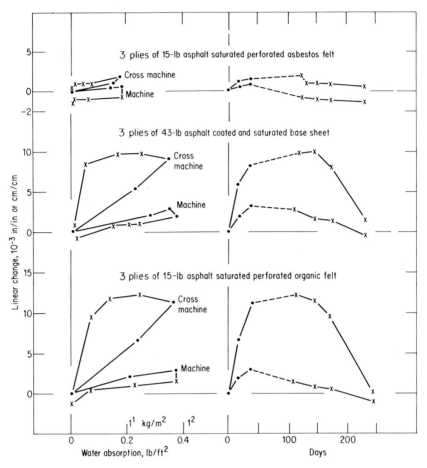

Fig. 12-23 Built-up membrane test samples, totally immersed in water for 115 days at 70°F and then dried in laboratory air, took more than 4 months to contract close to original dimension and still exhibited permanent lengthening. This slow-drying contraction, attributed to the lingering presence of interstitial water between the organic-felt fibers preventing the evaporation of intracellular water absorbed in organic-felt fibers, accounts for the persistence of membrane ridges, according to Prof. E. C. Shuman, who conducted the test. *Note:* All specimens 3 × 36 in. (75 × 915 mm), temperature 70°F (21°C). (From *Engineering Properties of Roofing Systems*, ASTM STP409, 1967, p. 54, American Society for Testing and Materials.)

- Double-layered insulation
- Joint taping

As its general response to widespread ridging problems back in the early 1960s, the roofing industry's solution was a coated base sheet, laid singly and followed by shingled saturated felts to complete the membrane. A coated base sheet is less permeable to water vapor than a saturated felt. Its greater stiffness also resists ridge formation more effectively than a membrane field-fabricated exclusively with saturated felts.

Orienting the felts perpendicular to the longitudinal insulation-board joints reduces the membrane's vulnerability to ridging because wetted felts expand less in the longitudinal than in the transverse direction.

Double-layered insulation reduces the risk of ridging because the bottom insulation layer blocks vapor flow from the interior to the membrane. With a highly impermeable insulation (e.g., foamglass), there should be virtually no upward vapor flow.

Double-layered insulation alleviates ridging in another way, i.e., by reducing the joint width between adjacent boards. By simple geometry, a 2-in.-thick insulation board with edges slanted 1° in from true vertical has twice the joint thickness of a 1-in.-thick board with the same edge-angle deviation. Moreover, improved overall stability (from superior anchorage potential) should also result in narrower joints.

Taped insulation joints can help prevent ridging, especially with single-layered insulation. They not only provide a barrier to upward water-vapor flow, they reduce splitting risks from stress concentrations over insulation joints.

MEMBRANE SLIPPAGE

Slippage is relative lateral movement between felt plies. A membrane plagued by slippage often assumes a randomly wrinkled appearance, like a badly laid carpet (see Fig. 12-24). Because it can expose the base sheet, slippage often reduces the membrane from a multi-ply to a single-ply covering, exposed to weather and other destructive forces. Although relatively uncommon, slippage is a costly mode of failure, extremely difficult to rectify.

Slippage failures usually occur within the first year or two of the roof's service life, with most failures on roofs from ½- to 1-in. slope. Slippage seldom occurs on roofs of less than ¼-in. slope, but roofs of ¼- to ½-in. slope are more vulnerable to slippage than is generally recognized. Membranes with slopes greater than 1 in. seldom slip because the roof designer, recognizing the risks of slippage on steeper slopes, usually takes the precaution of specifying backnailing or another positive anchorage technique.

Slippage-Promoting Factors

These findings are from an intense NBS laboratory investigation featuring slip-and-sag tests. This investigation followed field studies that

Fig. 12-24 Membrane slippage, a problem on sloped roofs, is caused by one or more of a complex of factors. (National Bureau of Standards.)

identified slippage problems and parameters. NBS materials expert W. C. Cullen identified six factors involved in slippage:

1. Slope

2. Bitumen

3. Felts

4. Climate

5. Substrate heat capacity

6. Surfacing

Slope is the prime factor in slippage (see Fig. 12-25). As a general rule, preventive backnailing of felts is recommended for Type III asphalt membranes over 1½-in. slope and for Type I asphalt or coal-tar-pitch roofs of ½-in. or greater slope (see Fig. 12-26). Even lesser slopes may require backnailing under unusual circumstances.

Low softening-point bitumen is often blamed for slippage. And low softening point is often attributed to softening-point fallback from overheating steep asphalt. Slippage from low softening-point asphalt has prompted several major roofing manufacturers to recommend steep asphalt (ASTM D312, Type III) for interply mopping on all roofs, regardless of slope.

But this remedy has its own special hazards. In many important respects, steep asphalt is inferior to dead-level asphalt as a plying cement. In fact, as pointed out in the discussion of equiviscous temperature (EVT), softening point, penetration, and ductility may not give a true picture of temperature susceptibility, or even flow characteristics. Viscosity is the most important index of flow. Heating asphalt above its blowing temperature in the absence of air produces a more fluid material at high roof-surface temperatures. For steep asphalt, softening-point fallback may reach 20 to 25°F. Steep asphalt is much more susceptible to softening-point fallback than dead-level asphalt.

Phased construction, when a base sheet is glazed and the shingled felts applied on a later date, with another interply mopping added to the glaze coat and thickening the asphalt film, is a prime cause of slippage. In phased membranes, slippage, if it does occur, is invariably at

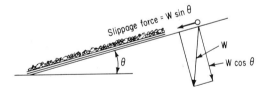

Fig. 12-25   Vector diagram shows how slippage force ($W \sin \theta$) increases with increasing slope and membrane weight.

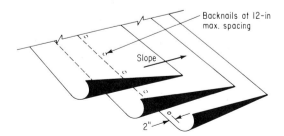

Fig. 12-26   Backnailing, to prevent slippage, is covered by overlapping felts.

the continuous, interply mopping plane at the base sheet's top surface. Eliminating phased construction, or shingling all felts, obviously eliminates this source of slippage.

Another flaw in felt application can cause a "wheeling effect" that promotes slippage. Where the membrane's felt plies are not properly staggered but instead terminated at the same vertical line, which may extend from eave to ridge, slippage potential may be especially high, with slippage exposure increasing as it nears this vertical line. The explanation for this wheeling effect, so-called because of its apparent rotation about two hubs located some distance back from the vertical line, is as follows. Slippage is aggravated along this line because of unbalanced, one-sided restraint (i.e., back toward the hub), whereas at other segments of the staggered felt lines there is generally two-sided restraint to slippage (i.e., tensile restraint from the felts on both sides of a given point from areas where the asphalt might be more stable, thus providing better anchorage for the felts). As another analogy, the line of spliced felt ends is like the ends of cantilevered beams, which have greater freedom to deflect because of their one-sided support. In fact, the slippage line may resemble the deflection curve of a continuous row of beams with cantilevered end spans.

Built-up membranes constructed of glass-fiber or perforated felts are most resistant to slippage. They let the interply bitumen flow through and form shear keys with the other interply moppings.[1]

Slippage risk increases with the thickness of interply moppings between felts. Adhesives should be as thin as practicable to provide waterproofing while still ensuring continuous, unbroken coverage and total separation of the felts. Increased adhesive thickness reduces horizontal shear resistance between the felt plies. Overweight moppings are likely to occur in phased construction because the phased construction plane may get two moppings: a glazed coat on top of the base ply and a final interply mopping to bond the top plies.

Climate and roof orientation affect slippage because they determine membrane temperature and the daily duration of high temperature. Obviously, the hotter the climate, the greater the slippage hazard. Of even greater importance is exposure to solar radiation, which explains the prevalence of slippage failures on south- and west-facing roof slopes.

Substrate heat capacity is another factor affecting the slippage risk, which varies *inversely* with substrate heat capacity. A structural con-

---

[1]W. C. Cullen, *Slippage of Builtup Roof Membranes—Causes and Prevention*, NBS Rep. 10950, p. 45.

crete deck substrate (150 lb/ft³) continually moderates membrane surface temperature—lowering it during a hot, sunny day, raising it during a cool night—through its high thermal conductivity and its high heat-storage capacity. In contrast, highly efficient, lightweight foamed insulation weighing 2 lb/ft³ isolates the membrane and raises surface temperature on a sunny day because it retards heat transmission through the roof sandwich.

Membrane surfacing affects the slippage hazard in two ways: color and surfacing aggregate weight. As discussed in Chapter 5, "Thermal Insulation," surface color has a tremendous effect on membrane temperature. A black surface increases the slippage risk because it promotes much higher temperatures than a light-colored roof surface.

Aggregate surfacing weight increases slippage risk by increasing the gravitational-force component acting parallel to the roof plane (see Fig. 12-18). Do not, however, try to alleviate this problem by eliminating aggregate surfacing in favor of a smooth-surfaced roof; this cure normally entails excessive sacrifice of other desirable membrane qualities.

Cullen has combined the previously described six factors into an equation relating slippage distance and time:

$$\frac{s}{t} = \frac{1}{\eta} \times \frac{FD}{A} \tag{12-4}$$

where s = slippage distance
   t = time
   F = force
   D = interply bitumen mopping thickness
   A = area
   η = bitumen viscosity

Multiplying through by t yields

$$s = \frac{1}{\eta} \times \frac{FDt}{A} \tag{12-5}$$

Examination of Eq. (12-5) shows that slippage distance varies *directly* with force (which depends on slope and surfacing weight), interply mopping thickness, and time (which depends on duration of high roof temperatures). Slippage distance varies *inversely* with viscosity. (This equation does not apply to membranes with backnailed felts.)

Poor field-application practices promote slippage, notably through overweight interply bitumen moppings and overheating of hot-mopped asphalt, which can result in softening-point fallback and consequent reduced bitumen viscosity. Improperly installing mechanically anchored base sheets obviously increases the membrane-slippage risk.

Slippage Prevention

Here are Cullen's recommendations for reducing membrane-slippage risk:

1. Set minimum viscosity limits for various asphalts at 140°F as well as 450°F. Cullen suggests a minimum viscosity of 5×10^6 P at 140°F as a guide value to reduce slippage potential.

2. Limit asphalt interply moppings to a minimum 15 lb/square, maximum 20 lb/square.

3. Limit coal-tar-pitch interply moppings to 20 lb/square minimum, 25 lb/square maximum.

4. Limit coal tar pitch to ¼-in. maximum slope without mechanically fastening the base sheet.

5. Avoid phased construction in climates subject to high slippage risk; substitute four- or five-ply shingled felt membranes for coated base sheet plus three or four shingled plies.

6. Design light, heat-reflective roof surfaces.

Backnailing is required generally for slopes of 1½ in. or greater for Type III asphalt, ½ in. for Type I asphalt (see Fig. 12-26).

For the owner who needs a cure for a currently developing slippage problem, Cullen advises sprinkling water on the roof during hot portions of the day or painting a smooth-surfaced roof with whitewash. Because slippage usually occurs within the first year or two (before natural aging has significantly increased bitumen viscosity), these inexpensive remedies may be enough to hold off the problem until the membrane's aging, hardening asphalt develops its own slippage resistance.

For a more drastic, and expensive remedy, consider mechanical fastening to the substrate. Fastener heads must be waterproofed with plastic cement (or substitute) and at least two layers of reinforcing felt or fabric.

DELAMINATION

A less common membrane failure, delamination, is promoted by

- Insufficient bitumen between felts
- Improper embedment of felts because of inadequate brooming or underweight mopping
- Application of inadequately heated bitumen

- Use of coal-tar-saturated felts with asphalt, or asphalt-saturated felts with coal tar pitch

- Water absorption during the winter, followed by evaporation in spring or summer

- Freeze-thaw cycling of water entrapped between plies

This last-named phenomenon can literally tear the membrane apart. In its milder aspects, delamination may produce wrinkling and then cracking, much like a wrinkled membrane.

Freeze-thaw cycling is a more serious problem than is generally assumed in cold climates with winter-long snow cover. Contrary to the general assumption that a snow-covered membrane does not freeze, Canadian researcher H. O. Laaly found that a roof membrane with shallow snow cover may undergo many freeze-thaw cycles in one winter.[1]

Laaly's research yielded other conclusions:

- It takes a minimum snow cover of 12 in. to maintain an essentially constant membrane temperature at the freezing level.

- Snow depth can vary widely—from a low of 2 in. to above 3 ft—at the same time in different parts of the same roof.

- A membrane (with or without snow cover) can freeze when moisture accumulates in the felts.

Laaly's temperature readings were from thermocouples embedded in the top surface and the center of the test membranes.

ALLIGATORING

Alligatoring occurs chiefly in exposed asphalt, especially steep asphalt. It occurs in smooth-surfaced roofs and sometimes in bare spots of aggregate-surfaced membranes. It consists of deep shrinkage cracks, progressing from the surface down, a result of continued photo-oxidation, aging, and embrittlement (see Figs. 12-27 and 1-2).

Alligator cracks can retain water, which threatens eventually to penetrate through to the felts, where it can work its familiar mischief: opening paths for leakage, wetting the felt fibers, reducing membrane tensile

[1]H. O. Laaly, "Temperature of Bituminous Roofing Membrane and Its Relation to Ambient Conditions," *Proc. Symp. Roofing Technol.*, NBS-NRCA, September 1977, p. 187.

Fig. 12-27   Advanced stage of alligatoring shows exposed felt deteriorating under attack of sunlight and moisture.

strength, and accelerating the membrane decay of both felts and bitumen.

The destructive, photo-oxidative process that culminates in alligatoring results from the ultraviolet acceleration of the normally slow oxidizing of the bitumen's hydrocarbon molecules into carbon dioxide and water molecules. The sun's high-energy ultraviolet radiation can actually break the bonds between carbon atoms, reversing the polymerization process that produced the long, chainlike molecules during the asphalt blowing process. Meanwhile, a secondary degradation process produces cross linkages of the smaller hydrocarbon molecules. The bitumen consequently becomes harder and more brittle.[1]

Alligatoring is aggravated by thick moppings, which raise the thermal coefficient of contraction-expansion and consequent contraction of the surface coating with falling temperatures. V-shaped in cross section, the progressing alligator cracks penetrate down to the felts, admitting water, which then invades the membrane by wicking along felt fibers. In advanced stages of alligatoring, curled segments of asphalt peel away, exposing the top felt (see Fig. 12-27).

Overly thick asphalt moppings result from several slovenly field practices, notably:

- Dumping the contents of hot-asphalt applicators

- Slow turning of hot-asphalt applicators when dispensing openings are not closed

[1]See H. E. Ashton, "Weathering of Organic Building Materials," *Can. Build. Dig.*, no. 117, 1975, p. 117-2. Also see Maxwell Baker, *Roofs: Design, Application and Maintenance,* Multisciences Publications, Ltd., Montreal, 1980, pp. 124, 125, 306, 308.

- Runoff from base-flashing mopping down onto the membrane
- Application of overly cooled asphalt, at temperatures below 300°F
- Overly thick interply mopping asphalt squeezed out beyond felt lap edges and then further thickened by top-coat moppings on a smooth-surfaced roof or by the flood coat on an aggregate-surfaced roof

A clay-based asphalt emulsion is a much more durable coating for a smooth-surfaced membrane because it is far less vulnerable to photooxidative degradation.

SURFACE EROSION

Displacement of the surface aggregate (and sometimes the flood-coat bitumen) often results in progressive failure. It may be followed by alligatoring, felt deterioration, blistering, or other failures promoted by water infiltrating the membrane or the insulation.

Underweight flood coat is probably the chief cause of aggregate-surfaced membranes' vulnerability to erosion. Erosive forces include:

- Wind suction, which sometimes attains negative pressures exceeding 100 psf at building corners for winds less than 100 mph
- Water flow or drip, especially where conductors from a roof above discharge onto a lower roof
- Rooftop traffic, especially on regular paths or where snow is shoveled and mechanical installation or repair work is done

Surface erosion on smooth-surfaced roofs, where it entails loss of surfacing asphalt, usually results from one or more of the following causes:

- Excessive foot traffic
- Weathering
- Discharge of corrosive or solvent-type fumes or liquids on the roof surface

THIRTEEN

Reroofing and Repair

A building owner's decision to tear off and replace a leaking roof system often follows this scenario: A prolonged series of futile attempts at repair, sometimes stretching out for years, harasses the owner with continuing irritation, inconvenience, functional handicaps, and sometimes with serious economic loss (e.g., lost rentals of motel rooms). Finally, these problems overwhelm any resistance to the cost of reroofing. The owner decides to reroof—often in the dead of a northern winter, when built-up roofing application problems oscillate between the difficult and the impossible.

So, as their first advice on reroofing, building owners should attempt to anticipate the need for reroofing. This does not mean that they must become soothsayers prophesying the fate of their roofs. It merely implies the exercise of greater prudence and control in recognizing a perennially troublesome roof leaking moderately in summer and autumn as an almost inevitably more severe problem in late winter, when it is much more difficult to repair, replace, or recover.

A recovered, or even totally replaced roof has less chance of lasting out a 20-year service life than a new roof built with comparable care and skill. This is a second important point for roof-troubled building owners. Despite its high cost, recovering or tearoff-replacement generally requires major compromise with roof-design principles. Working on the clean slate of a new building, the roof designer faces far fewer geometric constraints (discussed in detail later in this chapter). Tearoff is a slow, dismal job, requiring much greater care than new construction (see Fig. 13-1).

The difficulty of tearoff-replacement needs emphasis because building owners seldom comprehend the complexities of reroofing, and they are often shocked by the high cost. A reroofing project is almost always more difficult to carry out successfully than a new project of similar scope. Aggregate-surfaced roofs in particular are an extremely difficult

Fig. 13-1 This tearoff-replacement project started with gasoline-powered cutter slicing old aggregate-surfaced membrane into roughly 10-ft² segments (top), pried off roof (middle), and carted away (bottom). The project *should* have started (but did not) with removal of aggregate, including power vacuuming, to reduce risk of tracking aggregate into new, replacement roof areas.

problem. Keeping the old aggregate out of the new membrane, where it can form blister-forming voids and puncture felts, is always a difficult, sometimes a Herculean, task. It takes good planning and careful execution by the roofing contractor to solve this problem because the aggregate has a remarkable affinity for the workers' bitumen-covered shoes and equipment wheels.

RE-COVERING OR TEAROFF-REPLACEMENT?

After investigation and analysis, the roof designer faces the first major decision: whether to re-cover the existing membrane or go all the way with total tearoff-replacement. Tearoff-replacement offers two major opportunities:

- Inspection and repair of the deck

- Possible application of tapered insulation (depending on flashing heights) to slope an existing dead-level membrane that ponds water

Accompanying these two advantages are two disadvantages:

- Greater disruption of building operations

- Substantially higher cost

For some sensitive building occupancies—e.g., top-floor computer rooms, laboratories, or telephone equipment—tearoff-replacement may be an unacceptable risk. Removal of the entire existing roof system may expose the roof to intolerable leakage if the roofer is caught by a sudden, unanticipated rainstorm. Prefabricated decks—e.g., steel, plywood, and tongue-and-groove boards, precast concrete—are obviously a greater hazard than poured-in-place concrete. Joints between prefabricated deck units provide easy, open access for water entry into the building. But even poured-in-place concrete and gypsum decks will leak through cracks caused by shrinkage or thermal contraction.

Thus the decision to accept a tearoff may depend on building use. For most uses, the owner can accept some slight risk of leakage. (The owner has normally been doing that for some time prior to the decision to reroof.) But for some operations—e.g., sensitive production or laboratory operations—a tearoff-replacement may require protection of the contents from dust, dirt, and water. Substantial areas of deck replacement will definitely require a shut-down for several days (or weeks) for reroofing operations. The owner must weigh the immediate costs of a shutdown against the future benefits of owning a thoroughly rebuilt roof. So, as one early question, the owner must decide whether to accept a tearoff.

Against the difficulties and greater expense of a tearoff-replacement, the roof designer must consider the feasibility of re-covering. Re-covering requirements differ for different kinds of roof systems: conventional built-up bituminous membranes, protected membrane roofs, loose or adhered single-ply sheet synthetics, or sprayed polyurethane foam plus fluid-applied membrane coating. But re-covering does carry several general requirements:

1. The structural deck must be sound.

2. The existing roof system must be adequately anchored.

3. Existing insulation should be strong enough to resist traffic and normal impact loads, and it should be essentially dry.

4. The existing membrane must form a reasonably smooth surface or be economically repairable into a smooth surface.

Whether all the foregoing requirements are mandatory for all re-covering systems is debatable. Compromise is an inherent aspect of most reroofing projects. A loose-ballasted roof system requires a stronger deck than an adhered system to carry an additional 10 to 25 psf of loose aggregate or pavers. But a loose system, with a vapor-permeable membrane, can tolerate more moisture in the existing insulation than an adhered system susceptible to blistering. If wet insulation is accepted, the designer must decide how much insulating value to sacrifice from wet insulation. And the cost of each alternative method is an ubiquitous factor in any reroofing design decision.

In any event, the foregoing rules should be considered highly desirable for any type of re-covering roof system. Violation of any of these rules for re-covering should, at the least, prompt consideration of total tearoff-replacement. Planning a reroofing project is an exercise in running through the entire spectrum of possible solutions and balancing liabilities and advantages of systems A, B, C, and possibly D against each other. The solution to a reroofing problem will seldom, if ever, appear as definitely clear as the decision on a new project, where the designer has greater control of all the conditions.

INVESTIGATION AND ANALYSIS

In a paper titled "Technical Aspects of Retrofitting," R. L. Fricklas, Director, the Roofing Industry Educational Institute (RIEI), recommends that investigative analysis proceed through the following five stages:

- Nature and condition of existing roof-system components

- Roof-system history

- Roof-system environment (interior or relative humidity, exterior surface contaminants, structural loads, vibrations, and other unusual conditions)

- New performance criteria (especially for thermal insulation or heat-reflective coatings)

- Improved drainage (many existing roofs pond water, and, if at all practicable, reroofing must correct this condition to justify the owner's capital investment)

Preliminary Investigation

Start by examining original plans and specifications, if available, to identify roof-system components: deck, vapor retarder (if any), insulation, and built-up membrane (number of plies, type of felts, bitumen, and so on). Unfortunately, many building owners have lost the specifications; you are lucky to find a set of drawings. Check whether the architectural drawings call for slope. The structural roof framing plan can help determine the roof's load-carrying capacity. You should find the design live load on the roof framing plan.

Be forewarned that drawings and specifications often fail to show the final roof system. They are not always emended to reflect job-change orders.

Besides checking drawings and specifications, check for subsequent modifications to the roof. Even an oral record, from the maintenance person or building superintendent, is better than nothing. As part of this inquiry into modifications, also ask about possible changes in building use or additions of rooftop equipment. For example, a change in building use might raise interior relative humidity, thereby increasing water-vapor migration into the roof system.

Next, interview the building superintendent about the roof system's history. Successful reroofing requires proper identification of leak sources. Wall leaks are sometimes mistaken for roof leaks, as is dripping condensation.

Leak-Detection Technique

A leak-detection interview proceeds as an elimination process, designed to eliminate suspects, first in big batches, then in smaller ones as you converge onto the solution. For example, if the building leaks *only* when it rains, that fact indicates a true exterior leak. But if it leaks at other times—e.g., after high-humidity occupancy, followed by cold weather—the "leaks" could consist of condensation dripping down from the insulation. Leaks from certain wind directions can be identified as roof flashing leaks, or sometimes wall leaks, where wall flashing is defective or even totally omitted. Persistent leakage irrespective of rainfall may indicate water-saturated insulation, especially prevalent under ponded membranes.

When it is necessary to link a leak to a definite roof defect, local flood-testing may be required. A hose carried up to rooftop, or simply a bucket or two of water poured on a suspected flashing leak source, can simulate rain needed to check the interior for leakage. Note horizontal offsets from the leak source in prefabricated deck units. Steel decks will generally leak at end laps, not side laps, where the flute forms a trough capable of retaining substantial amounts of water.

Visual Inspection and Analysis

Interior inspection of the deck soffit through removal of a ceiling panel (or other access mode) should seek the following information:

> *Steel*   Rusting? Differential deflection at side or end laps? Excessive deformation? Sound welds? Do rooftop components—HVAC, other equipment, access hatches, and so forth— have their own structural angle supports?

> *Wood*   Rotting? Warped? Shrunk? Excessive joint gaps? Unanchored?

> *Structural concrete*   Cracks over ⅛ in.? Excessive deflection in evidence?

> *Precast concrete*   Excessive joint gaps? Differential deflection at adjacent units?

> *Poured gypsum*   Excessive deflection of subpurlin bulb tees? Cracking? Evidence of excess moisture?

> *Corrugated steel supporting lightweight insulating concrete*   Underside venting slots or side laps? Effluorescence on steel? (Make note to check deck surface carefully during topside inspection.)

> *Structural wood fiber*   Excessive deflection? Differential deflection between adjacent units? Excessive joint gaps?

During this interior inspection generally applicable to all deck types, note also the following:

- Evidence of foundation settlement (bearing walls' cracks and so on).

- Changes in deck type or span direction. (These lines should have an expansion-contraction joint.)

- Drain locations and drain leader accessibility.

- Location of rooftop HVAC units and their supply ducts or chiller pipes.

- Areas of most severe leakage (or condensation drippage).

Roof-surface inspection should start with a general observation of ponding. Obviously, the best time to observe the roof's ponding is shortly after rainfall. But the sedimentary deposits of darkened areas (algae in warm climates) indicate ponding. And the building superintendent or maintenance person can also report on ponding.

In your visual assessment of ponding, note particularly the following:

- Evidence of ponding along parapet walls, gravel-stop edges, expansion joints, penthouses, HVAC supports, and other flashed components.

- Location of drains. (Are they, in accordance with the generally diabolical law governing drain elevations, at the roof's high points? Are they blocked or constricted?)

- General flashing-height elevations

These observations can give you a quick, general impression of the scope of the problem: whether you can economically correct the roof's drainage problem, raising flashed components out of low spots and so forth.

Your attention should then turn to these specific items:

- Damaged, unadhered, wrinkled, or deteriorated flashings?

- Splits in membrane at gravel-stop strips?

- Damaged, inadequately sealed counterflashings?

- Membrane blisters, splits, ridges?

- Flashing conditions at HVAC units, vents, skylights, parapets, hatches, and so forth?

- Pitch-pan condition?

- Drains—intact lead seal: clamping bolts tight?

Field-Test Cuts

Have sample roof test cuts taken down to the deck for (1) visual field inspection and (2) laboratory analysis, if required.

A visual inspection should reveal the following information:

- How many membranes.
- Condition of deck—e.g., has lightweight insulating concrete suffered freeze-thaw disintegration? Is it wet?
- Is there a vapor retarder?
- Insulation well-adhered to deck (or to vapor retarder)?
- Type, condition (wet or deteriorated), thickness of insulation?

Laboratory analysis of sample test cuts can refine and confirm your visual observations. From a 40 × 6 in. sample cut across the felts, you can get a laboratory determination of the number of plies, laps, and so on for better assessment of the existing membrane's (or membranes') condition. From core samples, the laboratory can give you moisture contents of lightweight insulating concrete fill or other insulating material, or even the membrane. Photographs can provide a permanent visual record of conditions.

Large-Scale Moisture Surveys

On projects of sufficient size to justify the expense, recent advances in nondestructive techniques for locating wet insulation in built-up roof systems have made accurate surveys of large roof areas economically practicable (see Chapter 19, "Nondestructive Moisture Detection"). Regardless of the nondestructive testing technique—nuclear, capacitance, or infrared—core samples are required to verify results. By coordinating data from nondestructive roof surveys over large areas with core samples and construction data, you can get an accurate picture of the moisture content over the entire roof area.

ASSESSMENT FOR REMEDIAL ACTION

With the results of your investigation and test-cut sample reports, you should now have a general idea of what is economically practicable and technically feasible.

Tearoff-replacement is indicated if your investigation has turned up any of the following:

- Extensive ponding
- Deteriorated deck
- Wet and/or deteriorated insulation

- Poor anchorage between deck and insulation (or other roof-system component) and no practicable way to mechanically anchor them; i.e., the deck is structural concrete (cast-in-place or precast)

- An essentially irreparable membrane surface; blistered, wrinkled, and deteriorated beyond economically practicable repair into a smooth surface for re-covering

Re-covering the existing membrane may be practicable if the scope of the foregoing problem (or problems) is restricted to minor areas. If there are only a few rotted plywood panels, they can be replaced and reinsulated, and the remaining roof can be re-covered. If a full-scale moisture survey indicates wet insulation in 10 percent of the roof area, that segment can be removed and replaced and the remaining roof covered. Determining the break-even point for economical partial replacement vs. total tearoff is a complex question to be answered for each specific project. But whenever you start approaching one-quarter removal of random areas of wet insulation or membrane, you are probably moving close to an economic decision for total tearoff-replacement.

New Performance Criteria

Most reroofing projects require a decision about the practicability of two major improvements:

- Additional thermal-insulating quality

- Providing slope for positive drainage

Extensive ponding normally provides ample justification for tearoff-replacement instead of re-covering. If the rooftop geometry permits the raising of flashing heights, you can usually provide slope for positive drainage through the addition of tapered insulation or lightweight fill.

Improving thermal resistance, through addition of insulation, can similarly provide justification for a tearoff-replacement instead of re-covering. Most water-leaking roofs leak heating and cooling energy as well, and the vast majority of existing roofs, like walls, are drastically underinsulated. Consider the benefits of double-layered insulation.

Other improvements you may consider, especially in association with improved drainage, are the practicality of:

- Relocating or remounting rooftop equipment

- Adding new drains (or scuppers)

- Reconstructing (or adding) expansion joints, where required

- Changing the roof surfacing

If the existing membrane has aggregate surfacing, it may be wise to change to a smooth surface for industrial buildings. If cement, food, or other material are regularly discharged on the roof, a smooth-surfaced roof permits easier removal than an aggregate-surfaced or loose-ballasted roof. However, remember that a smooth-surfaced membrane is a *water-shedding*, rather than a *water-resistant* membrane, requiring minimum ½-in. slope. Consider adding wood walkways or special mineral-surfaced treads on paths that get the heaviest roof traffic (around HVAC units and so on).

When the existing slope is inadequate, it is seldom practicable to achieve ½-in. slope, or even the recommended ¼-in. minimum for aggregate-surfaced roofs. (With a drain just 24 ft away, it takes at least 7-in.-thick insulation to provide for ¼-in. slope: 6-in. allowance for slope, plus 1-in. minimum thickness at or near the drain.) You are lucky to get ⅛-in. slope with tapered insulation, and often you would settle for ¹⁄₁₆ in. And because most rooftop components are located without the slightest thought of placing them clear of low spots (much less at high spots where they should be), flashed components often sit within 2 ft or less of a drain, thus ensuring that they will always be exposed to the hazards of ponded or flowing rainwater.

As a possible solution when it is impracticable to use tapered insulation to provide slope, consider the following alternatives:

- Drainage via (1) electronically controlled siphon system or (2) solar-powered siphons.

- Use of a loose-ballasted elastomeric sheet membrane system. (Check manufacturer's guarantee to make sure it is not nullified by standing water on the roof.)

REROOFING SPECIFICATIONS

The architect should prepare detailed specifications and drawings for a reroofing project. The specifications should contain the following:

- Time limits for job completion

- Acceptable membranes (and surfacing)

- Acceptable insulation (and required thermal resistance)

- Provision for unit costs for insulation and/or deck replacement

- Provision for contractor proposals on alternative details
- Responsibility for disposal of debris and waste material

Isolating the New Membrane

On re-covering projects, never mop the new membrane to the existing membrane; always isolate it with either a venting base sheet or porous insulation (e.g., fiberglass). Venting base sheet is preferably nailed, but if nailing is impracticable, specify spot mopping. For a deck that will take mechanical fasteners, you can slash venting holes through the existing membrane, mechanically anchor a felt slip sheet and, on top of the felt, a venting base sheet or porous insulation, and then apply new insulation in hot-mopped Type III asphalt.

For re-covering projects in areas with high vapor-pressure gradient (interior relative humidity above 50%, average January temperature below 40°F), slash holes in the old membrane, so that it vents upward-migrating water vapor to the venting base sheet or porous insulation (see Fig. 13-2).

Make sure that the venting base sheet is above the dew-point temperature. (See Chapter 5, "Thermal Insulation," for computation technique for locating dew point in the roof cross section.)

Flashings

For all reroofing projects—re-covering as well as tearoff-replacement—replace flashings and gravel stops, for several reasons:

- Bituminous base flashings generally deteriorate faster than the membrane.

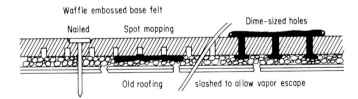

Fig. 13-2 On re-covering projects, with a high vapor-pressure gradient from a humid interior, holes should be slashed in the old membrane remaining in place, to let it vent. Diagram shows three different techniques for anchoring the base sheet, with its waffle-grooved vent ducts: nailing, the most dependable technique (left); spot mopping (center); dime-sized, grid-patterned holes (right), through which hot asphalt flows to adhere the base felt to the existing membrane below. (R. L. Fricklas, Roofing Industry Educational Institute.)

- Edge details and flashings are a major source of water entry.

- Roof-assembly-thickness change requires new wood nailer edge strips (thicker for overall tapered insulation or merely for tapered fiberboard strips designed to raise edges and flashed components out of water).

On reroofing projects lacking tapered insulation, specify tapered fiberboard strips to raise flashed components out of low areas that pond water.

For all re-covering projects, install pressure-relief vents at two levels:

- Lower level, to relieve vapor pressure under venting base sheet or insulation

- An upper vent system, to relieve vapor pressure in new insulation (see Fig. 13-3).

Surface Preparation of Existing Membrane

For re-covering projects, require a smoothly prepared, clean surface on the existing membrane for application of the new system, via the following steps:

- Power brooming and vacuuming to remove all loose aggregate from the surface

- Cutting out of ridges, blisters, loose felts, and other surface projections

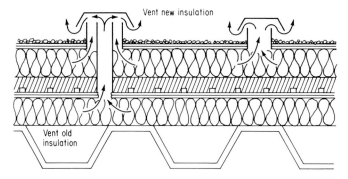

Fig. 13-3 Re-covering projects need pressure-relief vents at two levels, upper and lower. (R. L. Fricklas, Roofing Industry Educational Institute.)

- Airblasting or vacuuming to remove dirt, dust, and other loose materials

- Application of asphalt cutback primer for surfaces designed to receive hot-mopped asphalt (e.g., for application of a layer of porous insulation)

Existing aggregate-surfaced membranes require special attention in the specifications. Highlight the roofer's responsibility to keep this aggregate out of the new membrane. Aggregate sticks to the bitumen coating on workers' shoes and equipment wheels. Without special precautions by the roofing contractor, the old aggregate will be tracked into areas of new roof application and get trapped in the new roof system: at the deck-insulation interface, at the insulation-membrane interface, and between felt plies. Entrapped aggregate punctures felts and creates interply-mopping voids, thus posing a tremendous threat to membrane integrity. Warn the contractor in the specifications that the bid price must include the cost of necessary precautions to prevent aggregate entrapment. These precautions may include:

- Laying plywood walkways over old work areas to prevent workers and equipment from picking up aggregate

- Establishing "aggregate-removal" stations to keep workers and equipment "uncontaminated" when they enter new roof-application areas

- Arranging work patterns to avoid aggregate contamination

- Requiring special gravel-vacuuming equipment to ensure proper removal of aggregate from areas to be reroofed.

ALERTS

For New Built-Up Roof Membranes

1. Favor tearoff-replacement over re-covering if the existing roof system fails to satisfy (and cannot be repaired to satisfy) the following requirements:

- Sound structural deck

- Secure insulation-to-deck anchorage

- Reasonably dry, firm insulation

- Smooth membrane surface, suitable as substrate for new membrane

2. Consider upgrading of remedial roof system for (a) improved thermal performance and (b) improved drainage. (If impracticable to provide slope

through tapered insulation or fill, consider other methods of drainage: electronically controlled siphon or solar-powered siphon systems.)

3. Check for conformance of new system with Factory Mutual and Underwriters Laboratories requirements (which may be more stringent than when original roof system was applied).

4. Favor mechanical fastening for steel decks in tearoff-replacement projects because of (a) more dependable anchorage for wind-uplift and membrane-splitting resistance and (b) risk of excessive hot-mopped bitumen added to bitumen possibly on deck from original application.

5. On all reroofing projects (re-covering and tearoff-replacement), require removal and replacement of all flashing.

6. On tearoff-replacement projects, require removal of old systems down to the deck. (Leaving an old vapor retarder in place on a steel deck can (a) obscure water in flutes and (b) obscure defective side lap or end lap fastening.)

7. On re-covering projects:

   • Always isolate new membrane from existing membrane, either with a felt slip sheet or porous insulation layer.

   • Always provide for ventilation of the old system (to accommodate suspected and probable moisture migrating as vapor upward toward the new roof system).

   • Always reanchor existing insulation suspected of being inadequately adhered to the deck. (If you cannot fulfill this condition, tearoff-replacement becomes your rational choice.)

8. Take special precautions to prevent entrapment of old aggregate from existing membrane in the new membrane.

For Loose-Laid Synthetic Single-Ply Membranes

1. Check roof's structural framing and deck for additional gravel or paver ballast load for loose-laid, ballasted roof systems.

2. Before specifying a synthetic single-ply system as a remedial membrane over an existing bituminous membrane substrate, check for the synthetic material's compatibility with asphalt or coal tar pitch.

3. Require the following preparation of an existing built-up roof system as a substrate for a new synthetic membrane:
   (a). Remove *all* aggregate if new membrane is applied directly on top of the existing membrane. (If a protective insulation board is applied to the existing membrane as substrate for the new membrane, remove only *loose* aggregate.)
   (b). Repair defects in existing membrane (blisters, ridges, splits, fishmouths, and so on).

(c). Apply new materials (protective insulation or membrane) only after existing built-up substrate is clean, dry, smooth, and free of loose particles. (Consider application of a primer or a single-ply base sheet to improve existing substrate.)

(d). For loose-laid synthetic membranes, require smooth, firm substrate. (If protective insulation board is applied over existing roofs, it can be laid loose.)

For Sprayed Polyurethane Foam (Plus Fluid-Applied Coating)

Substrate Preparation (Replacement Roof after Tearoff Down to Existing Deck)

1. Remove moisture, grease, oil, loose particles, dust, and rust. Use special treatment—e.g., wire brush, commercial sandblasting, or chemical treatment—where required to prepare substrate for good adhesion.

2. Prime or seal all substrate surfaces to receive sprayed foam. Check primer manufacturer for specific primer to apply to substrate. Let primer dry and cure, per manufacturer's directions. Primer must be dry to touch before foam spraying starts. Avoid asphalt cutback primers because they contain high-boiling-point solvents that delay drying. Chlorinated rubber, which generally dries to a hard film within 30 min, is a generally suitable primer for structural decks (poured-in-place or precast, new or existing) and plywood. (Steel decks, with good factory coating, do not normally require a primer.)

Substrate Preparation (Reroofing over Existing Built-Up Membranes)

To qualify as a satisfactory substrate for application of sprayed foam insulation, an existing built-up membrane must be essentially dry and well-anchored. If investigation of the roof indicates only small, isolated areas with soft, wet insulation, remove these areas and replace with dry material. But do not risk placing a sprayed-foam system over wet insulation.

After investigating the roof system for entrapped moisture and adequate anchorage (taking test cuts, if necessary), proceed to prepare the existing membrane surface as follows:

1. Cut and patch all existing blisters, buckles, wrinkles, fishmouths, and soft-spot punctures. (Remove wet insulation from soft areas.)

2. Anchor loose sections to be sprayed.

3. Renail or otherwise secure all loose base and counter flashing, flanges, gravel stops, vent pipes, pitch pockets, scuppers.

4. Repair membrane splits, removing gravel and cleaning area 12 in. on each side of split. Embed 6-in.-wide woven glass fabric over split in plastic cement and trowel a top coating of plastic cement.

5. Vacuum and power broom all loose aggregate from surface.

6. Airblast the surface to remove dirt, dust, and other loose material. Repeat once or twice if required to ensure a clean substrate for foam application.

7. Apply a chlorinated rubber, if required by presence of residual dust remaining after airblasting.

Weather

Before starting foam-spraying operation, check weather report to ensure high probability of the following weather conditions:

- Low probability of rain (limit area planned for spraying to applicator's ability to protect substrate from wetting in event of sudden rain)

- Air temperature 40°F or above

- Maximum 80% RH

- Maximum 15-mph wind (or shielding precautions)

Take precautions to prevent overspray and coating of surrounding buildings, automobiles, shrubs, and so forth.

Spray Limits

Spray in minimum ½-in.-thick lifts. Allow maximum 4-h period between lifts. Discoloration—progressing from yellow to orange rust, sometimes accompanied by powder formation—may occur if foam surface remains exposed for a day or more. When such discoloration occurs, skin the foam surface and apply a new surface coating before applying either (1) another foam lift or (2) a membrane coating.

FOURTEEN
Wind Uplift

Dramatically rising wind-damage insurance claims over the past two decades attest to the rising incidence of wind-uplift failures. There are several explanations:

- Increased use of flexible steel decks

- Increased use of heavy equipment for roof-system application

- Advent of inferior anchorage techniques—notably cold adhesives—following the 1953 Livonia, Michigan, fire at a General Motors plant (which demonstrated the hazards of excessive bitumen under the insulation)

The hurricane belt, stretching from the Texas Gulf area eastward through Florida and up the Atlantic Coast to North Carolina, accounts for one-third of the nation's total wind-damage losses, although it contains less than one-tenth of the United States population. Gales and squalls occurring all over the country account for nearly 40 percent of total wind-damage losses. Tornadoes, although much more severe than hurricanes and occurring over a much larger area focused on Oklahoma, chiefly southern, southwestern and prairie states, account for about a quarter of total United States windstorm damage losses.[1]

Roofs can be practicably designed for anchorage that resists hurricane winds, which can attain speeds up to 130 mph. But for roofs on buildings caught in a tornado vortex, with wind speeds of 300 mph, there is no practicable design solution for the roof system. Buildings enveloped in a tornado vortex sometimes explode under the pressure differential created by the sudden atmospheric pressure drop.

Wind uplift stresses all the interfaces of the roof system—between framing members and deck, between deck and vapor retarder, between vapor retarder and insulation, and between insulation and membrane.

[1]Factory Mutual, *Loss Prevention Data Sheet 1-7*, June 1974, p. 1.

Roof-system components can be anchored via one of two basic techniques:

- Nailing, or other mechanical fastening
- Adhesives (hot-mopped asphalt or cold-applied emulsions or solutions)[1]

Faulty anchorage can both originate a wind blowoff and extend its area of damage. But as is discussed later in greater detail, most blowoffs start with a perimeter flashing failure and progress into the roof's interior. Perimeter anchorage is critical for both completed roofs in service and those under construction. Whenever flashing application lags days behind membrane application, temporary measures are required to protect the roof system from uplift pressures that can roll the membrane off the insulation or, more likely, pry the insulation off the deck.

The same organizations that set national fire-resistance standards, Factory Mutual Engineering Corporation and Association (FM) and Underwriters Laboratories, Inc. (UL) have established tests and ratings for wind resistance.

MECHANICS OF WIND-UPLIFT FORCES

To understand the nature of wind-uplift forces, first consider the positive pressures created by wind. Against a windward wall that stops its natural lateral movement, wind exerts a normal pressure dependent on its density and velocity squared (see Fig. 14-1). This positive pressure

[1]Factory Mutual, *Loss Prevention Data Sheet 1-49.*

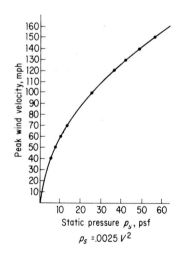

Static pressure p_s, psf

$p_s = .0025\,V^2$

Fig. 14-1 Wind-velocity-to-pressure conversion graph.

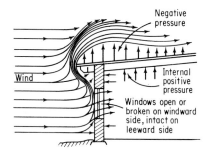

Fig. 14-2 Wind uplift is severe at the roof perimeter, especially at corners, where it may exceed the normal static pressure against the wall. (Factory Mutual Engineering Corporation.)

is known as static or stagnation pressure because it results from the wind's total loss of momentum—i.e., velocity loss from windspeed V to 0.

Uplift pressure is related to this basic positive stagnation reference pressure by negative coefficients, with the minus sign denoting uplift pressure. Thus an uplift pressure coefficient of -1.3 for a 90-mph wind velocity denotes an uplift pressure of -27 psf $(= -1.3 \times 0.00256 \times 90^2)$.

Wind blowing across a roof plane exerts uplift pressure because of a phenomenon known as the Bernoulli principle: The pressure exerted by a moving fluid drops with increased velocity and rises with decreased velocity to maximum pressure when the fluid is stationary. (As a simple demonstration of the Bernoulli principle, hold two vertically suspended sheets of paper in parallel planes about 2 in. apart and blow between them. The two sheets will come together, demonstrating the reduced pressure accompanying faster air flow.)

A building in a moving air mass behaves like an ungainly airfoil. A level roof is analogous to the upper surface of an airplane wing, which, largely through the negative pressure produced by the fast-flowing air, creates an uplift force that keeps the plane airborne. Unlike an airplane wing, however, a sharply angled building is not designed for smooth laminar flow; uplift pressures can vary widely because of irregular airflow patterns. At the windward roof edge, uplift forces can reach two times the static wind pressure. Over the remaining roof area, uplift may equal the static pressure. At the windward corners, the uplift pressure may be three times the static pressure.

Uplift forces depend basically on wind angle and slope. For a conventionally shaped building with rectangular plan, the worst wind direction is 45°, diagonally across the building. This 45° angle produces strong vortices along the roof's windward edges (see Fig. 14-2). A level roof also maximizes uplift force. As roof slope increases, suction over the windward roof plane decreases to zero at about 30° On steeper slopes, the wind exerts a positive pressure on windward roof planes and suction on leeward planes. Wind blowing perpendicular to a gable end wall produces suction over both sloping roof planes.

Less important factors affecting wind forces on roofs are height, size, and shape. Because wind velocity increases with elevation, the roofs of tall buildings are subjected to greater uplift forces than low buildings. And because wind gusts may engulf a small building but only partially engulf a large building, small buildings should be designed for larger wind pressures than large buildings.

Parapets introduce still another complication. Tall parapets can drastically reduce wind-uplift pressure—e.g., a 5-ft parapet can cut the uplift coefficient by a factor of nearly 3, according to wind-tunnel tests conducted at the University of Toronto.[1] A low parapet, however, in the 18-in. range, may increase uplift pressures. As a further complication, the parapet's effect depends on other building dimensions: height, length, and width.[2]

Hurricane wind-uplift forces may entail a loading reversal of two or even three times the normal gravity of roofs' live loads; they can rip away the structural deck. However, most wind-uplift failures occur within the built-up roofing system, where the weakest plane is normally at the interface between the deck and the insulation. The insulation-membrane interface is sometimes the weakest plane; wind may peel the membrane off the insulation, which remains in place on the deck. More often, however, failure occurs between deck and insulation. Sometimes both interfaces fail. Over a mechanically anchored substrate, the membrane base sheet may tear around the perimeters of the fastener heads.

LABORATORY WIND-UPLIFT TESTS

Both FM and UL conduct laboratory wind-uplift tests and publish approved roof-deck assemblies. FM publishes its approved list annually, in the *Approval Guide,* under "Building Materials and Construction." UL publishes a list of roof-deck assemblies rated Class 30, 60, or 90, depending on their successful resistance to 45-, 75-, or 105-psf total negative pressure in the UL uplift test. (See UL *Building Materials Directory,* "Roof Deck Constructions.")

The UL laboratory test features a sophisticated apparatus designed to simulate actual wind loading on a roof system. This UL apparatus

---

[1]W. A. Dalgliesh and W. R. Schriever, "Wind Pressures and Suctions on Roofs," *Can. Build. Dig.,* no. 68, May 1968, p. 68–4.

[2]H. J. Leutheusser, "The Effects of Wall Parapets on the Roof-Pressure-Coefficients of Block-Type and Cylindrical Structures," Department of Mechanical Engineering, University of Toronto, April 1964.

comprises three basic units: a lower chamber, where positive pressure is applied to the deck soffit; a 10 × 10 ft test specimen (deck-through-membrane); and an upper chamber, where negative pressure is applied to the roof membrane. Both upper and lower chambers contain glazed ports for viewing tests. A complete test cycle lasts 1 h and 20 min. The assembly passes if it withstands the full loading for its particular rating (i.e., Class 30, 60, or 90).[1]

FM tests steel decks only, in a steel pressure vessel with a hollow, rectangular prism 9 × 5 ft in plan and 2 in. deep. The test sample is clamped to the top angle of the vessel, with a rubber gasket controlling air leakage. Compressed air, entering through an opening in one side of the pressure vessel's frame, pumps up pressures in 15-psf increments from 30 to 75 psf, measured via a manometer connected to a ¼-in. opening on the opposite side. Each 15-psf pressure increment is held for 1 min. To qualify for an I-60 rating, the roof-deck assembly must withstand a minimum 60 psf for 1 min without showing evidence of any bond failure between components and no delamination of insulation.

The 60-psf value is derived as follows: Average wind velocity seldom exceeds 88 mph, which corresponds to 20-psf static pressure. Converted to uplift, that pressure produces about 30-psf uplift (−1.5 shape factor × 20 psf, static pressure). A safety factor of 2 raises the test load to 60 psf. Where winds exceed 88 mph, roofing components must meet I-90 uplift testing, as listed in the *Approval Guide* published annually by Factory Mutual.

ANCHORAGE TECHNIQUES

Drastic changes in the method of anchoring insulation boards to steel decks followed the Livonia fire in Michigan (see Chapter 15, "Fire Resistance"). It stimulated research into new cold-applied adhesives and mechanical fasteners to replace the hazardous, fire-feeding bitumens. Properly applied for such noncombustible insulations as foamglass, fiberglass, and perlite board, hot-mopped steep asphalt provides excellent uplift resistance, equal to about 2000 psf at normal ambient temperatures for a solid-mopped substrate. But it is highly dependent on weather conditions and careful, competent application. As a consequence, hot-mopped asphalt approximates all-or-nothing anchorage: extremely high resistance for properly applied asphalt, zero resistance for improperly applied asphalt allowed to cool before the insulation boards are placed.

Approved anchoring techniques satisfying both the fire-spread and wind-uplift requirements for Class I insulated steel-deck assemblies

[1]See UL 580, "Tests for Wind-Uplift Resistance of Roof Assemblies."

are listed in FM Engineering Associates' periodically updated *Loss Prevention Data Sheet 1-28.*

In 1978, there were six FM-approved mechanical anchors for Class I steel-deck assemblies, whereas in 1968 there were only two. One mallet-driven anchor has a series of locking tongues that spring open once they penetrate the steel deck. Depending on insulation thickness, one of the regularly spaced tongues bears against the deck soffit, resisting uplift. Other approved anchors have serrated shanks that key into the penetrated steel, providing friction resistance to uplift. Some anchors have large capped heads; others require plate washers. They are driven by hand, by pneumatic gun, or with a dry-wall-type screw gun. These fasteners provide from 200-lb up to 800-lb pullout resistance per fastener, depending largely on fastener size and deck thickness (the thicker the better). Note, however, that these fastener pullout values refer to fastener anchorage to the steel deck, not to insulation shear failure around the fastener head, the normal failure mode for mechanically fastened insulation boards.

Other mechanical fasteners supplement the stock of traditional large-headed roofing nails still used on wood plank decks. For low-density, preformed wood fiber, one manufacturer's fastener has a collapsible head driven against a cap located farther down the shank. As the head collapses, a locking foot rotates out from the shank to anchor the fastener. Another fastener, designed especially for gypsum decks, has a conical tubular shank that expands on driving to develop frictional resistance. And for lightweight insulating concrete, a hammered fastener has a slightly curved shank and a square, tubular cross section formed by two separate sheet metal pieces that separate under driving resistance to lock the fastener in place, like an expansion bolt (see Fig. 14-3).

For nails and other mechanical fasteners, the Asphalt Roofing Manufacturers' Association on Builtup Roofing recommends a minimum 40-lb pullout strength. One nailed plywood roof-deck assembly has qualified for UL Wind Uplift Class 90. UL construction requires 2-in.-wide masking tape over all plywood joints, to prevent air leakage from the underside. Nails are ring shank type, with 1-in.-square washers. Maximum spacing is 12 in., in any direction.

Improved wind-uplift resistance is another reason for completing insulation and membrane installation in one day's operation rather than delaying application of the whole membrane or its top plies. Uplift resistance of a nailed deck is greatly increased after mopping of the top plies. The mopped plies stiffen the whole membrane and thereby equalize stresses instead of permitting stress concentrations to start progressive failure through large areas.

Deck type	Fasteners for mechanically anchored decks		
Steel (lightgage)	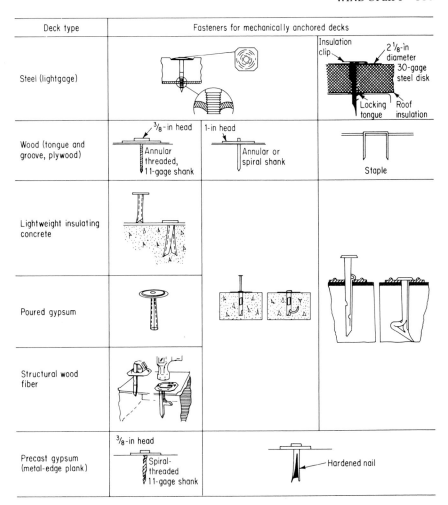		Insulation clip / 2 ⅛-in diameter / 30-gage steel disk / Locking tongue / Roof insulation
Wood (tongue and groove, plywood)	⅜-in head / Annular threaded, 11-gage shank	1-in head / Annular or spiral shank	Staple
Lightweight insulating concrete			
Poured gypsum			
Structural wood fiber			
Precast gypsum (metal-edge plank)	⅜-in head / Spiral-threaded 11-gage shank		Hardened nail

Fig. 14-3 Roofing nails and mechanical anchors come in assorted shapes and sizes, with varied anchorage devices—spiral or annular shanks, locking tongues and hollow shanks that expand when driven. Small-headed nails are driven through minimum 1-in.-diameter steel disk caps, minimum 30-gage thickness.

Determination of nail pullout strength is fairly straightforward for wood and preformed wood fiber (see Fig. 14-4). But for poured-in-place materials like gypsum or insulating concrete, it is more complex. Pullout strength of fasteners anchored in insulating concrete depends chiefly on the mix ratio, ultimate 28-day strength, and the deck's age when the fastener is pulled.

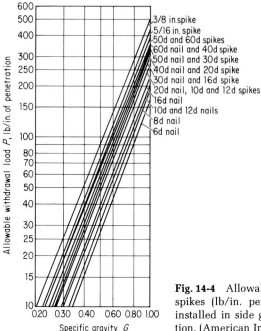

Fig. 14-4 Allowable withdrawal levels for nails and spikes (lb/in. penetration) for one nail or spike installed in side grain under normal loading duration. (American Institute of Timber Construction.)

ANCHORAGE FAILURE MODES

Although not without their own unique problems, fasteners are the most reliable of the three anchoring techniques. Under many field conditions, adhesive anchorage to steel decks with either hot-mopped asphalt or cold adhesives is highly unreliable. Field-measured temperature loss in hot, steep asphalt mopped directly onto steel deck in cold weather showed asphalt temperature dropping below 250°F within 6 s or less.[1] Such rapid dissipation of heat by the cold, heat-conductive deck makes it extremely difficult to get a good bond in cold weather because the asphalt may congeal before the insulation board is applied. Moreover, there is abundant evidence, in the sloppy adhesive lines on steel decks bared after the cutting of roof test samples, that hot asphalt is often poorly applied even in good weather, with consequently unreliable anchorage.

[1]R. M. Dupuis, J. W. Lee, and J. E. Johnson, "Field Measurements of Asphalt Temperatures During Cold Weather Construction of BUR Systems," *Proc. Symp. Roofing Technol.*, NBS-NRCA, September 1977, p. 279.

Cold adhesives cause so many problems that three major manufacturers have withdrawn approval for them for bonded roofs. Construction timing is less critical with cold adhesives than with hot asphalt. Slow evaporation of the volatiles in cold adhesives allows plenty of time to tamp insulation boards in place. Traffic deflections of the steel deck can break the bond between deck and insulation, and irregular deck surfaces can prevent proper contact between the adhesive and the components to be bonded. Moreover, their slow attainment of full adhesive strength makes cold adhesives vulnerable during construction, when traffic loads are at their heaviest. These loads may deflect the deck and break the bond between deck and insulation. (FM allows 4 days for a cold adhesive to attain half strength—i.e., to resist a 30-psf uplift—and 14 days to resist full 60-psf uplift.) Irregular deck surfaces can prevent proper contact and spreading of cold-adhesive ribbons. And careless use of a cold adhesive aged beyond its permissible shelf life can result in deficient bond strength. (A cold adhesive aged beyond its shelf life loses volatiles required to maintain its fluidity, leaving the material too viscous to attain proper contact with the bonded components and cure properly.)

Fastener problems begin with faulty driving—notably in driving the fastener over a deck flute where it fails to engage the flange. A thin deck, less than 22 gage, may also fail to engage the fastener's serrated shank, and a thick deck—18 gage or more—can break some fasteners. Overdriving mechanical fasteners can cup their disks, with consequent reduction in pullout strength. Locating fasteners too close to board edges—within 6 in. as a minimum recommended distance—can also weaken anchorage. And on fasteners with loose caps that can move downward on the shank, the shank can puncture the membrane unless the mechanical anchorage is for the bottom of a two-layered board insulation system.

As noted, the most common failure mode for mechanically fastened insulation boards is shear failure of the insulation around the fastener head rather than fastener pullout from the steel deck. Tests on mechanically fastened perlite boards' pullout resistance by R. L. Fricklas, Director, Roofing Industry Educational Institute, indicate the following tentative conclusions:

- Increasing insulation-board thickness increases shear pullout resistance by an exponent of roughly 2; that is, 1½-in. perlite board has nearly four times the shear pullout resistance of ¾-in. board.

- One layer of mechanically anchored, thick insulation has roughly twice the shear pullout resistance of two layers of equal total thickness, with the bottom layer nailed and the top layer hot-mopped.

- Mopping a coated felt to the insulation before installing the mechanical anchors significantly increases pullout strength.

CAUSES OF ROOF BLOWOFFS

Most roof blowoffs, about 80 percent, according to FM data, start with failure at the perimeter fascia-gravel stop (see Fig. 14-5). The metal fascia strip is bent outward and upward, exposing more of its area to the wind. Wind forces then pry off the fascia strip and/or the wood nailer-cant assembly. Failure of the cant-nailer assembly opens the roof assembly to peeling and suction forces, which can roll back the membrane sandwich. Positive pressure from wind entering steel-deck ribs supplements uplift forces at the roof surface. (See Fig. 14-6 for faulty fascia–gravel-stop details that led to wind-uplift failures investigated by FM engineers.)

Wind-uplift failure can also start in the membrane area as well as at edge flashings. In what is sometimes called "ballooning," negative pressures in high winds can lift up a poorly anchored membrane, or membrane-insulation unit, and gusting winds can set the membrane billowing in an undulating wave action at amplitudes observed and photographed at 1 ft or more above the deck. These wind-induced oscillations produce dynamic stresses much higher than corresponding static stresses. Because of this dynamic magnification of wind forces in

Fig. 14-5   Wind forces ripped this metal gravel stop from its faulty connection to the insulation—an expensive lesson on the need for anchoring to wood nailers around the perimeter of a nonnailable deck. (GAF Corporation.)

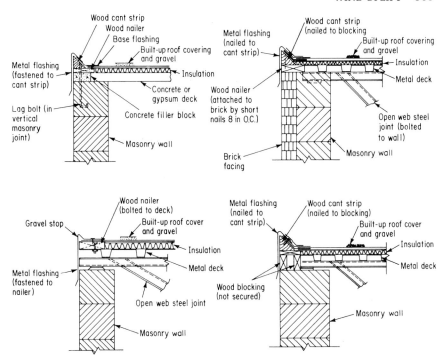

Fig. 14-6 These edge-fascia flashing details were found most vulnerable to wind-uplift failure by Factory Mutual engineers. Note that they all omit a continuous cleat, which provides bottom anchorage. (Factory Mutual Engineering Corporation.)

a vibrating system, suspension bridges have failed at relatively low wind speeds. Since this phenomenon applies also to roof membrane wind-induced oscillations, roof blowoffs may apparently occur at relatively low wind-uplift pressures, multiplied by the accidental resonance or near resonance between the wind impulses over the roof and the membrane's natural frequency. Poor anchorage at just a few locations may allow oscillations of increasing amplitude that could progressively break the deck-insulation bond in adjacent areas and finally result in a major blowoff. At the least, this phenomenon deserves further study and research.

The common practice of leaving roof coverings incomplete for several feet back from the roof edge, with the edge flashing and closing membrane installation performed later, can cause a wind-uplift blowoff during construction. This practice lets the wind get under the membrane edge and peel back the covering. When it is unavoidable, the incomplete membrane edge should be temporarily lined with weights until it is spliced.

Other factors in wind-uplift failure are

- Inadequate design

- Faulty practice in installing cold-applied adhesives

- Excessive longitudinal or transverse deflection (dishing) of lightgage steel decks with cold-applied adhesives

- Cold-weather application of hot-mopped bituminous adhesive

- Faulty application, or spacing, of mechanical fasteners

(See "Anchorage Techniques" section for further discussion.)

The wind-uplift failure in Fig. 14-7 shows a grossly faulty adhesive pattern, across instead of parallel to the deck flutes. The adhesive was applied in a random, overspaced pattern instead of the specified 6-in. spacing.

Because they are so flexible, the thinner metal roof-deck sections— 24 or 26 gage—can cause serious trouble for cold-applied adhesives. Transverse deflection (dishing) of light, springy decks can break contact between the steel surface and the rigid insulation above, thus precluding good bond. Some roofers, confronted with a specification calling for cold adhesive, have recommended and obtained authorization to substitute mechanical fasteners. To reduce the risk of excessive steel-deck deflection, FM Engineering Associates recommend the use of a minimum 22-gage steel deck.

Fig. 14-7 Random adhesive pattern on this wind-peeled steel deck reveals sloppy adhesive application. Adhesive ribbons should parallel the deck flutes at maximum 6-in. spacing. (Factory Mutual Engineering Corporation.)

BUILDING CODE REQUIREMENTS

The widely used American National Standards Institute (ANSI) *Building Code Requirements for Minimum Design Loads in Buildings and Other Structures* (ANSI A58.1-1972) contains wind-uplift design data that relate more to structures than to the roof system. Generally, the ANSI requirements, which are the basis for three of the four national model building codes' wind-load provisions, are embodied in the FM wind-uplift design procedure, explained in detail in the following section. This section, accordingly, concludes merely with a reminder to the roof designer to satisfy any local code requirements. By following the FM design procedure, the designer will usually, if not always, fulfill this responsibility.

Note, however, that ANSI A58.1-1972 was undergoing extensive revision while this manual was in publication. In the draft revision, wind-pressure coefficients are substantially reduced. Roof size becomes a factor, with the pressure coefficients higher on smaller roofs. (The latest available ANSI standards should always be consulted.)

COMPUTING WIND UPLIFT

Many architects responsible for roof design do not understand the principles of wind-uplift design, and some are even unaware of FM and UL requirements. FM's recent research into this subject has increased design wind loadings to surprisingly large values. In some parts of the United States (notably the hurricane belt), these recently published wind-design loadings will greatly exceed the gravity loadings of light roof assemblies. Because it requires only a single occurrence of these large wind loadings to rip off a portion of the roof assembly, the costly hazards of inadequate design are obvious.

FM's *Loss Prevention Data Sheet 1-7*, issued June 1974 (under revision in 1981), sets forth a design procedure for determining wind-uplift forces on a roof. Following this procedure should generally ensure safe wind design. But there are exceptions—e.g., for a building in an unusual location where the surrounding topography, or even the presence of surrounding structures, drastically alters normal wind conditions. In such instances the designer may require wind-tunnel testing or some other extraordinary measures beyond the scope of this manual. FM's procedure for determining wind-uplift pressure recognizes the following major variables:

- Geographical location
- Local topography
- Building height

- Building size and shape

- Roof slope

- Percentage of wall openings

Geographical location establishes basic wind speed to be used in wind-uplift design. In the continental United States this wind speed varies from 130 mph in the Florida Keys and 120 mph in Miami to 70 mph for Los Angeles and San Diego, and even as low as 60 mph in Fairbanks, Alaska. Figure 14-8 has this basic design for wind speed; the figure shows wind isotachs (i.e., curves of equal maximum wind speed) for 100-year mean recurrence interval.

Topography and *building height* affect wind pressure in obvious ways made explicit by the wind-speed formula:

$$V_h = V_{30} \left[\frac{h}{30} \right]^{1/n} \qquad (14\text{-}1)$$

where V_{30} = wind speed at 30-ft above ground
h = height (ft)
V_h = windspeed at height h
n = ground roughness factor

As Eq. (14-1) indicates, wind speed increases with increasing elevation and decreases with increasing ground roughness factor (an index of the obstacles to wind flow).

FM considers three types of ground roughness:

A $(n = 3)$ = rugged or mountainous terrain; central business districts of cities with many tall buildings
B $(n = 4.5)$ = suburban areas, towns, city outskirts, wooded areas, wood terrain
C $(n = 7)$ = Flat, open country; flat, open coastal plains, grassland

(FM recommends elimination of ground roughness A from consideration for FM-insured locations. Another ground-roughness type, D, will appear in FM's revised *Loss Prevention Data Sheet 1-7.*)

Velocity pressure is calculated from Eq. (14-2):

$$p_h = 0.00256 V_h^2 \times G_f \qquad (14\text{-}2)$$

where p_h = static pressure on vertical surface
G_f = gust factor

Gust factor decreases with increasing ground roughness and elevation. It equals 1.59 at 30 ft for ground roughness B, 1.32 at 30 ft for ground roughness C. It approaches 1 at elevations 1500 ft and higher. (These gust factors will be reduced in the revised *Loss Prevention Data Sheet 1-7*.)

Gust factor is included because weather station instruments usually measure the "fastest-mile" of wind, not peak gust velocity. The fastest-mile gives the average wind speed over the time required for the wind to travel 1 mi; thus the time for a fastest-mile wind speed of 90 mph is 40 s. Gust factor also varies with building size.

Roof slope is one of three parameters (along with building size and shape) that determine pressure coefficient c_p, which relates velocity pressure to the pressure actually exerted at different parts of the structure. It depends on building shape and size and roof slope. A minus sign signifies vacuum pressure, or uplift.

Percentage of wall openings is a fourth parameter affecting pressure coefficient c_p. In buildings with large openings—e.g., an aircraft hangar or a warehouse with a long loading dock—wind entry can produce positive pressure at the underside of the roof assembly in addition to the negative topside pressure, all contributing to the net uplift. FM's design technique recognizes this factor.

Illustrative Example

The following example problem demonstrates the FM procedure for computing wind-uplift pressures on a roof. Determine the wind uplift on the roof of a level-roofed, 50-ft-high building, 100 × 150 ft in plan, in Des Moines, Iowa.

1. From the isotach map of Fig. 14-8, pick a design wind speed equal to 90 mph (100-year mean recurrence interval).

2. Choosing ground roughness B (for outskirts of cities) from Table 14-1, read a p_h (velocity pressure) value of 21 psf.

3. Entering Table 14-2, pick c_p (pressure coefficient) values for three conditions on the roof: (a) perimeter, (b) corners, (c) interior.
   (a). For flat roofs, along a 4-ft perimeter strip, $c_p = -2.4$.
   (b). 4-ft-square corner $c_p = -5.0$.
   (c). With wall openings less than 0.2 wall area, interior $c_p = -1.3$.

4. With these c_p values, compute uplift as follows:
   (a). For a 4-ft perimeter strip, wind uplift $= -2.4 \times 21 = -50$ psf.
   (b). For a 4-ft square at each building corner, wind uplift $= -5.0 \times 21 = -105$ psf.

Table 14-1 Velocity Pressure p_h (psf) for Ground Roughness B* (Left) and Ground Roughness C† (Right)

Height above ground, ft	Wind isotach (from Map Fig. 14-8), mph						
	70	80	90	100	110	120	130
0– 30	10	13	18	22	26		
30– 50	12	17	21	26	32		
50– 100	15	20	25	31	38		
100– 200	19	25	31	39	47		
200– 300	22	29	37	45	55		

Height above ground, ft	Wind isotach (from Map Fig. 14-8), mph						
	70	80	90	100	110	120	130
0– 30	18	22	28	34	42	50	58
30– 50	20	25	32	40	47	56	65
50– 100	22	28	36	44	53	63	75
100– 200	25	32	42	51	62	74	86
200– 300	28	36	45	56	67	80	94

300– 400	24	32	40	49	60
400– 500	26	34	43	53	64
500– 600	28	36	46	56	68
600– 700	29	38	48	59	71
700– 800	31	40	51	62	75
800– 900	32	42	53	65	80
900–1000	34	44	55	68	83

(Zone markers: "2" between columns near 32/40 and "3" near 59/71)

300– 400	30	38	48	60	72	86	100
400– 500	31	40	51	63	76	90	105
500– 600	32	42	53	65	79	93	109
600– 700	33	44	55	68	82	97	114
700– 800	34	45	56	70	84	100	117
800– 900	35	46	58	73	86	102	120
900–1000	36	47	59	75	89	104	123

(Zone markers: "2" near 42, "3" near 65/79)

*Ground roughness B describes city suburban areas, towns, wooded and rolling terrain.

†Ground roughness C, for flat, open country, flat, open coastal belts and grassland, is associated with higher-level wind velocities and thus higher pressure.

Notes: In Zone 1, roofing components must pass the Factory Mutual 60-psf wind uplift test (Data Sheet 1-28). In Zone 2, roofing components must pass the Factory Mutual 90-psf wind uplift test (Data Sheet 1-28). In Zone 3, steel deck is not recommended.

Table 14-2   Wind-Pressure Coefficients for Building Roof Surfaces up to 10° Slope

Roofs, flat, 0.1d perimeter strip along eave* including flashing (where d = least dimension)	Upward	−2.4 (1)
Roofs, gabled, 0.1d perimeter strip at eave, including flashing, and at ridge	Upward	
0°–30°		−2.4 (1)
30°–40°		−1.7
Roofs, flat, corners (0.1d)² including flashing, wind on diagonal	Upward	−5.0 (1)
Roofs, gabled, corners (0.1d)² including flashing (where 0 = angle of slope)	Upward	
0°–30°		0.10–5.0 (1)
30°–40°		−2.0
Roofs, flat or 0°–10° slope; exterior wall openings ≥ 0.2 wall area (area within perimeter strips)	Upward	−1.3 (1)
Roofs, flat or 0–10° slope; exterior wall openings > 0.2 wall area (area within perimeter strips)	Upward	−1.5 (1)
Roofs, gabled slope > 10° (area within perimeter and ridge strips)	Upward or downward	See Table 9 (FM Data Sheet 1-7)

*4-ft-wide roof perimeter and ridge strips used instead of 0.1d.

Source: Factory Mutual, *Loss Prevention Data Sheet 1-7*, June 1974, "Wind Forces on Buildings and Other Structures."

(c). For the roof interior (i.e., inside the perimeter strip), wind uplift = −1.3 × 21 = −27 psf. (Use −30 psf, the minimum design load acceptable to FM.)

With these figures we now choose our anchorage technique. Assume that because of cold-weather construction, which threatens the bonding of hot asphalt, we decide to use mechanical anchors throughout the entire roof-deck area. Assume a design withdrawal strength of 200 lb per anchor. (Withdrawal strength can be determined by test or from manufacturer's literature.)

Multiplying each psf uplift factor by 2 (for safety factor), we get (1) −100 psf for 4-ft perimeter strip, (2) −210 psf for 4-ft-square corner areas; (3) −60 psf (FM's minimum recommended uplift value) for interior area.

Accordingly, required fastener spacing is as follows:

1. 4-ft perimeter strip: one fastener per 2 ft², to resist 100-psf uplift

2. 4-ft² square corners: one fastener per ft², to resist 210-psf uplift

3. Interior roof area: three fasteners per board, to resist 60-psf uplift

For spacing and pattern of mechanical anchors, see the latest edition of FM *Loss Prevention Data Sheet 1-28*. Despite design requirement of only three fasteners per board for 2 × 4 ft insulation boards, specify minimum four fasteners per board. But note that FM may require six fasteners for an approved 2 × 4 ft board, depending on the type of fastener. For this information, consult the latest edition of the annually published FM *Approval Guide*. Also see Fig. 14-9 for fastener patterns for 2 × 4 and 3 × 4 ft insulation boards.

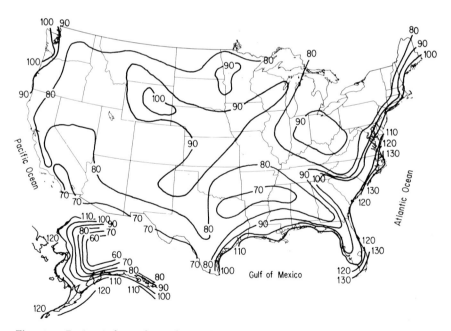

Fig. 14-8 Basic wind speed in miles per hour. Annual extreme-mile 30 ft above ground, 100-year mean recurrence interval. (Factory Mutual Engineering Corporation.)

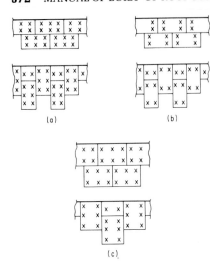

Fig. 14-9 Factory Mutual-approved fastener patterns for (a) 2 × 4 ft (0.61 × 1.22 m) insulation boards, six fasteners per board; (b) 2 × 4-ft boards, four fasteners per board; and (c) 3 × 4-ft (0.91 × 1.22 m) insulation boards. (Factory Mutual Engineering Corp.)

WIND-UPLIFT MECHANICS ON BALLASTED ROOFS

Ballasted, loose-laid roof systems are a radically different wind-uplift problem from adhered or anchored roof systems. The normal 10- to 15-psf aggregate ballast weight on loose-laid systems falls far short of the 30- to 60-psf wind-uplift design pressures often required for adhered roof systems. Judged by field experience, however, ballasted, loose-laid systems resist wind-uplift pressure by an entirely different mechanism than adhered systems. In an adhered system, an area of lost adhesion no longer contributes to the roof's uplift resistance. A small area of lost adhesion can expand rapidly because of the low peeling resistance of the remaining adhered sections. An adhered system usually suffers a blowoff from local failure, like the tensile failure of the weakest link in a chain. However, in a ballasted system any wind-lifted area shifts its ballast to an adjacent area, whose uplift resistance is consequently increased.

GRAVEL BLOWOFF AND SCOUR

The problem of loose gravel aggregate scour and blowoff has grown with the rising popularity of loose-laid membranes for both conventional and protected membrane (i.e., "upside-down") roofs. Blowing aggregate threatens surrounding buildings' windows, which have been broken by these flying refugees from neighboring roofs. Added to the more obvious hazards to property and persons from wind-launched gravel missiles is the direct exposure of protected membrane roof (PMR) insulation or a conventional system's membrane to sunlight, traffic punctures, and other forces against which aggregate surfacing provides protection.

Two Canadian researchers, Prof. R. J. Kind of Ottawa's Carleton University and R. L. Wardlaw, of the National Research Council, published a design procedure for preventing loose aggregate blowoff and scour.[1]

Basic design strategy against aggregate blowoff and scour comprises one or more of the following techniques:

- Increasing stone size

- Increasing parapet height

- Substituting concrete pavers for gravel

Increased stone size increases the aerodynamic force required to move or lift the loose stones. Wind exerts an aerodynamic drag force on an exposed stone. At a certain critical value, it topples the stone away from its nested position. Through a complex interaction of forces, upward deflection of a rolling stone combined with increased aerodynamic wind force, faster wind speed can blow some stones off the roof.

Critical wind speed increases with particle size because the weight of the particle varies with its diameter cubed, whereas the frontal area exposed to wind force increases with the particle's diameter squared. Aerodynamic force rises proportionately with wind velocity squared (i.e., a doubling of wind velocity quadruples aerodynamic force). It follows that the ratio of aerodynamic force to gravitational force is proportional to $V^2 d^2 / d^3 = V^2 / d$ (see Fig. 14-10). Critical wind velocity is thus expressed by Eq. (14-3):

$$V_c = k \sqrt{d} \qquad (14\text{-}3)$$

where V_c = critical wind velocity
d = stone aggregate size
k = proportionality constant

Parapet height is a major design variable. Parapets shield the stones from the wind; increased parapet height reduces aerodynamic forces on the stones. When the wind angle is roughly 45° to a rectangular building's axes (the worst angle for gravel blowoff), a parapet elevates the cores of the vortices, thus reducing rooftop suction (see Fig. 14-11).

[1] R. J. Kind and R. L. Wardlaw, *Design of Rooftops Against Gravel Blowoff*, National Research Council of Canada, NRC no. 15544, September 1976.

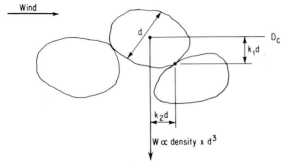

Fig. 14-10   Critical wind velocities, at which loose aggregate unit rolls over or flies off roof, vary with the square root of the unit diameters. Sand grains would obviously blow off easier than boulders, and the tendency to do so is a continuum between these two extremes (and even beyond them). The aerodynamic lifting force is a function of wind velocity squared, V^2, and d^2. Resisting force is a function of d^3. Thus

$$\frac{\text{Aerodynamic force}}{\text{Gravitational force}} = \frac{V^2 d^2}{d^3} = \frac{V^2}{d}$$
$$V^2 \propto d$$
$$V_c = k\sqrt{d}$$

where V_c = critical wind velocity. Note: k_1 and k_2 are proportionality constants depending on stone shape. (R. L. Wardlaw, National Research Council of Canada.)

Concrete pavers substituted for aggregate in the critical roof corners can prevent scour and blowoff in these most vulnerable areas. In plan, these corner areas are either square- or L-shaped, designed to match scour patterns.

Kind and Wardlaw derived their design data from wind-tunnel testing of scale-model buildings, with model dimensions, including gravel size, one-tenth full scale. Models and prototype buildings are related by the so-called Froude number ($V\sqrt{hg}$; V = wind velocity, h = building height, g = gravitational acceleration). Because g is constant, prototype and model wind speeds V_p and V_m are related as in Eq. (14-3):

$$V_p = \sqrt{\frac{h_p}{h_m}} \; V_m = \sqrt{10} \; V_m = 3.16 V_m \qquad (14\text{-}4)$$

Thus a 25-mph windspeed in the wind tunnel corresponds to a full-scale windspeed of $3.16 \times 25 = 79$ mph.

In the Kind-Wardlaw design procedure, you first select a design wind speed for 30-, 50-, or 100-year return period; pick an exposure-gust factor off a graph; and, finally, compute two critical wind speeds:

1. Gust speed at which scouring would continue if wind speed persisted

2. Gust speed at which several stones (say, six or more) fly off the roof (over the upstream gravel stop or parapet)

In an illustrative example, for a 20-ft high, 100 × 150-ft building in Washington, D.C., with design wind speed = 79 mph (50-year return period), you could raise the critical windspeeds for scour and blowoff to the required 79-mph value by one of two methods:

- Increasing gravel size from ¾ to 1 in., increasing parapet height to 2 ft, and omitting paver blocks

- Installing 6-ft-square paver block arrays at corners, maintaining ¾-in. aggregate size and gravel-stop edge detail (i.e., no parapet) elsewhere on the roof

A further variable in the Kind-Wardlaw design procedure is building type: (1) low; (2) medium; and (3) highrise, for which they offer appropriate criteria on height, length, and width ratios.

In PMR roof assemblies (see Chapter 10), permeable fabric sandwiched between the light plastic foam insulation and the gravel ballast reduces wind-scouring damage and limits the gravel and insulation blowoff at any given wind velocity. With increasing wind velocity, the impermeable fabric induces three staged failure modes:

- In F1, gravel scour at upwind corners leaves most of the displaced gravel in deep piles near the perimeter parapet (although some leaves the roof).

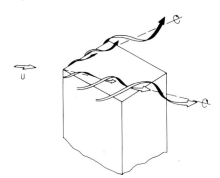

Fig. 14-11 Diagonal wind sweeping across roof creates worst uplift and scouring condition at windward corner. (R. L. Wardlaw, National Research Council of Canada.)

- In F2, at a higher velocity, the plastic foam insulation boards in the bare gravel-scoured portions of the roof oscillate vertically, but remain trapped under the fabric. Increased wind velocity pulls the fabric out from under the accumulated gravel piles and the now unrestrained insulation boards blow away.

- In F3, at still higher wind velocity, the fabric peels back farther until a new stabilized condition is reached, with gravel accumulated in the fold between lifted and unlifted fabric.[1]

FIELD WIND-UPLIFT TESTS

Field wind-uplift tests are generally applicable only to roof assemblies with adhered, not mechanically anchored, components. Because of the small field test area, 25 ft² at most, the loading of mechanical anchors will almost inevitably be eccentric, with consequent distortion of the test result.

For new steel-deck roof assemblies, or those of doubtful wind-uplift resistance—e.g., unpeeled areas in a roof system that has already experienced a blowoff in another area—FM recommends one of two field uplift tests:

- The Negative Pressure Test
- The Pull Test

As its chief advantage, the recently published Negative Pressure Test method is nondestructive (unless, of course, the roof assembly fails). The Negative Pressure Test apparatus (shown in Fig. 14-12) comprises three basic components:

- A 5-ft-square, transparent acrylic plastic dome, fabricated in four segments and field-bolted through flanges sealed with rubber gaskets and sealed at the roof surface with a flexible PVC foam strip

- An electrically powered vacuum pump, mounted on the vacuum chamber dome

- A manometer indicating vacuum pressure created within the chamber

[1]R. J. Kind and R. L. Wardlaw, "Model Studies of the Wind Resistance of Two Loose-Laid Roof-Insulation Systems," *Nat. Res. Counc. Can., Lab. Tech. Rep., LTR-LA-234,* May 1979.

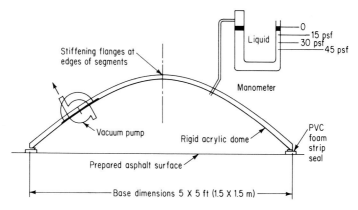

Fig. 14-12   Vacuum wind-uplift test apparatus can test uplift resistance up to 45 psf. (Factory Mutual Engineering Corporation.)

Choose a test area between supporting beams or joists, if practicable, at least 10 ft from the nailed perimeter. Sweep loose aggregate away from a 12-in.-wide perimeter strip and level this strip with poured hot asphalt, ½ in. maximum thickness, to create a smooth surface for sealing the vacuum chamber's gasketed bearing flange. A water film under the foam gasket can facilitate sealing.

A calculation to determine FM-required wind-uplift resistance in psf must precede the uplift test (see FM *Loss Prevention Data Sheet 1-28*, tables 1 and 2). For example, if Class I-60 rating is required, the test should proceed as follows: Run the vacuum pressure to 15 psf, indicated by the manometer in equivalent water pressure (= 5.2 psf/in.), and hold for 1 min. Then raise the pressure in 7.5-psf increments, hold for 1 min at each pressure, until the vacuum pressure reaches 30 psf, and hold for 1 min at that pressure. If I-90 rating is required, continue for two more 7.5-psf increments, holding each for 1 min, until 45-psf vacuum pressure (= 8.7 in. water) is attained. (For further details, see FM *Loss Prevention Data Sheet 1-52.*)

Failure is evidenced by sudden upward ballooning of the membrane. Moreover, the roof deck will normally deflect upward under the vacuum pressure. For an adhered roof system, upward deck deflection should not exceed 1/240 of the span.

Test areas differ for existing and new roofs. Test existing roofs wherever inadequate wind-uplift resistance is suspected. Test new roofs at the rate of one test per 10,000-ft² area—i.e., conduct seven tests on a roof of 65,000-ft² total area.

The older, destructive FM Pull Test apparatus comprises a tripod supporting a 200-lb capacity spring scale connected with a hand-chain hoist, or block and tackle, to a plywood stiffener bonded to the membrane. A 2-in. strip is cut around the 2-ft-square test section down to the steel deck, to isolate the test section from the rest of the roof. Uplift force is applied by a ratchet lever in the same prescribed increments used in the Vacuum Pressure Test. However, an adjustment is needed to account for the 15-lb weight of the plywood stiffener. Thus total force in an I-90 rating requires a total 195-lb force, for net 180-lb uplift (45-psf) over the 4-ft² test segment.

Smooth-surfaced roofs require only brooming of the test area, but aggregate-surfaced roofs must be heated with a torch to melt the asphalt, to permit shovel removal of the aggregate. This operation involves care, to avoid disturbing the adhesion between insulation and membrane. An indentation cut into the center of the test section, to match a centrally located ⅝-in.-diameter hole drilled in the 1½-in.-thick, 2-ft-square plywood stiffener, accommodates a washer-and-but assembly connected to a ½-in.-diameter eyebolt.[1]

REPAIR PROCEDURES AFTER WIND-UPLIFT FAILURE

Wind-uplift damage occurs in one or more of the following modes:

- Loosened (or ruptured) edge flashings and gravel stops
- Membrane and insulation peeled from deck
- Membrane peeled from insulation
- Membrane and insulation broken loose from deck but finally dropped back in place

The last condition obviously constitutes a grave threat of future peeling. Unless eyewitnesses have seen the ballooning (sometimes observed as high as 3 ft above the deck), it may totally escape notice. If you have any suspicion that ballooning (or "floating") has occurred, you should cut samples to examine the deck-insulation bond.

Replace damaged flashing and gravel stops with new metal. For flashing that remains in place, additional nailing (if practicable) may suffice. If such repair is impracticable, consider replacing all flashing to conform with FM requirements in *Loss Prevention Data Sheet 1-49*.

Replace peeled insulation boards with new insulation and anchor it to the deck with asphalt or mechanical fasteners, *not* cold adhesives. Be careful to examine insulation and reject damaged boards with crushed corners or edges or fracture lines resulting from the wind roll-off. Exercise the same care required for approval of new materials. Limit hot asphalt application to temperatures above 50°F.

---

[1]See FM *Loss Prevention Data Sheet 1-52* for more details on these FM field uplift tests.

A Class I roof with asphalt remaining on its deck is a special problem. If not removed, the old asphalt plus new asphalt adhesive may exceed the allowable weight for Class I rating. If it is impracticable to remove the old asphalt, anchor the insulation with FM-approved mechanical fasteners. (Note, however, that for a sprinklered building, retention of Class I rating may be unimportant.) Before readhering the insulation, remove any delaminated insulation layers stuck to the deck surface.

When the membrane is peeled from insulation that remains in place, test the insulation for solid anchorage before replacing the peeled membrane with a new membrane. Apply a minimum, three-ply, built-up membrane solidly mopped to the insulation, observing all other rules of good membrane application. (Solid mopping does not apply where the underlying insulation is poured lightweight concrete or gypsum.)

For ballooned insulation broken loose from the deck, first determine the extent of the affected area. Cut 12-in.-square test samples through the insulation, if necessary. Anchor the membrane and insulation with FM-approved mechanical fasteners. If the membrane has three or more plies, you can omit disk washers with the fasteners.

With serrated, nail-type fasteners, prepunch holes through the membrane to the insulation. (One manufacturer has a special prepunching hammer.) If you drive serrated fasteners through a bituminous membrane, bitumen will adhere to the serrations and reduce its holding strength at its engagement with the steel deck.

Apply a minimum two shingled plies (solid mopping of felt with a solid mopping of felt surfacing) above the existing membrane. Trowel asphalt-based plastic roofer's cement over the mechanical fastener heads.

ALERTS

General

1. Before specifying a roof assembly, consult with local building official and insurance agent or rating bureau to make sure that your proposed design satisfies local wind requirements.

2. Before specifying an anchoring technique:
   (a). Request data on wind-uplift resistance from manufacturer.
   (b). Check UL or FM listings for wind-uplift approval of proprietary products.

Design

1. Beware of substituting even a single component within a roof assembly rated for wind uplift. Such changes require restudy of the resulting new roof

assembly. Wind-uplift ratings apply to an entire roof assembly, not to its individual components.

2. Specify mechanical fastening instead of cold-applied or hot-mopped adhesive whenever practicable, for any type of nailable deck.

3. Specify minimum 22-gage thickness for steel decks.

4. Specify minimum 18-gage for metal gravel stops or fascia strips.

5. Specify a continuous cleat or hook strip for stabilizing the bottom of fascia strips.

Field

1. Require weighting of all membrane edges left incomplete before splicing with other sections of membrane.

2. Exercise extreme caution about high winds and/or low temperature when hot-mopped adhesives are specified to anchor insulation or other roof components to a substrate.

FIFTEEN

Fire Resistance

The roof designer must often satisfy two fire-resistive requirements: (1) building code criteria and (2) insurance company requirements for qualifying the building for fire coverage. Building code requirements are of course mandatory, and from a practical viewpoint, so are insurance company requirements. Use of a combustible instead of a fire-resistive roof assembly will usually raise the owner's total building cost by making a costly sprinkler system mandatory.

The most important standard-setting organizations for fire and wind-uplift resistance are the Underwriters' Laboratories, Inc. (UL) of Northbrook, Illinois, and Factory Mutual Research Corporation (FM), of Norwood, Massachusetts. Both organizations classify roof assemblies for fire and wind-uplift resistance for many of the nation's insurance companies. UL and FM maintain laboratories for testing and listing manufacturers' building products that satisfy their standards, and insurance companies use these construction listings to develop specifications and recommendations for their insureds. UL standards are also the basis for building code requirements for fire resistance of built-up roof assemblies.

NATURE OF FIRE HAZARDS

Fire hazards that concern building code officials and insurance companies are broadly classified as follows:

- *External, above-deck fire exposure* from flying brands or burning debris blown over from neighboring buildings on fire

- *Internal, below-deck fire exposure* from interior inventory or equipment fires

Criteria for testing and certifying these two classes of fire hazard fall into three categories. A set of UL criteria and tests is generally the basis for external fire resistance. Criteria for internal fire resistance comprise the following:

- Limitation of flame spread along the roof-assembly soffit

- Time-temperature rating, per ASTM E119 furnace test

Because it is simpler, the less serious external fire hazard, with its evaluating tests and standards, is discussed before the more serious internal fire hazard.

EXTERNAL FIRE RESISTANCE

External fire presents the risk of fire spreading from a burning, nearby building. The membrane's chief fire-resistive function is to reduce this risk. Membranes are accordingly tested and rated for their resistance to external fire. A membrane should not spread flame rapidly, produce flying bands endangering adjacent buildings, or permit ignition of its supporting roof deck.

According to the industry standard for rating roof coverings, "Test Methods for Fire Resistance of Roof Covering Materials," UL790, classified roof coverings "are not readily flammable, do not slip from position, and possess no flying brand hazard." Their performance is rated for different fire intensities:

- Class A roof coverings "are effective against *severe* fire exposure."

- Class B roof coverings "are effective against *moderate* fire exposure."

- Class C roof coverings "are effective against *light* fire exposure."

Still another distinction among roof-deck coverings is the degree of protection they give the roof deck. Under "severe fire exposure," a Class A roof covering affords "a *fairly high degree* of fire protection"; under moderate fire exposure, a Class B roof covering affords "a *moderate degree* of protection"; and under light fire exposure, a Class C deck affords "a *measurable degree* of fire protection."

Aggregate-surfaced membranes generally qualify as Class A roof coverings, regardless of deck type, combustible or noncombustible. Mineral-surfaced cap sheets sometimes qualify as Class A, but smooth-surfaced membranes seldom rise above Class B rating.

To determine roof-covering fire classification, UL has three tests:

- Flame spread
- Flame exposure
- Burning brand

The testing apparatus consists of a 3-ft-long gas burner placed between a large air duct and the roof-covering specimen, which is mounted on a standard timber deck and set at the maximum slope for which the tested covering is recommended by the manufacturer (see Fig.15-1). In all three tests, a 12-mph air current blows across the test specimen, simulating wind.

The *flame-spread test* is conducted in all roof-covering testing programs regardless of deck type (i.e., combustible or noncombustible). Flame spread depends chiefly on the following parameters:

- Roof slope
- Nature and quantity of surfacing
- Membrane composition

Roof slope affects roof-covering rating because hot, burning gases tend to rise from a horizontal surface, thus lengthening the flame-spread time. A vertical surface, representing the limiting case, obviously lengthens the dimension of flame spread up the surface. Depending on slope, a given roof covering can have several ratings. (The rating naturally declines with increasing slope.)

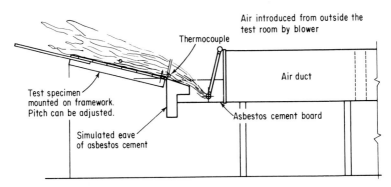

Fig. 15-1 Testing apparatus for qualifying roof coverings through flame exposure, flame spread, and burning tests. (Underwriters' Laboratories, Inc.)

Both the nature and quantity of surfacing and membrane composition affect surface combustibility. Clay-based asphalt emulsions have better ratings than asphalt cutbacks on smooth-surfaced roofs because the presence of noncombustible clay tends to retard flame spread.

The flame-spread test subjects a 13-ft-long test specimen to flames around 1400°F for either 10 min (Class A and B), 4 min (Class C), or until the flame recedes from the point of maximum spread. For Class A rating, the extent of burning, measured from the bottom of the deck, must not exceed 6 ft; for Class B, it must not exceed 8 ft; and for Class C, 13 ft.

Application of a maintenance coating later in the roof's service life could increase roof-surface combustibility and reduce its roof-covering classification. If the particular coating is not UL-listed for the tested membrane, its application could nullify UL classification.

The two flame-penetration tests—*flame exposure* and *burning brand*—are conducted only on roof assemblies with combustible decks (wood or plywood), not on noncombustible decks (concrete, gypsum, steel, preformed mineralized wood fiber).

The *flame-exposure test* subjects the specimens to intermittent flame exposures of 1400°F for Classes A and B, 1300°F for Class C. This test evaluates the roof covering's protection of the underlying combustible deck. Cracking or charring during this test's cooling cycles can reduce this protection.

The *burning-brand test* specifies different sizes of test brands simulating the firebrands hurled by wind or by uprushing gases of an actual fire onto the roof. Test-brand size varies from a 12-in.-square lumber lattice (4.4-lb weight) for Class A down to a 1½-in.-square block (0.02-lb weight) for Class C (see Fig. 15-2). The brands are ignited in a gas flame at temperatures over 1600°F for varying periods of time, depending on the sought classification, placed on the tested roof membrane, and allowed to burn until consumed.

To qualify for classification, the roof covering must withstand the three tests (*flame exposure, flame spread,* and *burning brand*) without (1) any portion of the roof covering material blowing or falling off in glowing brands; (2) exposing the roof deck by breaking, sliding, cracking, or warping; and (3) permitting ignition or collapse of any portion of the roof deck. (Other conditions of varying severity are established in "Test Methods for Fire Resistance of Roof Covering Materials," UL790 or ASTM E108.)

The classifications of built-up roofing membranes are published in the *Building Materials Directory* (Underwriters' Laboratories, Inc.) under "Roof Covering Materials." Felts used in these UL-listed built-up membranes carry the UL label.

Fig. 15-2 Test brands burned to establish UL Class A, B, or C roof covering range from a 12-in. square saw-cut block for Class A to a 1½-in. square block for Class C. (Underwriters' Laboratories, Inc.)

INTERNAL FIRE HAZARDS

Research on acceptable fire spread from internal (i.e., below-deck) fire has focused chiefly on steel-deck roof assemblies, which are especially important because of steel's predominance among deck materials. Because of its extremely high thermal conductivity, steel deck quickly transmits the heat energy of an interior fire to the above-deck roof components, which must be carefully selected to ensure adequate fire resistance.

Industry concern with steel-deck roof assemblies dates from the historic fire in 1953 at a huge General Motors plant in Livonia, Michigan. Representing the nation's greatest fire-insurance loss (until the 1966 McCormick Place fire), the Livonia fire gave the industry a stark lesson on the fire-feeding hazards of hot-mopped bitumen applied to a steel deck. Melted by the fire, the asphalt used to adhere the combustible insulation to the deck melted and dripped through the steel-deck joints, drastically prolonging the fire and advancing it 100 ft ahead of the burning contents. As a consequence, less hazardous, cold-applied adhesives and mechanical fasteners replaced the heavy bituminous moppings formerly applied directly to steel decks with combustible insulation. Note, however, that some of these fire-resistive improvements—especially cold adhesives and plastic sheet vapor retarders—have exacted sacrifices in other aspects of roof-system quality: notably wind-uplift resistance and vapor-retarding effectiveness.

Below-Deck Fire Tests and Standards

Following the Livonia fire, tests conducted in a 20 × 100-ft standard test building by FM and UL confirmed the hazards of placing large quantities of bitumen directly on a steel deck with combustible wood-fiber insulation. The bitumen and wood fibers spread test fires through the 100-ft-long building within 10 to 12 min, regardless of variations in the number of moppings (from one to three) between the deck and the insulation. The full-scale building tests conclusively demonstrated that no hot bituminous mopping of sufficient thickness to bond the insulation could be part of a fire-resistive steel-deck roof system with wood-fiber or similar insulation.

Noncombustible insulation—foamed glass, glass-fiber, perlite board—proved satisfactory with limited amounts of hot-mopped bitumen applied directly to the steel deck.

Flame-Spread Test

To satisfy building code and insurance requirements, a roof system must always resist internal fire (fire within the building). The chief safeguard required against internal fire is limitation of flame spread along the underside of the roof assembly.

As a result of the Livonia fire, a steel-deck roof assembly with 1-in., mechanically anchored, plain vegetable fiberboard insulation and a

Fig. 15-3 To qualify as a Class I steel-deck roof assembly, a test specimen, installed as the test tunnel roof, must not spread flame from twin gas burners more than 10 ft in the first 10 min, or 14 ft during the next 20 min. (Underwriters' Laboratories, Inc.)

four-ply, aggregate-surfaced built-up membrane became the standard roof construction for both UL and FM, the criterion for evaluating other roof-deck assemblies. To qualify for UL listing as "acceptable," a roof-deck assembly must not spread the flame of a modified UL723 flame-spread test farther than the standard steel-deck assembly described above.

In this modified UL723, ASTM E84, "Test Method for Fire Hazard Classification of Building Materials," the test roof assembly forms the top of a test tunnel, with twin gas burners delivering flames against its soffit. Gas supply and other variables are adjusted until the furnace produces a flame-spread rate of 19½ ft in 5½ min on select-grade red-oak flooring. (In this flame-spread rating spectrum, asbestos = O, red oak = 100.) To qualify as "acceptable" in this tunnel test, the roof assembly must not spread flame on the underside more than 10 ft during the first 10 min, 14 ft during the next 20 min (see Fig. 15-3).

In the more expensive 20 × 100-ft test building, an acceptable roof assembly must not spread flame more than 60 ft from the fire end of the test structure during the 30-min test.

FM Fire Classification

Factory Mutual's classification of resistance to interior fire divides roof assemblies into two basic categories: sprinklered and unsprinklered. (These classes refer to the roof construction for nonhazardous occupancies because some interiors containing combustible materials require sprinklers *regardless of roof-system fire resistance.*)

Roof assemblies *not* requiring sprinklers include:

- Class I steel-deck assemblies

- Noncombustible decks—concrete, gypsum, asbestos cement, and preformed structural mineralized wood fiber

- Wood decks treated with fire-retardant, inorganic salts limiting flame spread to 25 ft or less

Roof assemblies requiring sprinklers include:

- Class II steel-deck assemblies

- Combustible decks (untreated wood)

FM Research Corporation's calorimeter test for Class I steel-deck assemblies measures the fuel contributed to combustion by a 4 × 5-ft roof-assembly test specimen placed in the test-furnace roof. As the basis of comparison, a noncombustible panel undergoes the same 30-min fire exposure, with auxiliary fuel added to match the time-temperature curve recorded for the test specimen. The metered auxiliary fuel thus equals the fuel contributed by the test specimen. To qualify as a Class I assembly, the test specimen's average fuel contribution must not exceed certain tabulated values [in Btu/(ft^2·min)] for specified time intervals. Large-scale fire tests, conducted in a 100 × 20-ft building, have demonstrated the acceptability of assemblies that satisfy the tabulated limits.

The distinction between Class I and Class II steel-deck assemblies depends upon the heat release of the above-deck components.

The use of hot-mopped bitumen on a steel-deck surface to form a vapor seal or to bond combustible, organic insulation to a built-up membrane disqualifies a steel deck for Class I rating. One exception to this rule occurs when a deck soffit is sprayed with a noncombustible insulation. Otherwise, only FM-approved adhesives or approved mechanical fasteners can qualify a steel-deck assembly for a Class I rating. (For approved manufacturers' adhesives and fasteners, see Factory Mutual *Approval Directory*, latest edition.)

Application of foamed plastic insulation (polystyrene or urethane) directly to its top surface generally disqualifies a steel-deck roof assembly for Class I rating. Subjected to heat from an interior fire, most foam plastics disintegrate and expose the bitumen of the built-up membrane to heat conducted by the steel deck. Plastic foam thus creates the same fire-feeding hazard as a bituminous mopping.

To qualify as an FM Class I steel-deck roof assembly, most plastic foams require a lower layer of noncombustible insulation—e.g., glass-fiber or perlite board—between the steel deck and the foam. Nonetheless, some foamed plastics, notably glass-fiber-reinforced isocyanurate board with asphalt-saturated asbestos facer sheets, qualify as Class I without the interposition of a fire-barrier material.

Note the irony of the systems concept in roof design in regard to urethane. To improve the steel-deck roof system's fire resistance requires another insulation (e.g., perlite board) *below* the urethane. But to prevent urethane-promoted blistering may require another insulation board *above* the urethane (see Chapter 12, "Premature Membrane Failures").

Time-Temperature Ratings

Time-temperature rating is a second index of internal fire resistance, complementing the previously discussed index of flame spread along the deck soffit. Given in time units—hours and fractions—a roof assembly's time-temperature rating is established by its performance in a standard ASTM E119 furnace test, which subjects the tested assembly to a constantly rising temperature (see Fig. 15-4).

A time-temperature rating is required by many building codes. For example, the National Building Code, recommended by the American Insurance Association, requires a 2-h fire rating for roof assemblies in fire-resistive Type A construction. To qualify for this rating, a tested roof assembly (or any other tested system or component) must endure a test fire of progressively rising temperature—from 1000°F at 5 min to 1700°F at 1 h and 1850°F at 2 h. This standard test is promulgated by four major organizations: Underwriters' Laboratories (UL263), American Society of Testing and Materials (ASTM E119), National Fire Protection Association (NFPA 251), and the American National Standards Institute (ANSI A2.1).

The fire-endurance test measures roof-assembly performance in carrying loads and confining fire. To qualify for a given fire rating, the tested assembly (minimum 180 ft² with minimum 12-ft lateral dimension) must (1) sustain the applied design load, (2) permit no passage of flame or gases hot enough to ignite cotton waste on the unexposed surface, and (3) limit the average temperature rise of the unexposed surface to a maximum 250°F above its initial temperature or a 325°F rise at any one point.

(For steel assemblies with structural steel, prestressed, or reinforced concrete beams spaced more than 4 ft on centers, there are several other complex requirements, set forth in "Standard for Safety: Fire Tests of Building Construction and Materials," UL263, Underwriters' Laboratories, Inc., December 1976.)

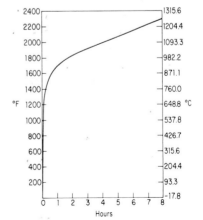

Fig. 15-4 Standard time-temperature curve (ASTM E119) shows rising temperatures that construction assemblies must endure to qualify for fire-resistance ratings, in hours. (American Society for Testing and Materials.)

Fire Ratings vs. Insulating Efficiency

With the advent of skyrocketing energy costs and the trend toward more thermally efficient roof insulation, a serious conflict has developed between fire resistance and energy conservation. Although some directory-listed, fire-rated roof assemblies have unlimited insulation thickness, many still contain only the minimal, usually 1-in.-thick insulations generally specified before the mid-1970s' impact of the energy crisis. These minimal insulations fall far below current standards of thermal-insulating efficiency. Thus the designer may be forced to choose between a listed fire-resistance rating and a thermally efficient roof system.

The problem arises because of the hazards of tampering with any component in a rated fire-resistive roof assembly. As noted in Chapter 2, "The Roof As a System," thickening the insulation of a fire-rated roof-ceiling assembly beneficially reduces roof-surface temperature. But in retarding heat loss *through* the roof assembly, this added insulation increases the ceiling-space temperature, and this higher ceiling-space temperature could buckle steel joists or accelerate the burning of combustible structural members that would otherwise continue carrying their loads. Thus increased insulation thickness could reduce the roof assembly's time-temperature rating.

Resolution of this conflict between fire-resistive standards and roof-system thermal efficiency requires more research sponsored by manufacturers marketing efficient new insulating materials.

BUILDING CODE PROVISIONS

Building code provisions for roofs, as well as other building components, almost universally follow FM or UL standards. All four model codes, the basis for most of the nation's local codes, classify buildings by degree of fire resistance and occupancy.

Under the Basic Building Code, promulgated by the Building Officials Conference of America (BOCA), the degree of fire resistance is defined by four basic types of construction:

1. Fireproof

2. Noncombustible

3. Exterior masonry walls

4. Frame

Each type has subclasses, e.g., Type 1, Fireproof Construction, 1A and 1B, and each subclass carries its own required fire rating for the roof assembly, for example, ¾ h for Type 2A, noncombustible, protected construction.

For any given use, the larger the building, the more stringent the fire requirements. Conversely, the more fire-resistive the construction, the larger the permitted building. For example, any residential hotel more than nine stories high or 22,800 ft² in area must have at least Type 1B roof assembly (1½-h fire rating) when the top floor ceiling dimension is 15 ft or less.

The National Building Code similarly establishes roof-construction requirements—"fire resistive," "ordinary," "wood frame," and so on. These classes similarly range from a maximum required 2-h fire rating down to zero fire-rating requirement for wood frame.

In accordance with the lesson of the Livonia fire, Section 705.3 of the National Building Code bans the use of "a metal roof deck with any material applied directly to its upper surface which represents the hazard of propagation of fire on the underside of the metal roof deck" for roofs of unsprinklered buildings that exceed 9000 ft² in area.

The National Building Code rates roof coverings in conformance with the UL classification: Class A, "effective against *severe* fire exposure"; Class B, "effective against *moderate* fire exposure"; and Class C, "effective against *light* fire exposure."

The code requires Class A or B roof covering on every building except

- Dwellings

- Wood-frame buildings

- Buildings located outside a fire district and qualified, based on height and area limits, for wood-frame construction

Any building exempted from the requirement for Class A or B roof covering, must, however, have at least a Class C covering. (Under the National Building Code there is no distinction between the requirements for a Class A or B roof covering.)

ALERTS

1. Before specifying a roof assembly, consult with the local building official and insurance company representative to make sure that your design satisfies local fire requirements.

2. Beware of substituting even a single component, especially insulation, within a fire-rated roof-ceiling assembly. Such changes require restudy of the new roof assembly thus created. A fire rating applies to an entire roof-ceiling system, not to its individual components.

SIXTEEN

Roof-System Specifications

Roof-system specifiers frequently take refuge in brief, vague roofing specifications that in effect simply require a 20-year bondable roof by manufacturer X, or equal, in an attempt to shunt responsibility from themselves onto the roofing contractor or manufacturer. These ineffectual, incomplete specifications often lie at the root of a roof-system failure. Indiscriminate substitutions are allowed—an inferior, cheaper system for a superior, more expensive system—and from then on everything slides downhill, with the contending parties ultimately battling to avert legal responsibility for a leaking roof doomed before its application to premature failure.

"Roofing may be coal, tar, pitch, or specially processed asphalt," says one architect's roofing specification. Judging from this slovenly edited specification, you could expect a slovenly designed roof, and without disappointment. The school-building roof constructed from this design was not made of coal, but it lacked slope for positive drainage and had improperly flashed, flat skylights, only 2 in. above the roof and located near the maximum deflection areas of long-span metal roof deck, where ponded water soon leaked into the school classrooms.

A common specification states: "Builtup roofs and flashing shall be a complete system for 20-year bond as specified. However, manufacturers considered equal will be considered by the Architect for approval."

Skimpy membrane specifications have launched many roofing failures. Many of the failed two-ply, coated-felt membranes were substituted in specifications like the one above. The specifier who writes a specification so devoid of technical content is effectively abdicating responsibility. The brief, off-handed specification expresses the designer's disdainful attitude toward the roof. It is the kind of specification you usually find on prematurely failed roofs.

SPECIFICATIONS AND DRAWINGS

Both words and pictures are indispensable for describing a roof system. It is impracticable to convey an accurate description of roof construction details in words alone.

Ideally, the drawings show the *form* of the construction, whereas the specifications establish construction *quality*. Drawings show what is to be done; specifications show how it is to be accomplished. Specifications should amplify but not repeat information shown on drawings. The architect's drawings graphically portray design details, location, and dimensions. The specifications establish standards for material quality and work. To serve their complementary functions, drawings and specifications should dovetail like a snugly fitted, two-part jigsaw puzzle, with no overlaps and no gaps. When information is erroneously duplicated on drawings and specifications, possible conflicts can cause incorrect construction, higher costs—at the least, confusion and delay.

Drawing Requirements

The drawings (minimum scale = ¼ in.) should cover the following information:

1. Scope—location and types of roofing, change of roof levels on plans, elevations and details

2. Slopes and drains

3. Walkways

4. Insulation location and thickness, lateral ventilation of insulation, stack vents

5. Details of cants, openings, crickets, curbs, expansion joints, eaves, and stack and edge vents, inside and outside corners

6. Ventilators, skylights, scuttles, and equipment

7. Wood nailers to receive flashing (both composition and metal)

Specification Requirements

The specifications should provide the following information:

1. Scope of the work should combine roofing membrane, roofing sheet metal, accessories, insulation, and vapor retarder.

2. Selection of a roofing system compatible with a reputable manufacturer's published specification.

3. Coordination of roofing specification with other relevant specifications.

4. Responsibilities of the different members of the construction team—owner, architect, manufacturer, general contractor, roofing and sheet metal contractor, mechanical contractor, plumber.

5. Description of vapor retarder, insulation, felt, bitumen, and flashing materials, with tests and standards for acceptance.

6. Maximum and minimum roof slopes (coordinated with deck specifications and structural drawings).

7. Methods of applying all components, with limiting conditions, e.g., heating of bitumen.

8. Provisions for inspection; requirements for bitumen; anchorage requirements for vapor retarder, insulation, and membrane.

9. References to (but not repetition of) mandatory provisions already stated in the general conditions.

WRITING STYLE

Roofing specifications must be written in clear, concise prose. Here are several rules:

1. Favor short, simple sentence structure over long, complex structure.

2. Give directions, not suggestions, in the imperative mood whenever practicable.

3. Do not parrot legal jargon ("said" and "same" as identifying pronouns) or long-winded, all-purpose safety clauses copied from old specifications. (Cluttering a specification with needless words obscures the essential information and reduces chances that the specifications will be read.)

4. Avoid indefinite expressions—e.g., "reasonable," "best quality," "or equal."

SPECIFYING METHODS

There are two basic classes of specifications: *prescriptive* and *performance*. A prescriptive specification describes the means for achieving desired but normally unstated ends. On the other hand, performance specification explicitly states the results to be achieved: e.g., a weathertight, fire-resistant, wind-resistant roof system with specified thermal resistance, with appropriate numbers and referenced test standards.

There are three varieties of prescriptive specifications: proprietary, reference standards, and descriptive. Normally, the specifier chooses between the first two, leaving descriptive specifications for exotic roofing projects.

Proprietary specifications are easily written, short, and precise. Their major disadvantage concerns the difficulty of defining "equal" products for substitutions. The notorious "or-equal" provision has spawned a host of roofing failures, entangled in a corresponding host of disputes or lawsuits resulting from substitution of inferior roofing membrane materials for superior, originally specified products. Obviously, the specifier relying on a proprietary specification should know the product. If the contractor requests a substitution, the specifier should demand evidence that this substituted product will meet the same performance standards met by the specified product—e.g., thermal shock factor (TSF) in a northern climate where splitting resistance is important. Architects and owners have suffered in the past from substitution of inferior coated-felt membranes with two or three plies for four-ply, saturated-felt membranes.

Specifying by performance criteria is the ideal goal toward which specifications for roof systems and other buildings are evolving. But, as the discussion of performance standards in previous chapters has indicated, complete performance standards for roof systems remain a long way in the future. Performance standards for particular requirements— e.g., fire and wind resistance—are well established, and performance requirements for other attributes can be incorporated into specifications, in conjunction with reference standards. The specifier can thus call for a membrane that meets the performance requirements of a UL Class A roof covering, a minimum TSF of 100°F, maximum linear coefficient of thermal expansion-contraction of $40 \times 10^{-6}/°F$, and other attributes listed in the "Preliminary Performance Criteria for Bituminous Membrane Roofing," *Build. Sci. Ser.* 55, NBS. Along with these performance criteria, the specifier can specify ASTM (or other standards)—e.g., Type I asphalt, per ASTM D312, and so on. And for the roof system as a whole, the specifier can incorporate the requirement of UL Class 90 wind-uplift resistance and 1- or 2-h fire time-temperature rating, per ASTM standard furnace test E119.

MASTER SPECIFICATION

MASTERSPEC, a national computerized master-specification system available to the design professions from the AIA-sponsored Production Systems for Architects and Engineers (PSAE), places special emphasis

on roofing because of the high incidence of roofing failures. MASTER-
SPEC follows the standard CSI format, a standardized procedure that
not only reduces the probability of a specifier's omitting a required item
but offers greater convenience to the contractor preparing a bid.

There are two basic specifying methods in the MASTERSPEC system: *broadscope* and *narrowscope*. Broadscope is a proprietary-speci-
fication format. Narrowscope is basically a reference-standards specification that allows inclusion of performance criteria.

Narrowscope is generally the more satisfactory choice for well-
informed designers who know what they want. It is best for larger,
more complex projects because roof size is itself a complicating factor,
aggravating splitting hazards from poorly anchored insulation, making
drainage design more difficult over large expanses. Broadscope is for
less well-prepared specifiers relying on the manufacturer's expertise.

Whichever alternative architects select, they must make a definite
choice, without attempting to mix the two.

DIVISION OF RESPONSIBILITY

When assigning responsibility, the specification writer should follow
these general principles:

1. The *architect* bears ultimate responsibility for the design of
   all components in the built-up roof system and its final
   inspection and acceptance. This person is ultimately responsible for use of a specified material and its compatibility with
   adjoining materials, regardless of information received from
   manufacturers, suppliers, or applicators.

2. The *manufacturer* (or manufacturers) of the various roofing
   components is responsible for furnishing materials that conform with the specifications and for furnishing accurate technical data about the physical properties, chemical composition, and other information pertaining to the material and its
   compatibility with adjoining materials.

3. The *general contractor* is responsible for:
   (a). Coordinating the work of the roofing contractor with that
   of other subcontractors penetrating, adjoining, or working
   on the roof
   (b). Ensuring that the roofing subcontractor follows the
   specifications
   (c). Ensuring the protection of stored roofing materials and
   components from moisture and other hazards before and
   after installation

(d). Providing walkways, where required, to protect the roof from traffic damage

(e). Installing a satisfactory deck

(f). Ensuring that all roof-penetrating elements and perimeter walls are in place and that perimeter nailing strips, cants, curbs, and similar accessories are installed *before roofing work starts*

4. The roofing contractor is responsible for:

(a). Following the specifications for all roof-system components: vapor retarder (if specified), insulation, membrane, and flashing. (If the roofing contractor finds any specified material or procedure impracticable or contrary to good roofing practice, the architect should be so informed. No material or procedural substitutions are acceptable without a formal change order.)

(b). Inspecting the roof deck surface and either (1) accepting it as suitable for application of the roofing system or (2) notifying the general contractor about deficiencies requiring correction before the roof-system application can proceed.

(c). Upon completion of work, giving to the general contractor an "alert" report recommending steps (e.g., walkway protection) that the general contractor should take to protect the roof system from abuse during the remainder of the project.

TEMPORARY ROOFING

Omitting specification provision for temporary roofing can create colossal roofing problems, particularly on projects requiring roof protection during cold winter weather. As discussed in Chapter 7, "Elements of the Built-Up Roof Membrane," a temporary roof (two-ply built-up membrane or single-ply coated-felt) is advisable when the job is hampered by any of the following:

- Mandatory in-the-dry work within the building before the weather permits safe application of a permanent roof system

- Prolonged rainy, snowy, or cold weather

- Necessity of storing building materials on the roof deck

- Large volume of work on roof deck by trades other than the roofing contractor

The specification should require removal of temporary roofing (including insulation boards required to provide a substrate over steel deck). Some manufacturers permit repair of temporary roofs left in place as vapor retarders. But the safest rule on temporary roofs is to remove them, including their price in roofing contractors' bids. A vapor retarder subjected to the rigors of construction traffic and weathering before it is covered with insulation may lose its vapor-retarding ability.

Here is the Associated General Contractors (AGC)-National Roofing Contractors Association (NRCA) advice to remove a temporary roof:[1]

> In general no attempt should be made to retain a temporary roof as the base for the permanent roofing system. The fact that permanent roofing could not be properly installed at the time is evidence that the temporary roof is probably imperfect. Its imperfections may contribute to failure of the permanent roofing at a later date. Temporary roofing should be removed before installing permanent roofing. In general "phased construction" of roofing application cannot be recommended.

SPECIAL REQUIREMENTS

Some roofing projects—notably tearoff-replacement—create special problems. When these special problems are anticipated, the specifier should make special note of them in the specification. For example, when an old aggregate-surfaced membrane is to be removed, a prudent specifier concerned about the potential damage to the new membrane (mopping voids and felt punctures) can insert into the specification the following clause:

> Contractor shall furnish plywood walkways and take any other precautions required to prevent tracking of aggregate from existing membrane to be removed into new work areas where aggregate pieces can be trapped within the new membrane. Contractor shall instruct and police his workmen to assure that aggregate is not tracked into new work areas on workmen's shoes or equipment wheels. Discovery of entrapped aggregate within the new membrane is sufficient cause for its rejection.

Such clauses can answer possible claims by the contractor that it is impracticable to prevent the tracking of aggregate, that the cost of preventing it was not included in the bid, and that it does not make any difference anyhow. If the specification contains an explicit clause warning the contractor to beware of tracking aggregate, the discovery

[1]AGC-NRCA, *Roofing Highlights*, p. 7.

of entrapped aggregate in the new membrane will constitute an incontrovertible specification violation.

When other serious problems can be anticipated, the specifier should insert similar clauses highlighting the problem and forcing the contractor's attention to it. These clauses can also help the owner if the project winds up in litigation. Judges and juries, in their vast ignorance about roofing technology, need all the help the specifier can give in identifying good roofing practice.

SPECIFYING NEW ROOFING PRODUCTS

In today's hazardous legal environment, architects specifying new products of any kind run heavy malpractice risks, and in the roofing area, these risks rise exponentially. Although it may be legally safer to stick with the old methods, the poor performance of the traditional practices is itself a goad to architects/engineers to try new materials and methods. Ultraconservatism in specifying materials is no answer for a designer who wants to maintain professional utility.

The ultimate answer to the design professional's dilemma in dealing with new products is the systems-building approach, with its more rational apportionment of responsibilities among designer, contractor, and manufacturer and its reliance on performance standards with practical testing methods and certification procedures attesting to the building subsystem's—e.g., roof system's—satisfactory performance. The advent of performance standards for roof systems is, however, probably a decade or two in the future. Meanwhile, the contemporary roof specifier must cope with today's roofing industry in its glorious fragmentation and divided responsibilities.

When specifying a new roof-system product, the architect should judge a new product guilty until proven innocent. It is the manufacturer's responsibility to establish the fitness of this product for its intended use, and the architect's responsibility to document research of this product every step of the way. From a longer list drawn up by attorney John S. Martel comes the following list of investigatory steps:

1. Obtain from the manufacturer a list of projects on which the product has been previously used, including the owner, architect/engineer, and roofing contractor. Interviewing these principals, you can ascertain whether the conditions on their projects are similar to your project's.

2. Ask the manufacturer's representative about conditions in which the product is not recommended, about its failures as well as its successes. (This advice is not so naive as it might

appear. Perhaps the manufacturer will not candidly inform you about the product's shortcomings or failures. But later, in possible litigation involving the product's failure, the architect/engineer may be questioned in cross-examination whether there was at least an attempt to get this information. An admission that there was not cannot help but weaken the defense.)

3. Notify the manufacturer in writing how you intend to use the product.

4. Request relevant technical data from the manufacturer. (If it cannot be supplied, you have, ipso facto, an excellent reason for not specifying the product.)

5. Insofar as practicable, investigate the manufacturer's past performance, seeking answers to such questions as these:
   (a). Does the manufacturer warrant the product's performance?
   (b). What is the company's past record for manufacturing and marketing reliable roofing?
   (c). Has the manufacturer produced previous roofing products?

Before specifying a new product, the architect should inform the client (preferably in writing) about possible risks, as well as the advantages, of using the new product. If the client is unwilling to accept these risks, specify a more conventional material.

QUALITY CONTROL

On most building projects there is no attempt to check the conformance of built-up roof systems with the specifications. Checking of materials should start with laboratory analysis of the bitumen, to make sure that specified material is used. A simple softening-point (ring-and-ball) test, per ASTM D2398, is an inexpensive precaution to ensure the owner that the roofer is not substituting Type III (steep) asphalt for the normally required Type I (dead-level) asphalt in the flood coat of low-sloped and even dead-level roofs destined to pond water during their service lives. Use of Type III asphalt for the entire project is convenient for the roofer; it is common practice in some regions of the United States, notably south Florida. But Type III asphalt lacks Type I asphalt's cold-flow property and durability. Thus its substitution can shorten a membrane's service life. (See Chapter 7, "Elements of the Built-Up Membrane," for further discussion of the deleterious effects of Type III asphalt on built-up roof construction.) On large projects the architect may want to check other properties—e.g., penetration, ductility—to ensure conformance of specified asphalt with ASTM D312 provisions.

TEST CUTS

Some roofing consultants periodically take test cuts to determine approximate quantities of roof-system components and to check for membrane quality—presence of entrapped moisture, interply mopping voids, bare felt spots, and so on. Taken during roofing-application operations, the test sample is examined and weighed in the field, replaced if practicable, and the roof repaired with at least an equal number of felts. These samples are normally taken *before* flood coating and aggregate surfacing are applied, per ASTM Standard D3617, "Standard Recommended Practice for Sampling and Analysis of New Built-up Roof Membranes." This standard recommended at least one specimen per roof, plus one per 400 m² (4300 ft²).

Field-test sampling shows whether the roofer's application is meeting specified requirements: e.g., *average* interply mopping weight of ±15 percent of specified weight. If mopping weights are found out of tolerance or voids discovered in the interply moppings, these findings signal a warning to both the roofer and the owner that the system has built-in potential trouble. The field report is a helpful document to the owner in event of future litigation, a seldom improbable destiny for a roofing project.

Roof test cuts are opposed by some roofing contractors. Test cuts weaken the membrane, and their results are often misinterpreted by ignorant owners or architects who demand harmfully overweight bitumen moppings and force contractors into applying excessive bitumen. So go the arguments against test cuts during application.

These two criticisms are easily answered. Small test cuts can be replaced and patched. And the solution to erroneous interpretation of test cuts is not abolition of test cuts, but correct interpretation. Critics of test cuts generally advocate a "trust-us" policy that emphasizes experience in observing application operations as the best guide to a good membrane. Experience *plus* test-cut findings are better than experience alone.

Opposition to test cuts extends to the post-construction phase, when test samples are often needed to determine the quality of work or materials. Obscurantist lawyers typically question whether it is not possible that the roof is better than the test samples indicate. That is of course a possibility. It is also possible that the roof is *worse* than the samples indicate.

The basic argument against the validity of random test sampling focuses on the statistical validity of extrapolating from, say, four 1-ft^2 test cuts on a 100,000-ft^2 roof. How, ask these critics, can you infer the quality based on a test sampling of only 0.004 percent of the roof's surface area?

To answer this question, I present the following statistical example. Assume that you find four test samples with no adhesion of insulation to deck, and that a satisfactory roof system must have at least 90 percent satisfactory samples (thus indicating that 90 percent of the roof is well-adhered). Under these conditions, the probability that you would get four unsatisfactory (and no satisfactory) samples = $(1 - 0.9)^4 = 0.0001$. In other words, the odds are about 10,000 to 1 against such an improbable combination. They are, moreover, about 270 to 1 against finding three or more bad samples and 18 to 1 against finding two or more bad samples out of four in a roof where, as postulated, 90 percent of the samples are satisfactory.[1]

Moral: You do not have to test a major portion of a roof to make virtually certain inferences about its general condition.

For test cuts through existing roof systems, you can have a laboratory analysis of several properties of both insulation and membrane. A laboratory-analyzed test cut can yield the following information:

- Number of felt plies

- Weight and defects in interply bituminous moppings (e.g., mopping voids that grow into blisters)

- Evidence of entrapped foreign materials—e.g., pieces of aggregate from a previously removed membrane on a tearoff-replacement project)

- Flood-coat weight

- Percentage of surfacing aggregate embedment

Test-cut size presents another decision. A 12 × 12-in. test cut is normally sufficient, a convenient size for easy packaging in a plastic bag.

---

[1]The probability of two or more bad samples is calculated by the binomial law, as the sum of the coefficients for the terms containing p^2, p^3, and p^4 representing, respectively, 2, 3, and 4 bad samples in which $p = 0.1$ probability of a bad sample and $(1 - p) = 0.9$ represents the probability of a good sample. Then

$$P = (6 \times 0.9^2 \times 0.1^2) + (4 \times 0.9 \times 0.1^3) + 0.1^4$$
$$= 0.0486 \qquad + 0.0036 \qquad + 0.0001$$
$$= 0.0523$$

With a 6-in.-wide × 40-in.-long sample, cut across the 36-in.-wide felt widths, you can check for correct 2-in. felt laps and lap exposures. But poor felt laps constitute a relatively uncommon violation of good roofing practice. The greater convenience of the smaller samples usually justifies the 12 × 12 in.-size. Moreover, a 12 × 12-in. sample also can be analyzed for number of felt plies, material, and weight—everything, in fact, except the felt laps.

ALERTS

General

1. Specify roofing materials and components preferably to conform with ASTM standards or federal specifications as a minimum, with any additional requirements appended to the selected standards. Using standards from a single source simplifies the task of checking specifications and promotes consistency. Make sure that the specifications are understood and available for reference.

2. Avoid duplication of material properties covered by specifications, e.g., "aggregate shall be ¼ to ⅝ in. in size, clean, and free from dust and foreign matter." Instead, require conformance with ASTM Specification D1863.

3. Check other specification sections for inclusion of information necessary for other roof system components. For example, Section 0610—"Rough Carpentry:"

   • Include wood nailers to secure metal flanges and devices requiring connection with bituminous flashing.

   • Include openings in wood members at edges of insulation to allow venting, if not provided by the design of the insulation.

4. Require manufacturer of flashing materials to approve compatibility with roofing materials in coefficient of thermal expansion, adhesion, elasticity, resistance to sunlight, and other relevant properties.

New Products

1. Commit the manufacturer to a specification—either its own or *written* approval of a specification written by the architect/engineer. (This procedure helps establish an implied warranty of the product's suitability, undermining the manufacturer's possible argument that the product was improperly used.)

2. Include in the general or supplemental conditions a clause requiring the roofing contractor to become familiar with all specified products and, if the contractor disapproves of any, to submit objections in writing.

General Field

1. Require a joint job inspection and conference by the roofing contractor, general contractor, architect's representative, manufacturer's representative, and building inspector before application starts.

2. Require the general contractor to have perimeter walls and roof-penetrating building components in place *before roofing work starts.* (Installation of these elements after roofing work starts requires needless patching and repairs that multiply the chances for leaks and other modes of premature failure.)

3. Require the roofing contractor to approve the deck substrate as satisfactory *before* work is started.

4. Require the general contractor to give the architect ample notice of intent to start roofing operations (usually 2 days or more).

5. Require the general contractor to coordinate the work of the roofing subcontractor, mechanical subcontractor, or other subcontractors, with each assuming full responsibility for any damage that may be inflicted on another's work.

6. Require subcontractors installing rooftop equipment to notify the architect and general contractor of any damage they cause to the membrane. Require offending subcontractors to finance the necessary repairs made by the roofing subcontractor.

Technical

For technical recommendations for the various roofing components, consult the Alerts in the various technical chapters: Chapter 4, "Structural Deck," Chapter 5, "Thermal Insulation," Chapter 6, "Vapor Control," Chapter 7, "Elements of the Built-Up Roof Membrane," and so forth.

As a guide to writing the specification, use the specification work sheets provided by Production Systems for Architects and Engineers (PSAE) or other suitable master-specification system.

SEVENTEEN

Bonds and Guarantees

The high proportion of premature roofing failures makes financial responsibility for repairing or replacing defective roofing systems a major problem for building owners. This financial responsibility can be negotiated via one, or more, of the following:

- Manufacturer's bond: for 10, 15, or 20 years

- Manufacturer's guarantee: generally for 5 years, renewable to 10

- Roofer's guarantee: for 1, 2, and, less frequently, 5 years

Disillusionment with the manufacturer's bond has virtually wiped out this formerly popular method of negotiating for financial responsibility as building owners slowly learned, through bitter experience, about the lean protection offered by these often misunderstood contracts. Nonetheless, because of their historical importance as the first available method of contracting for financial responsibility, the manufacturer's bond is discussed before roofers' and manufacturers' guarantees that, by 1980, had virtually replaced the manufacturer's bond.

MANUFACTURER'S BOND

Formerly the most popular method of negotiating financial responsibility for repairing a defective roof system, the manufacturer's roofing bond has steadily declined, to a tiny fraction of today's built-up roof projects. Its chief influence today is probably more psychological than financial. Diffident specifiers insert the phrase "20-year bondable roof" into their specifications in a vague attempt to cover themselves as having specified a high-quality roofing membrane capable of lasting 20 years. Only a minor fraction of specified "bondable" roofs are actually bonded.

Historically, the manufacturer's bond appeared in response to a challenge created within the roofing industry. In the early days of built-up roofing, the manufacturer doubled as roof applicator and exercised total control over the whole roof construction process—from factory production in the factory to field installation. As the volume of built-up roofing increased during the latter part of the nineteenth century and the early part of the twentieth, manufacturers found their dual role as fabricator-applicator economically impractical. Thus was born the independent roofing contractor, and what was once an integrated responsibility was split in two.

Under the new arrangement, roofing quality frequently declined. Lured by the newly opened opportunities, inexperienced (and often unethical) roofers entered the business and lowered the previous standards set by the manufacturer-applicators. These new roofers sometimes moved from place to place, leaving behind a trail of leaking roofs.

In response to growing roofing troubles, materials manufacturers established standards for manufacturing quality, membrane design, and field practices. In 1905, the old Barrett Company originated the specification roof, with its prescription for number of plies, quantity of bitumen, and best application procedures. In 1916, Barrett established a network of approved roofers and, under prescribed conditions, began guaranteeing roofs applied by these roofers. This original roofing bond guaranteed the built-up membrane for 10 years against leaks attributable to material failure or faulty application.

Roofing bonds are sometimes mistakenly considered insurance policies for roofs in case of total failure. This gross error can be corrected simply by reading the bond's conditions. Moreover, under a manufacturer's bond (as opposed to a guarantee), the top coverage available in 1980 was about $60/square, a minor fraction of the cost of tearoff-replacement, which may exceed $350/square.

How the Manufacturer's Bond Works

The mechanics of the manufacturer's bond is as follows: The roofing materials manufacturer issues the bond, which is backed by a surety company pledged to assume the manufacturer's potential liability for a failed roof. The bond is designed to assure the owner that:

1. The manufacturer's materials were used

2. The membrane and base flashing (if included in a supplementary flashing bond) are installed by a manufacturer-approved roofer

3. The manufacturer's representative has inspected the installation, during application and after completion

Especially during periods of high inflation, a roofing bond can be an extremely poor investment. Consider the following example: a 20-year manufacturer's bond with \$10/square liability limit, \$6/square premium. Invested at 12 percent compound interest, this \$6 premium would take less than 5 years to exceed the \$10 liability limit.[1] Thus a 12 percent interest rate makes this roofing bond a poor investment unless you collect the full amount before the roof reaches its fifth year. If you own 100 roofs, you would have to recover payment on the 100 bonds' total liability limit at an average rate of 4.5 years for the bond premium investment to prove more profitable than investing the total premium at 12 percent interest. This example is an extreme one, but it nonetheless shows the need for rational economic analysis. A 5-year bond with unlimited liability at a slightly larger premium makes far more economic sense than the type of bond just analyzed.

The bond coverage is narrow as well as shallow. Bonds are normally nullified by violations of any of the following partial list of exclusions:

- Roof areas with standing water (except for coal-tar-pitch membranes)

- Performance of repair work without the manufacturer's supervision or approval

- Roof damage from "natural disasters" (e.g., windstorms, hail, hurricanes, tornados, lightning, earthquake)

MANUFACTURERS' GUARANTEES

Offered by both manufacturers of single-ply synthetics and conventional built-up systems' manufacturers, guarantees, backed by the company itself instead of by a surety company, have largely replaced the old roof-bonding system. Because most major roofing manufacturers have been in business for decades, the risk of bankruptcy is too slight to merit consideration of the added protection provided by the independent surety company. (Concern about bankruptcy may not, however, be misplaced with some of the newer, smaller, single-ply synthetic roofing manufacturers.)

Manufacturers' guarantees generally offer better protection than bonds. Their higher liability limits—unlimited in some instances—

---

[1] $6(1 + 0.12)^n = 10$

$$1.12^n = 10/6 = 1.667$$

$$\log 1.12^n = \log 1.667$$

$$n = \frac{\log 1.667}{\log 1.12} = 4.5 \text{ yrs}$$

more than compensate for their shorter coverage periods. Most roofs that perform satisfactorily throughout the first 5 years of service life continue to do so for many more years. Thus the guarantees' shorter period, coupled with higher liability, focuses the owner's protection on the period when it is most needed. The typical 5-year guarantee, with a renewable option for another 5 for a total of 10, should normally offer adequate protection (see Fig. 17-1). Some guarantees run as long as 15 years.

Note too that the second half of a 20-year guarantee is worth far less than the first half (which is equivalent to a 10-year guarantee). On a Present-Worth basis, the only rational way to calculate such value, a 10-year guarantee for $20 liability limit carries a $6.44 Present-Worth value at the end of its 10-year period at 12 percent interest. At the end of 20 years, this value drops to $2.07. With double-digit inflation apparently destined to continue through the 1980s, fixed long term liability limits grow increasingly unattractive as money continues depreciating at a rapid rate.

TYPICAL GUARANTEE TERMS

Guarantees offered for conventional built-up membranes and the new single-ply membranes differ in two major respects:

1. The single-ply manufacturers' guarantees are generally better than guarantees offered by conventional built-up roof materials manufacturers.

2. The range of guarantees offered for single-ply membranes is narrower than the range of guarantees offered for conventional, multi-ply built-up membranes.

The contrasting treatment of ponded water illustrates the superiority of the single-ply manufacturers' guarantees over the conventional built-up manufacturers' bonds and guarantees. Most built-up membrane bonds and guarantees are nullified in areas subjected to ponding. Few, if any, single-ply manufacturers' guarantees contain this exclusion.

Both bonds and guarantees generally exclude:

• Damage from (1) natural disasters (lightning, hurricanes, tornados); (2) vandalism; (3) abuse; or (4) negligent maintenance

• Consequential damages (i.e., damage to building contents from leaking water)

A guarantee or bond is also nullified by unauthorized repairs.

190 FAIRFIELD AVENUE, WEST CALDWELL, NJ 07006

Telephone: 201-226-2136   Telex: 130-223

GUARANTEE

Barra Corporation of America (hereinafter Barra) guarantees that Braas Rhenofol® plastic roof sheets, roof covers and Braas Rhenofol® roofing system accessories will maintain the roof of the following structure:

PROJECT
LOCATION
SQUARE FOOTAGE AND SYSTEM

In a water tight condition at Barra's expense for TERM from the date hereof. Barra's obligations under this guarantee are conditional upon the owner of the above structure giving Barra written notice at the above address of any leak immediately upon discovery, however, no later than thirty (30) days of discovery of such defect or defects.

This guarantee covers only those defects occurring during the term of the guarantee and which defects are due to defective Braas Rhenofol® material and/or Braas Rhenofol® roofing systems accessories, or which defects originate from defective workmanship. It shall be Barra's obligation to repair such defects or, at Barra's option, replace such defective material at no charge to the owner for labor and material. Any such repair or replacement shall be done by a licensed Barra applicator.

Barra shall not be liable for:

1. Any damage to other components of the roof or the building

2. Any incidental or consequential damages flowing from any defect

3. Any failure of the Braas Rhenofol® material or leakage due to hurricanes, lightning, full gales, wind launched debris or other substances, or other similar acts of God or natural causes

4. Any failure or leakage due to any condition or defect caused by any deliberate act or negligence in the maintenance of machinery situated on the roof or structural integrity of the parapet and deck.

This guarantee shall be void in the event any repair or other work on or through the roof is done without Barra's prior written approval of the methods, materials and contractor to be employed.

Barra, its agent or employees shall have free access to the roof during normal business hours throughout the term of this guarantee.

Upon full and satisfactory payment of the invoices and bills for installation, supplies and services, this guarantee shall become effective.

Barra's failure at any time to enforce any of the terms or conditions stated herein shall not be construed to be a waiver of such provision.

This guarantee is in lieu of all other warranties or guarantees whether expressed or implied.

For and on behalf of
Barra Corporation of America Inc.

_____
PRESIDENT, Barra Corporation of America Inc.

                                                        (Seal)

_____   _____
Effective date   Registration No.

Fig. 17-1 Typical guarantee by single-ply synthetic manufacturer does not exclude coverage because of ponding, an exclusion that nullifies most guarantees issued by manufacturers of conventional bituminous built-up membranes. (Barra Corporation of America.)

gle-ply and conventional built-up roof system materials illustrate the superiority of single-ply roof guarantees. For a 5-year guarantee with *unlimited liability* for the single-ply membrane, the premium costs one-third less than the identical guarantee for a four-ply bituminous membrane.

Both single-ply and conventional built-up multi-ply guarantees generally cover membrane and base flashing for 5 years. (Single-ply coverage may be a little wider because at least for PVC systems, PVC-coated metal counterflashing is also covered.) Some manufacturers write guarantees for 10 years. Others offer a 5-year renewable option that requires manufacturer's inspection of the roof on expiration of the original 5-year period.

Liability limits also vary. One manufacturer limits its liability to original material cost: i.e., repair or replacement cost (labor and materials) cannot exceed the original material cost. Others provide unlimited coverage: i.e., whatever it costs to make the roof watertight.

Premiums also differ. Some manufacturers offer no-premium guarantees for the first 5 years, and one manufacturer charges no premium for 15 years' coverage.

The typical single-ply 5- or 10-year period extends the guarantee well beyond the normal 2-year roofer's guarantee offered with conventional built-up membranes. The 5-year guarantee covers quality of work (formerly omitted from some roofing bonds). Thus the 5-year (or 10-year) guarantee offers wider coverage than the common combination of 2-year roofer's guarantee and a manufacturer's bond that excludes coverage for leaks resulting from poor work.

One major manufacturer offers a roof system guarantee with these major features:

1. Coverage for entire roof system (i.e., everything above the deck—insulation, membrane) and base flashing (but not cap flashing)

2. 20-year coverage

3. Unlimited liability for first 10 years, $25/square liability for the next 10

Premium for this coverage is obviously high: $7.50/square (see Fig. 17-2).

ROOFER'S GUARANTEE

The normal 1- or 2-year guarantee may supplement or, in some cases, replace a long-term manufacturer's bond or guarantee. Manufacturers often require a 2-year guarantee from the roofing subcontractor. After an 18-month inspection by the manufacturer, defects uncovered by this inspection must be corrected by the roofing subcontractor before the manufacturer's bond or guarantee takes effect. Failure of the roofing

JM

Johns-Manville
BLUE CHIP
ROOFING SYSTEM GUARANTEE

Guarantee No:

Date:

Building Name:
Building Address:
Building Owner:
Owner Address:
Applied By:

COPY

Roof Area:　　　　　　　　　Roof Specification:
Completion Date:
Lineal Feet of Flashing:　　　　　Flashing Specification No.:
Lineal Feet of Expand-O-Flash Expansion Joint Covers:
Roof Insulation: Type:　　　　　Manufacturer:

Johns-Manville Sales Corporation, a Delaware Corporation with its principal office at Ken-Caryl Ranch, Denver, Colorado 80217 (sometimes hereinafter called "J-M"), guarantees to the original owner named above that, subject to all the provisions of this document, for a period of ten (10) years from the date of the completion of the roofing, J-M will at its expense, make or cause to be made any repairs necessary to maintain the J-M roof, J-M composition flashings (if listed above) and J-M insulation (if listed above) in a watertight condition.

J-M makes this guarantee subject to the following conditions:

1. The owner's only recourse to J-M or its affiliated companies is this Guarantee. THERE ARE NO OTHER WARRANTIES OF ANY KIND, INCLUDING WITHOUT LIMITATION, WARRANTIES OF MERCHANTABILITY OR FITNESS FOR A PARTICULAR PURPOSE.

2. J-M has no responsibility or liability under this Guarantee other than the repairs that may be necessary to maintain the roof in a watertight condition subject to all the conditions of this document. The owner recognizes that J-M and its affiliates shall have no other liability for incidental or consequential damages of any nature whatsoever, including without limitation, loss of profits.

3. The roof owner will notify J-M if repairs that are covered by this Guarantee are required. The notice shall be in writing and delivered to a J-M office by certified mail within sixty (60) days after the necessity of such repairs is or should have been discovered. Failure to give notice shall terminate J-M's responsibility under this Guarantee.

4. If any of the following events occur, J-M's responsibility under this Guarantee shall end:

A. If a sprinkler system, water or air-cooling equipment, radio or television aerial, framework for a sign, watertower or other addition shall be installed on the roof after the completion of the roof, unless J-M shall first be given notice of the proposed installation and give written approval to the making of the necessary roofing application and the materials to be used to join the proposed installation to the roof and the method of application of such materials and structures;

B. If any repair work, except temporary emergency repairs, is done other than under the supervision of or subject to the inspection and approval of J-M;

C. If the roof is damaged by natural disasters; i.e., windstorm, hail, flood, hurricane, lightning, tornado, earthquake, or other phenomena of the elements;

D. If the roof has been subject to misuse, neglect, or accident in any way; or if the roofing system or the building is used for a purpose for which it was not originally designed or designated;

E. Owner fails to give notice or to perform strictly all of the conditions specified in this document that are within his responsibility;

F. If the roof has been subject to test cuts not approved by J-M;

G. If the owner fails to make repairs for which he is responsible within a ninety (90) day period after discovery of need for repair is made or should have been made in the exercise of reasonable care.

5. The owner shall not bring any action under this Guarantee later than one year after the need for repairs is or should have been discovered.

Leaks from the following causes, except where caused by those exclusions set forth below, are covered by this Guarantee:
A. Natural deterioration of membrane;
B. Natural deterioration of base flashing (only if endorsed);
C. Natural deterioration of the roof insulation if J-M supplied the insulation;
D. Bare spots;
E. Blisters;
F. Fishmouths;
G. Ridges;
H. Splits, unless due to movement or failure of substrate;
I. Buckles and wrinkles;
J. Workmanship in application of roofing membrane;
K. Workmanship in application of base flashing (only if endorsed);
L. Workmanship in application of roof insulation if insulation supplied by J-M;
M. Slippage of roofing membrane; and,
N. Slippage of base flashing (only if endorsed).

EXCLUSIONS:

By way of illustration and example and not limitation, the following types of damage and repairs are excluded from this Guarantee:
A. Roof maintenance for corrections of conditions other than leaks;
B. Natural disasters; i.e., windstorm, hail, flood, hurricane, lightning, tornado, earthquake, or other phenomena of the elements;
C. Structural defects or failures;
D. Damage to building or its contents;
E. Changes in building usage unless approved in writing in advance;
F. Damage resulting from any new installation, on, through, or adjacent to the roofing membrane;
G. Repairs or other applications to the membrane or base flashing after date of completion unless performed in a manner acceptable to J-M;
H. (Damage to, or resulting from,) any material used as a roof or wall base over which a J-M roof system is applied;
I. (Damage to, or resulting from,) any material used as insulation unless the insulation is manufactured or supplied by J-M;
J. (Damage resulting from) traffic or storage of material on the roof surface;
K. (Damage resulting from) infiltration or condensation of moisture in, through, or around walls, copings, building structure or underlying or surrounding areas;
L. Damage due to lack of proper drainage (standing water); and,
M. Damage due to movement or deterioration of metal components adjacent to the roof.

RENEWAL OPTION

At the end of the initial ten (10) year period, the owner shall have the option to renew this contract for an additional ten (10) years under the following conditions:

During the tenth year of this Guarantee, and prior to its expiration, if the owner of the building so requests, J-M will make an inspection of the roof and issue to the owner a report on the condition of the roof outlining any and all maintenance work that should be done. This inspection by J-M is free of charge and without obligation.

If the owner elects to exercise his option to renew this contract, he shall perform or arrange to have performed the maintenance work described in the report at his cost by a roofing contractor acceptable to J-M and will notify J-M upon the completion of this work. Maintenance work required must be completed no later than ninety (90) days after expiration date of this Guarantee.

Upon payment of a charge which shall not exceed J-M's then current initial Guarantee charge, the roof will be reinspected by J-M and, if found to be acceptable, this Guarantee will be extended for an additional ten (10) year period.

All conditions of this Guarantee shall apply for the renewal period, except that Johns-Manville's total liability shall not exceed $25.00 per square, for

a total of $_____

In Witness Whereof: Johns-Manville has caused this Guarantee to be duly executed the date set forth above.

ATTEST:　　　　　　　　　　JOHNS-MANVILLE SALES CORPORATION

_____　　By:_____
　　Vice President　　　　　　　　　Attorney-in-Fact

Fig. 17-2 This guarantee, unique in the industry, covers most of the system components—membrane, flashing, and insulation—if supplied by the guaranteeing manufacturer. (Johns-Manville Corp.)

413

subcontractor's guarantee to the manufacturer does not relieve the manufacturer of liability under the bond or guarantee.

The typical roofer's guarantee requires the roofer to repair leaks resulting "solely from faults or defects in material or workmanship

---

Roofing Guarantee

Whereas ....,

of...,

herein called "the Contractor," has completed application of the following roof:

Owner:..

Address of owner:...

Type and name of building:..

Location:...

Area of roof:...

Date of completion:...

Date guarantee expires:..

Whereas, *at the inception of such work the Contractor agreed to guarantee the aforesaid roof against faulty materials or workmanship for a limited period and subject to the conditions herein set forth;*

Now, Therefore, *the Contractor hereby Guarantees, subject to the conditions herein set forth, that during a period of Two (2) years from the date of completion of said roof, it will, at its own cost and expense, make or cause to be made such repairs to said roof and composition flashing resulting solely from faults or defects in materials or workmanship applied by or through the Contractor as may be necessary to maintain said roof in watertight condition.*

This guarantee is made subject to the following conditions:

1. Specifically excluded from this guarantee is any and all damage to said roof, the building or contents caused by the acts or omissions of other trades or contractors; lightning, windstorm, hailstorm, or other unusual phenomena of the elements; foundation settlement; failure or cracking of the roof deck; defects or failure of material used as a roof base over which the roof is applied, faulty construction of parapet walls, copings, chimneys, skylights, vents, supports, or other parts of the building; vapor condensation beneath the roof; penetrations for pitch boxes; or fire. If the roof is damaged by reason of any of the foregoing this guarantee shall thereupon become null and void for the balance of the guarantee period unless such damage is repaired by the Contractor at the expense of the party requesting such repairs.

2. The Contractor is not liable for consequential damages to the building or contents resulting from any defects in said roof or composition flashing.

3. No work shall be done on said roof, including, but without limitation, work in connection with flues, vents, drains, sign braces, railings, platforms or other equipment fastened to or set on the roof, and no repairs or alterations shall be made to said roof, unless the Contractor shall be first notified, shall be given the opportunity to make the necessary roofing application recommendations with respect thereto, and such recommendations are complied with. Failure to observe this condition shall render this guarantee null and void. The Contractor shall be paid for time and material expended in making recommendations or repairs occasioned by the work of others on said roof.

4. This guarantee shall become null and void if the roof is used as a promenade or work deck or is sprayed or flooded, unless such use was originally specified and the specification is noted in paragraph 8 below. Areas that pond water shall not be covered by this guarantee.

5. This guarantee shall not be or become effective unless and until the Contractor has been paid in full for said roof in accordance with the agreement pursuant to which such roof was applied.

6. This guarantee shall become null and void unless the Contractor is promptly notified of any alleged defect in materials or workmanship and provided an opportunity to inspect the roof.

7. This guarantee is in lieu of all other guarantees or warranties, express or implied. THERE ARE NO WARRANTIES OR GUARANTEES WHICH EXTEND BEYOND THE DESCRIPTION ON THE FACE HEREOF.

8. Additional conditions or exclusions..

....

In Witness Whereof, this instrument has been duly executed this .... *day of*.............................,
198.......:

MEMBER
MIDWEST
ROOFING
CONTRACTORS
ASSOCIATION

By...

Approved Guarantee Form No. 1970A, Midwest Roofing Contractors Association, Inc.

serving qualified roofing contractors in

ARKANSAS * COLORADO * ILLINOIS * INDIANA * IOWA * KANSAS * KENTUCKY * MICHIGAN * MINNESOTA * MISSOURI * MONTANA
NEBRASKA * NORTH DAKOTA * OHIO * OKLAHOMA * SOUTH DAKOTA * TENNESSEE * TEXAS * WISCONSIN

Fig. 17-3 Typical roofing contractor's guarantee. (Midwest Roofing Contractors' Association.)

Roof Maintenance Program

As a part of the specification requirements, complete and deliver to Owner the following proposal. The Proposal is conditioned on acceptance by the Owner within 30 calendar days of receipt of Proposal.

Form of Proposal

Project: Date of Roofing Completion:

Roof Section(s):   Roofing Contractor:

Owner: Name:

Address: Address:

Date of Proposal: Phone:

1. The undersigned Contractor agrees, beginning at the termination of the 2-year guarantee on the roofing installation covered in the construction contract, to provide the following services at the listed rates:

    a. Inspect the entire roof area twice each year; once during the months of April or May, and once during the months of September or October. Following each inspection visit, the contractor shall prepare and deliver to the Owner a Report of Condition, including recommendations for needed repair or maintenance work. Include an estimate of a not-to-exceed cost of recommended work.

        Cost of Inspection & Reports: $ Each.

    b. Upon authorization by the Owner, to complete the repair and maintenance work on a Time & Material basis as follows, such material and labor costs to include overhead, equipment, employee benefits, transportation, etc.

        Material: Cost plus   %

        Labor: Cost plus   %

2. Terms of the Agreement shall be 10 (or other) years, and shall be cancellable by either party upon providing written notice to the other party.

3. Payment for services included — Inspection & Reports and authorized repair/maintenance work — shall be upon completion and acceptance of the work. Terms shall be net 30 days from receipt of invoice.

Contractor: _____   Accepted:

    By: _____   Owner: _____

                                                     By: _____

                                                   Date: _____

Fig. 17-4 Roof-maintenance program can be negotiated via this form. (A. L. "Pete" Simmons, Roofing Consultants, Inc.)

applied by or through the roofer." It excludes the following: all damage attributable to lightning, windstorm, hailstorm, or other unusual phenomena of the elements; foundation settlement; failure or cracking of the roof deck; defects or failure of substrate; vapor condensation under the membrane; faulty construction of parapets, copings, chimneys, skylights, and so on; fire; or clogging of drains. Like the manufacturer's bond or guarantee, this roofer's guarantee excludes liability for damage to building contents or other parts of the structure (see Fig. 17-3).

Occasionally, owners negotiate broader coverage and longer terms in contractors' guarantees—5 years on insulation, vapor retarders, and roof sumps as well as the roofing membrane.

ROOF-MAINTENANCE PROGRAM

Complementing the roofer's guarantee and starting with its expiration date, a roof-maintenance program can be negotiated between owner and roofing contractor. As part of the proposal for roof application, the contractor includes a proposal for semiannual (spring and fall) inspection of the roof, with a report to the owner on the roof's condition, plus recommendations for repair or maintenance work. (See Fig. 17-4 for sample roof-maintenance proposal form and conditions.)

EIGHTEEN

Field Inspection

Inspection is one of the weaker links in the roof construction chain. Many owners are unaware of the need for rigorous field inspection, and some are reluctant to pay the price for such apparently nonproductive work. Good inspection is, however, essential to good construction, neglected at an owner's peril. One careless or ignorant worker can nullify the most painstaking efforts of the design team. Guarantees or bonds lull many owners into a false sense of security; because of the difficulty of proving legal responsibility, they are of limited value. Rigorous inspection is the strongest line of defense against poor-quality work.

The basic purpose of inspection is to secure the construction of a good roof. Inspection also has a secondary purpose: to eliminate poor work as a factor in the analysis of any subsequent roof problems. Thus the roofer should welcome inspection as a possible means of escaping involvement in future litigation. However, neither the roofer nor the general contractor should mistakenly assume that inspection in any way relieves them of contractual responsibility to follow the plans and specifications.

For the architect, inspection can be a valuable legal safeguard. If the architect is sued for faulty design of a failing roof, complete inspection records can avert the risk of taking legal responsibility for poor work or possibly nonconforming materials.

Yet good field records are the rare exception rather than the general rule. On many failed roof projects under litigation, the architect has failed to even note dates of roofing application and weather conditions. Rain recorded on the date of membrane application can help explain the presence of membrane blisters on the roof. Cold temperatures could also explain blistering or membrane-splitting failure (from rapid cooling of hot-mopped asphalt adhesive and consequent lack of adhesion between membrane felts or between insulation boards and steel deck). To be deprived of such valuable weather data can jeopardize the architect's prospects of escaping responsibility in litigation, which occurs with such relatively high frequency on roofing projects that it should never be ignored.

INSPECTION ARRANGEMENTS

There are four basic arrangements for roof inspection:

1. Manufacturer's representative (on roofs covered by a manufacturer's bond)

2. Architect's or owner's representative

3. Roofing consultant

4. Nonprofit industry-sponsored roofing inspection agency

Manufacturer's Representative

Under the terms of most manufacturer's bonds, the manufacturer's representative inspects the installation of flashing and membrane. As part of their bonded roof inspection program, some manufacturers require the inspecting representative to certify that the required inspections have been made—by filing with the manufacturer a certificate signed at the jobsite by the architect, the architect's or owner's representative, or the general contractor's superintendent.

Manufacturer's inspection has a built-in conflict of interest. The manufacturer's inspector, often a sales representative, is in the awkward spot of rejecting the work of a customer whose continued good will is necessary for future material sales.

The flaws in manufacturer's inspections are sometimes evidenced by the actions of manufacturers' representatives during litigation. A built-up membrane hopelessly corrugated with random wrinkles produced by water saturation of felts may become acceptable to a manufacturer's representative unwilling to offend a roofing materials' customer, despite the conflict between this approval and the manufacturer's published instructions.

Architect or Owner's Representative

Inspection by either the architect's or owner's representative or by a roofing consultant eliminates the conflict of interest. On large projects, a full-time architect's or owner's inspector can inspect roofing application. If not, a roving inspector, retained by the architect, may make periodic site visits. As jack-of-all-inspection trades, however, the architect's inspector is unlikely to be master of roofing.

Roofing Consultant

The roofing consultant offers inspection as part of a service that usually includes advising on roof design and specifications. Consultants' services range from this professional service, by consultants who are professional engineers or architects, to consultants functioning more like a conventional testing laboratory, inspecting and reporting on the roofer's work.

Inspection Agencies

Nonprofit, industry-sponsored roofing inspection services originated on the West Coast in response to widespread premature roofing failures. In addition to jobsite inspection, the service offered by the prototype organization parallels that of the roof consultant. It includes preliminary consultation with the architect on the roof-system specification, laboratory-analyzed test cuts, and recommendation as to whether the architect should accept or reject the roof after its completion.

PRINCIPLES OF GOOD INSPECTION

A roofing inspector probably has the most demanding job of construction inspection. The built-up roofing system is the most problem-prone building subsystem manufactured on the jobsite. The most intricate work on a welded steel truss or a complex set of elevator controls is done in a fabricating shop or factory. But the roof system is custom-built in the field. Regardless of the quality of the component materials, the roof's integrity depends totally on good fieldwork.

The lack of widely accepted standards and test methods handicaps the roofing inspector. Concrete and welding inspectors, for example, can follow well-established and accepted tolerances, standards, and tests. Determining the bar-replacing accuracy in a flat slab, where depth tolerance is ¼ in., is simple and direct. Inspecting built-up roof construction, however, often requires judgment that cannot be simply resolved with a rule. Are the fishmouths bad enough to require repair? (See Fig. 18-1.) Are the felts "wet?" Is the substrate "smooth?" Is the aggregate uniformly spread and well-embedded in the flood coat?

A good start is crucial. Firmness at the start establishes the right attitude, with one-tenth the effort required to rectify a bad start. Once poor work has been tolerated even for a short time, it becomes difficult to correct.

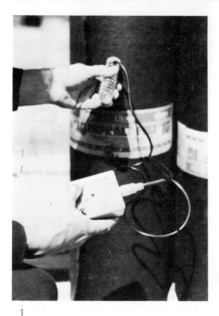

Fig. 18-1 Inspector checks moisture content of roofing felt with moisture meter.

The inspector should obviously know the specifications and details before setting foot on the roof deck. Nothing will destroy a roofer's respect for an inspector quicker than an early display of naivete or ignorance. The overeager inspector who acts on uncertain knowledge and then has to back down has an uphill battle for the rest of that job.

A successful inspection program generally entails the following:

- A preapplication conference, with inspector (owner's or architect's representative), general contractor, roofer, sheet-metal contractor, and materials manufacturer or agent, held at least 2 days before application begins

- Notification to the inspection agency at least 2 days before application of vapor retarder, insulation, or membrane

- Intense inspection as each component is installed, especially at the start

- Formal issuance of inspection reports, promptly distributed to all parties concerned, on the progress of the work and its conformance with, or violation of the specifications

- An irregular schedule of inspector appearances

- When ready for final inspection, the contractor should issue formal notice

Documents and Equipment

At the jobsite, the inspector should have access to the following:

- Complete contract documents—specifications and drawings
- Standards (especially ASTM) referred to in the specifications
- List of approved subcontractors and material suppliers
- Copies of approved shop drawings and submittals
- Copies of correspondence on roof-system components

Jobsite equipment should include the following:

- Infrared thermometer (to check mopping bitumen temperature)
- Camera (for field photos of poor construction)
- Long level
- Straightedge
- Whisk broom

Probably the most important piece of equipment of hot-applied built-up membranes is the hand-held infrared thermometer. As dis-

Fig. 18-2 Great care and rigorous inspection are required to get a clean steel-deck surface for application of new insulation on a tearoff-replacement project.

Fig. 18-3 Sprinkle mopping, an unreliable technique for anchoring insulation boards, is shown on a reroofing project.

cussed elsewhere in this manual, cooled, congealed bitumen is a major source of roofing problems. Knowing that 390°F is the lowest generally acceptable mopping temperature for Type III asphalt and 345°F for Type I asphalt, the inspector can anticipate membrane problems from low-temperature bitumen. Low-temperature and consequently excessively viscous bitumen normally produces mopping voids and overweight interply moppings, which increase the risks of membrane blistering and splitting. A record of low interply mopping temperature can alert responsible parties to the need for corrective action or, at least, provide invaluable information during possible subsequent litigation over the failed membrane.

The inspector should follow a formal routine when making the inspection, using a detailed checklist. On projects of ample size and complexity, preparation of a special project checklist can facilitate the inspector's work.

Material Storage Checklist

1. Check labels for conformance with specifications:
   (a). Bitumen: Correct asphalt type? Softening point? Equiviscous temperature (EVT)?

(b). Felts: Correct manufacturer? Type? Weight?

(c). Insulation: Correct manufacturer? Type? Thickness? R factor?

(d). Flashing (or plastic) cement: As specified?

(e). Mechanical fasteners: Correct manufacturer? Type? Length?

(f). Cold adhesive: Correct manufacturer? Type? Shelf-life exceeded?

(g). Cant strips: Correct material? Correct size?

(h). Flashing materials: As specified?

2. Storage conditions:

(a). Are all materials stored above ground?

(b). Require pallets over laminated kraft paper-covered concrete slabs, pallets over plywood or tarpaulin-covered dry earth as floor for storing insulation or felt rolls.

(c). Are materials covered with tarpaulins? (Roofer must cut polyethylene wrappings around materials and substitute tarpaulins.)

(d). Check felts, insulation with moisture meter for moisture content (see Fig. 18-1 and Table 18-1).

Table 18-1 Equilibrium Moisture Contents of Roof-System Materials

Material	Equilibrium moisture content, % by wt., @ 75°F, 90% RH
Fiberboard	8.5–10.0
Gypsum	1.0–2.5
Steel	0
Wood	7.5–18.0
Concrete	1.0–2.0
Asbestos formboard	6.0–7.0
Cellular glass	0.1–0.5
Styrene	0.1–0.5
Urethane	2.0–4.0
Insulating concrete	5.0–6.0
Perlite board	2.5 –3.5
Organic felt, asphalt-impregnated	2.5–4.0
Asbestos felt, asphalt-impregnated	1.0–2.0
Glass felt, asphalt-impregnated	0.1–1.0

Source: H. W. Busching, R. G. Mathey, W. J. Rossiter, and W. C. Cullen, "Effects of Moisture in Builtup Roofing—A State-of-the-Art Literature Survey," NBS Techn. Note 965, July 1978, p. 24.

(e). Felt rolls stored vertically, not horizontally? (Reject felt rolls compressed into oval cross section.)

(f). Reject insulation boards with crushed or otherwise damaged edges or corners.

Note: Reject defective material *before* application starts, if possible (to give roofer more time to replace rejected material).

Deck Preparation

1. Drains: Correct location? (Guard especially against high elevation.)

2. Nailers: Correct elevation? Correct cross-section dimensions? Treated? Solidly bolted?

3. Cant strips: Correct material? Correct size? Firmly adhered or nailed?

4. *Before* roofer starts application, check deck for slope, smoothness, and cleanliness.

5. Check prefabricated decks for joint tolerances of ⅛-in. vertical gap, ¼-in. horizontal gap plus the following. For *steel decks:*

   (a). Dishing = $\frac{1}{16}$-in. maximum measured over three adjacent flange surfaces (i.e., two deck flutes).

   (b). Unequal deflection at sidelaps? (Minimum one side lap fastener for spans less than 6 ft, two side lap fasteners at one-third points for deck spans over 6 ft.)

Fig. 18-4 A broken corner of insulation board (left) was trimmed and plugged with a triangular fragment cut from scrap (right). The repair was better than nothing, but the damaged board should have been replaced. It should have been saved until a roof interruption required a worker to carve a full-size board for fitting around an opening or other roof penetration.

(c). End laps, properly crimped to ensure maximum ⅟₁₆-in. break at overlap? End laps located at supports (i.e., not cantilevered over supports, which allows two units to deflect differentially)?

6. Deck structurally supported at openings (HVAC equipment, skylights, scuttle access openings, drains, large vent openings)?

For *wood decks:*

(a). Plywood joints stripped with felt? H clips at unsupported joints (to prevent differential deflection)?

(b). Board sheathing or plank: covered with 5-lb rosin-sized paper, 15-lb felt, or base sheet, kraft-paper laminate? Cover sheet correctly nailed (minimum 11-gage shank, ⅜-in.-diameter minimum nail heads, nailed through 1-in.-diameter disks? Spacing at 9 in. along edges, two interior rows at 18 in. staggered)?

For *precast concrete:* If 1/8-in. vertical misalignment tolerance is exceeded at joints between units, require leveling of the deck surface with cement mortar.

For *lightweight insulating concrete fill:* Coated base sheet in place with proper laps, fasteners, and spacing?

For *poured gypsum deck:* Check method of anchoring insulation—mechanical anchors or hot-mopped asphalt?

APPLICATION CHECKLISTS

Before application starts, check the official weather forecast—tuning in the local cable TV weather report or calling the local weather bureau. If the forecast predicts rain, warn the contractor. If it predicts temperatures below 40°F, warn the contractor to prepare for cold-weather application. Other special weather conditions, e.g., high winds, may require added precautions—for example, in laying plastic sheets if they happen to be specified.

As a final preapplication item, check for a thermometer on the bitumen kettle or tanker, after it arrives. You are now presumably ready for application on a clean, properly prepared deck.

Underlayment Application

1. Prohibit application of light, flexible, plastic vapor retarders on windy days.

2. Check underlayment in specification: laps, fastening method (hot-mopped Type III asphalt, mechanical fasteners, spacing).

3. Is underlayment a vapor retarder? If so, it must be flashed at openings and edges.

Insulation Application

1. Check for specified mechanical fasteners: proper length, pattern, and spacing. Check FM spacing requirements for
    (a). Building corners
    (b). Perimeter strip
    (c). Interior roof area

2. Check for insulation-board joint pattern: staggered or aligned (for taped joints).

3. Check insulation boards bearing (1½-in. minimum) on steel-deck flanges. Prohibit cantilevering of boards over flutes.

4. Before hot mopping wet decks (poured-in-place concrete, lightweight insulating concrete, or poured gypsum), test for dryness as follows: Pour a small amount of hot bitumen on the deck. If it froths or bubbles, the deck is too wet. As a second test, try to remove it with your fingers. If you can manually remove it, the deck is too wet.

5. Check to see that deck surface to be hot-mopped is primed with asphalt primer before hot mopping begins.

6. Check specification for hot mopping of decks.

7. Spot-check insulation-board anchorage by manually prying a few random boards with exposed edges. Reject boards that come loose under a few pounds of manual uplift force.

8. Check for clean preparation of steel decks on reroofing projects (see Figs. 18-2 to 18-5).

Membrane Application

1. Check membrane materials: bitumen (asphalt type, coal tar pitch); felt type; number of felt plies and pattern (base sheet plus shingled plies, all shingled plies, felt-lap dimension).

2. Require working thermometer on all kettles.

Fig. 18-5   This mopped-felt cutoff detail has been placed prematurely. To protect the entire exposed insulation surface from water infiltration, it should have been applied after completion of felt laying over the insulation. (See Fig. 5-12 for correct method of installing cutoff.) (Acoustical and Insulating Materials Institute.)

3. Require fire extinguisher on all kettles.

4. Require draining of steep asphalt from kettles before dead-level asphalt is introduced. Prohibit change of kettle use from coal tar pitch to asphalt, or vice versa.

5. To reduce traffic over newly applied membrane, roofer should start application at far points on deck and work toward area where roof materials are hoisted.

6. Felt-laying order: On slopes 2 in. or less, apply felts perpendicular to slope, if practicable, starting at low point.

7. Ensure that water flow is over or parallel to felt lap, never against lap. (If flow against any lap is unavoidable, require roofer to mop in 6 in. felt stripping over the joint.)

8. Require backnailing of base sheet of steep (Type III) asphalt membranes of 1½ in. or greater slope and ½ in. or greater dead-level (Type I) asphalt or coal-tar-pitch membranes.

9. Hot bitumen temperature ranges at time of application:

Steep (ASTM Type III) asphalt	375°F–450°F
Dead-level (ASTM Type I) asphalt	350°F–425°F
Coal tar pitch	250°F–325°F

With EVT available, require hot bitumen temperature at application time within $\pm 25°F$ of the EVT (see Fig. 18-6).

10. When an asphalt tanker is on the site, maintain asphalt temperature *below* the *finished blowing temperature* (FBT) and feed asphalt into a kettle for further heating before pumping it to the roof.

11. In cold weather, require double or insulated lines and insulated asphalt carriers and mop buckets.

12. Check for continuous, uniform interply hot moppings, with no contact between adhered felts.

13. Require manual brooming following 6-ft maximum distance behind unrolling felt (see Fig. 18-7).

14. Require immediate repair of felt-laying defects: fishmouths, blisters, ridges, splits, omitted envelopes, and so on (see Figs. 18-7 to 18-12).

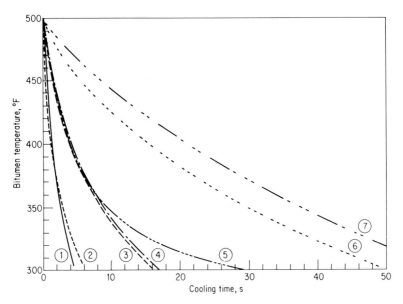

Fig. 18-6 Hot-asphalt cooling curves (calculated by NBS researchers) show the tremendous variation in cooling time from 500 to 300°F for hot asphalt mopped to substrates of varying thermal resistance and heat capacity. Air and substrate temperatures are 70°F; wind speed is 0 mph. Substrates are (1) concrete, (2) steel, (3) insulating concrete, (4) plywood, (5) felt/PUF insulation, (6) fiberglass insulation, and (7) PUF insulation. Concrete, which promotes the fastest cooling, has low thermal resistance and high heat capacity. Urethane, which promotes slow cooling, has high thermal resistance and low heat capacity. (National Bureau of Standards.)

Fig. 18-7 Felt-laying caravan, complete with broomer to promote uniform mopping adhesion, applies felt with strong, machine-direction perpendicular to short direction of insulation boards, as recommended in Fig. 5-11. Center worker, unrolling felt, is eliminated with felt-laying, asphalt-dispensing equipment instead of asphalt dispenser shown.

Fig. 18-8 Finger wrinkling in this asphalt-saturated asbestos felt, left exposed to weather for 3 months with no protective asphalt glaze coating, resulted from moisture expansion, predominantly across the felt.

15. Terminate day's work with complete glaze-coated seal stripping, preferably coated with plastic cement.

16. Aggregate-surfaced membrane may be glaze-coated with flood coat and aggregate surfacing delayed until later, if necessary. Limit delay between felt application and flood-coat, aggregate surfacing to a maximum 4 days.

Fig. 18-9 Suspecting this roof of containing water entrapped between the insulation and felts, the inspector had the felts cut to investigate. The bitumen rolled easily off the glass-fiber insulation, confirming his suspicions. (The bitumen would have adhered tightly to a dry insulation surface.) Subsequently, the roofer removed an 80-ft² area of membrane. He repaired it with two coated base sheets, amply overlapped, and three saturated felts, more than equaling the basic three-ply membrane. Where the lower repair felts overlapped the roofing felts, "bucking" water on the slope, an asphalt-saturated fabric tape covered the joint.

17. Roofer must move dry aggregate to roof at rate of application, without stockpiling. If unavoidable, roofer may stockpile in small mounds or rows on completed areas, but *not on unsurfaced felts.* Do not let roofer leave stockpiled aggregate overnight. Reject wet, dusty, or snow-covered aggregate.

18. Permit no traffic across completed membrane except where the membrane is protected with minimum ½-in. plywood.

Flashings and Accessories

1. When hot-mopped, felted base flashings are specified, check to ensure use of steep (Type III or IV) asphalt, to avoid threat of sagging from erroneous use of dead-level (Type I) asphalt.

2. Check for use of *flashing* cement, not *plastic* or ordinary roofer's cement, when cold-process trowelings are specified for flashings instead of hot-mopped steep asphalt.

3. Require installation of base flashing and counterflashing, if practicable, as membrane application proceeds. If delay in installing counterflashing is unavoidable, apply three-course stripping seal (one ply of fabric flashing between two trowelings of flashing cement), as a temporary joint seal.

Fig. 18-10 Folded edge of base felt (envelope), nailed to edge strip, prevents bitumen drippage from interply moppings (top). Lack of such a folded edge, or envelope, permitted staining of the walls (bottom). (GAF Corporation.)

4. Check for cant strips at curbs, walls, and other right angle intersections.

5. Check for specified treatment of wood nailers and cants.

6. Check for proper installation of gravel stops—e.g., metal

Fig. 18-11   A result of laying felt sloppily, the 3-in.-wide, ½-in.-high fishmouth in the smooth-surfaced membrane shown provided a conduit for chronically ponded water on the poorly drained roof surface to invade the membrane and ultimately saturate the insulation.

Fig. 18-12   Sweeping this wet coated base sheet (note shimmering ladder reflection in water) is not enough to dry it for application of remaining felt plies. Roof operations should have been (but were not) canceled until the surface had dried.

Fig. 18-13 Pitch pan, a bad detail at best, is close to its worst in this detail. Instead of placing it on the top of the membrane and stripping it in with two plies of felt, the roofers simply placed it on top of the aggregate surfacing.

flanges set on *top* of the membrane, stripped in with two felts or fabric plies (also, for proper installation of pitch pans, if used; see Fig. 18-13).

7. Check for proper priming of walls and gravel stop metal before applying base flashing.

8. Check for 4-in. asbestos felt vertical sealer strips over end laps of exposed base-flashing felts.

9. Check for proper molding and adhesion of base flashing to backing.

10. Check for flashing toe-sealer strips at horizontal joint with membrane for smooth-surfaced membranes.

11. Check for use of specified nails at specified spacing (maximum 8 in.).

12. Check base flashings for maximum 12-ft length, staggered end laps (2-ft minimum offset), 3-in. minimum end laps.

13. Check vertical intersection flashing heights for proper dimensions: minimum 8 in., maximum 12 in., 4-in. minimum vertical counterflashing lap.

NINETEEN

Nondestructive Moisture Detection

Nondestructive moisture-detection techniques have been developed over the past decade in response to the problem of wet insulation. These advanced techniques—notably infrared (IR) thermography and nuclear moistures—have made accurate surveys for locating wet insulation economically practicable. (In 1981, these surveys cost about $0.05 to $0.10 psf, ranging down to $0.03 psf.) Before the advent of these nondestructive techniques, the problem of accurately locating and defining the limits of wet areas and corresponding decisions on whether to tear off and replace the roof, to re-cover it, or simply to live with the problem and rely on ad hoc repairs were necessarily based on limited information.

Periodic nondestructive moisture surveys are like periodic medical checkups for a human being, with early detection of moisture problems analogous to early cancer detection, according to Wayne Tobiasson, research engineer with the U.S. Army Cold Regions Research and Engineering Laboratory (CRREL), Hanover, New Hampshire, and a pioneer in thermographic detection of roof moisture.[1] A cancer caught in its early stages, like a roof system gaining moisture, may require only minor surgery for its removal and restoration of the patient's health. Allowed to gain moisture undetected, however, a water-saturated roof can result in premature death of a roof system, just as a neglected cancer kills its human victim.

Illustrating the benefits of nondestructive moisture surveys, a $62,000 repair program instead of a $210,000 tearoff-replacement project was credited to nuclear moisture-meter survey by a major building owner.[2]

---

[1]Wayne Tobiasson, Charles Korhonen, and Alan Van den Berg, "Hand-Held Infrared Systems for Detecting Roof Moisture," *Proc. Symp. Roofing Technol.*, NBS-NRCA, September 1977, p. 265.

[2]Paul Geffert, "Diagnosing Roof Problems with a Nuclear Moisture Meter," *Plant Eng.* Apr. 28, 1977, pp. 213ff.

With its many leaks, plus a widely blistered membrane and other defects, the 14-year-old, 70,000-ft² roof system was apparently headed for total replacement at a cost of $3 psf before the owner bought the idea of periodic moisture surveys. Plotted from readings on a grid and then computer-analyzed for moisture content, the survey revealed 13 percent of the roof area wet. Spot repairs with removal of identified wet areas, plus replaced flashings and overall general repairs, can prolong the roof's service life for many years, postponing tearoff-replacement expense to some remote future date at a $148,000 saving.

Leak-source location is a potential ancillary benefit of nondestructive moisture detection, although it should not be promoted as a consistent leak-detection method. After several futile visual searches plus remedial repair designed to correct a roof leak that had shortcircuited a 1000-A busway, the same building owner cited in the foregoing paragraph had the problem solved by a nuclear moisture-meter survey. It traced a long path of wet insulation from the previously undiscovered leak source (a counterflashing gap between the roof and a penthouse) to the leak point. Rainwater flowed laterally from its entry point at the counterflashing gap to the leak point, a considerable distance away.

BENEFITS OF NONDESTRUCTIVE MOISTURE DETECTION

Accurate detection and location of roof areas containing wet insulation can alert the owner to the need for remedial action to avert the following problems:

- Leakage fed by water-soaked insulation
- Cooling and heating-energy waste
- Decay of roof-system materials
- Expansion of wet insulation into ever-increasing areas

Water trapped in insulation can be a voluminous source of leak water. In a porous insulation like fiberglass over a poured concrete deck, entrapped water can feed slow leaks through small cracks in the concrete for months, even years. If continually replenished by ponded rainwater over an imperfectly watertight membrane, this entrapped water can feed slow leaks perpetually.

Wasted heating and cooling energy is a direct result of wet insulation. Especially on large roofs covering low sprawling buildings, these energy-leaking losses can be extremely costly. Heat loss through wet roof areas has been measured by heat-flux sensors at three times the heat-loss rate through dry roof areas.[1]

Decay of roof-system materials, organic roofing felts and insulation, is caused by water absorption. Deterioration results not only from fungal rot of organic materials but from dissolution of phenolic binders in some organic materials.

Expansion of areas with wet insulation is a result of neglecting this problem. Regardless of whether the moisture source is liquid leak water from above or condensing water from below, the process of extending the areas of wet insulation will continue. Extension of the wet insulation areas by condensing water vapor is promoted by the following process: Areas with wet insulation are cooler and consequently at lower pressure than dry areas when the sun bakes the roof surface. This situation creates a horizontal dry-to-wet vapor-pressure gradient, which impels water vapor flow toward cooler, lower-pressure, wet areas. Upon reaching the cooler, wet insulation, the migrating water vapor condenses, thus extending the wetted areas. This process can be especially serious with porous insulation, which facilitates lateral vapor flow.

Nondestructive testing thus enhances the benefits of more commonplace, routine maintenance and repair programs. It can effect early detection of entrapped moisture, possibly averting the subsequent replacement of a roof system just starting to experience serious moisture problems. These nondestructive testing techniques can delineate existing roof areas with entrapped moisture or even signal impending moisture problems 9 or 10 months later.

LIMITATIONS

The new moisture-detection techniques have several limitations:

- Core samples should be taken to validate survey results, regardless of which nondestructive detection technique is used.

- Nondestructive surveys are limited to certain economies of scale, with higher unit costs for small roofs. Survey cost (1981) can nonetheless range below $1000.

[1]Charles Korhonen and Wayne Tobiasson, "Detecting Wet Roof Insulation With a Hand-Held Infrared Camera," U.S. Army Corps of Engineers, Cold Regions Research and Engineering Laboratory, Hanover, N.H., 1978, p. A14. See also Wayne Tobiasson and John Ricard, "Moisture Gain and Its Thermal Consequence for Common Roof Insulations," *Proc. Symp. Roofing Technol.*, NBS-NRCA, April 1979.

Nondestructive moisture-detection survey data can translate into benefit-cost ratios of 10 or more—in energy-cost savings and averted replacement costs.

PRINCIPLES OF NONDESTRUCTIVE MOISTURE DETECTION

As stated, the most important new moisture-detection techniques are infrared (IR) thermography and nuclear moisture meters. Infrared thermography comes in two versions: (1) airborne thermal IR and (2) hand-held thermal IR detectors.

IR thermography, or "imaging," depends on the emanation of IR, heat radiation from all bodies at an intensity varying, in accordance with the Stefan-Boltzmann equation, with the fourth power of the absolute temperature. As temperature rises, IR radiation gains intensity, with higher frequency and shorter wavelength. A "red-hot" body radiates energy in the red portion of the visible spectrum—about 0.7 μm = 0.0007 mm. In the normal range of nocturnal terrestrial temperatures—say, between 32 and 86°F—the emitted IR wavelengths are much longer, ranging around 10 μm (= 0.01 mm). Because the emitted IR heat energy occurs in the invisible part of the electromagnetic spectrum, it can be detected only by sensitive media: semiconductor crystals, such as mercury-doped germanium or indium antimonide as the thermograph sensor. Moreover, to attain the required sensitivity, these detectors must be cryogenically cooled, usually by liquid nitrogen, to temperatures below −300°F.

The thermograph indicates areas of wet insulation by showing contrasting temperatures between different areas—a darker tone for dry areas at lower temperature and thus emitting less IR radiation at longer wavelength, and a lighter tone for wet areas at higher temperatures. Because of the radiative subcooling of roof surfaces on clear nights, with little or no cloud cover to absorb and reradiate escaping IR radiation beamed out into space, such clear, cool nights provide the biggest contrast between wet- and dry-roof area temperatures.

An IR camera depicting these temperature differences is basically a TV camera, with the IR radiation sensed by the camera and converted into a video signal that displays the warm, wet areas in lighter shades, cooler areas as darker shades (see Fig. 19-1).

Because thermography depends on locating roof areas with wet insulation by their temperature differences from roof areas with dry insulation, the sharpest depiction occurs when the temperature difference between wet and dry areas is greatest. Judged from extensive experience, a clear, cool night after a clear, sunny day, several hours after sunset to several hours before sunrise, is the best time to run a thermographic moisture-detection survey. But thermographic surveys

Fig. 19-1 Hand-held thermographic camera system comprises four components: (1) camera, (2) display unit, (3) Polaroid camera and mount; (4) battery pack. (U.S. Army Cold Regions Research and Engineering Laboratory.)

are practicable year-round in all parts of the globe. In summer, transient solar heating raises roof-surface temperature and consequent roof-temperature differences. But steady-state heat flow during the winter-heating season is equally satisfactory because the interior heat energy creates temperature differentials between wet and dry areas. (During the heating season, wet-insulation areas gain more heat than dry areas because of the wet insulation's loss of thermal resistance.)

Nighttime thermography offers advantages over daytime because of the added complexity in the roof's daytime thermal behavior. The key problem is reflected solar radiation, which distorts surface-temperature readings. Another problem is the thermal effects of moving shadows cast onto the roof by nearby structures. Nighttime surveying eliminates these complicating factors (see Fig. 19-2).

The *nuclear moisture meter* works on a totally different principle. A neutron generator with a radioactive source emits high-energy ("fast") neutrons aimed at the target area. From their collisions with atoms in this material, some neutrons are reflected back to the vicinity of the neutron emitter. Neutrons that hit hydrogen atoms are slowed and can be counted by the instrument. The number of returning "slow" neutrons indicates the number of hydrogen atoms in the tested material. And the number of hydrogen atoms, which constitute two-thirds of the atoms in water, becomes an index of the quantity of water in the tested area.

Fig. 19-2 Nighttime is best for thermographic survey because of complicating thermal effects produced by moving shadows cast on the roof by parapets, chimneys, and so on, variable solar radiation, and other factors not present at night. (Tremco.)

Because hydrogen atoms abound in hydrocarbon bituminous roofing materials, this additional source of hydrogen atoms complicates the correlation of hydrogen atom count with the quantity of water. In more or less uniform cross sections, however, a datum level of hydrogen atoms can be established for dry areas, and the excess count assumed to indicate water. The kind of error arising from this complicating factor is illustrated in early uses of nuclear moisture meters, with flashing areas erroneously identified as high-moisture areas. The users had failed to allow for the increased quantities of bitumen and consequent hydrocarbon atoms in the vicinity of flashings.[1]

With the nuclear moisture meter, readings are taken on a 10-ft, sometimes a 5-ft, grid. A pattern of wet insulation can be inferred from the readings and plotted as shown in Fig. 19-3. The nuclear meter is a portable instrument, readily transported and handled on the roof (see Fig. 19-4).

[1]H. W. Busching, R. G. Mathey, W. J. Rossiter, and W. C. Cullen, *Effects of Moisture in Builtup Roofing—A State-of-the-Art Literature Survey*, NBS Tech. Note 965, July 1978, p. 38.

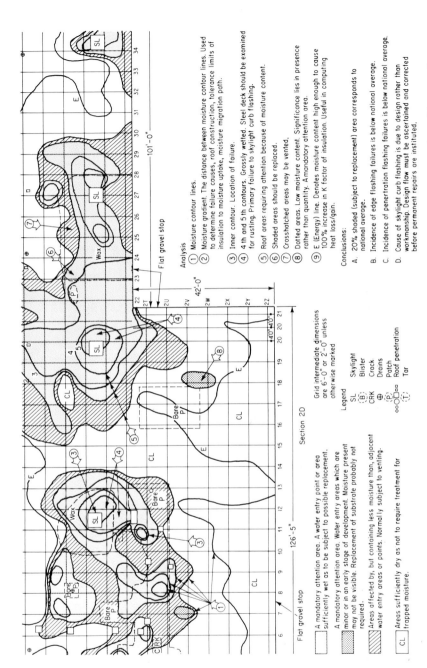

Analysis

① Moisture contour lines.

② Moisture gradient. The distance between moisture contour lines. Used to determine failure causes, roof construction, tolerance limits of insulation to moisture uptake, moisture migration path.

③ Inner contour. Location of failure.

④ 4th and 5th contours. Grossly wetted. Steel deck should be examined for rusting. Primary failure to skylight curb flashing.

⑤ Roof areas requiring attention because of moisture content.

⑥ Shaded areas should be replaced.

⑦ Crosshatched areas may be vented.

⑧ Dotted areas. Low moisture content. Significance lies in presence rather than quantity. A mandatory attention area.

⑨ E (Energy) line. Denotes moisture content high enough to cause 100% increase in K factor of insulation. Useful in computing heat loss/gain.

Conclusions:

A. 20% shaded (subject to replacement) area corresponds to national average.

B. Incidence of edge flashing failures is below national average.

C. Incidence of penetration flashing failures is below national average.

D. Cause of skylight curb flashing is due to design rather than workmanship. Design flow must be ascertained and corrected before permanent repairs are instituted.

Section 2D

Grid intermediate dimensions are 6'-0" or 2'-0" unless otherwise marked

Legend

SL   Skylight
(B)   Blister
CRK   Crack
⊕   Drains
(P)   Patch
Roof penetration
(T)   Tar

A mandatory attention area. A water entry point or area sufficiently wet as to be subject to possible replacement.

A mandatory attention area. Water entry areas which are minor or in an early stage of development. Moisture present may not be visible. Replacement of substrate probably not required.

Areas affected by, but containing less moisture than, adjacent water entry areas or points. Normally subject to venting.

CL   Areas sufficiently dry as not to require treatment for trapped moisture.

Flat gravel stop

Fig. 19-3 Roof-surface grid, at 6-ft spacing in this example, is plotted with insulation moisture contents inferred from nuclear meter readings. Shaded areas indicate areas of wet insulation recommended for removal and replacement. (Gammie Nuclear Service Co.)

441

Fig. 19-4 Nuclear moisture meter is easily wheeled around the roof. (Gammie Nuclear Service Co.)

IR THERMOGRAPHY VS. NUCLEAR MOISTURE METERS

As noted, nuclear moisture detection can be economically justified on a smaller scale than thermography. Hardware cost for IR thermographic equipment runs upward from $12,000, about 5 times the cost for a nuclear meter. But on a large-scale survey, IR thermography may be less expensive, and it has a technical advantage because it can cover the roof more thoroughly. A nuclear meter "sees" only about 2 ft² of surface area at each grid point. On a 10-ft-square grid, nuclear detection thus covers only 2 percent of the roof area. Wet spots can occur between grid points. Closing the grid to 6-ft-squares reduces the chances of missing wet spots, but it triples the number of required readings.

As a possibly offsetting practical advantage, a nuclear moisture-meter survey can be carried out in the daytime, a generally more convenient schedule than the nighttime one required for good thermographic readings.

OTHER MOISTURE-DETECTION TECHNIQUES

Another nondestructive moisture-detection technique in general use is the *capacitance meter*, a technique borrowed from the paper industry. Like nuclear detection, it requires a grid layout for measurements. Moisture-detection readings depend on the tremendous difference between the dielectric constant of water (80) vs. the constants of most roofing materials, which range from 1 to 4. This technique resembles nuclear detection, with sensors utilizing capacitance circuits roughly paralleling the nuclear backscatter gages. Variations in dielectric constant from a norm obtained in the dry roof areas indicate varying quantities of entrapped water.

In a study comparing IR thermography, nuclear backscatter meters, and capacitance meters, an NBS report indicated inferior performance for capacitance meters. The three techniques were judged on the basis of two performance criteria:

- Minimum moisture content detected by the instrument (i.e., "threshold moisture")

- Quantitative accuracy in measuring moisture content beyond threshold moisture

A slight superiority indicated for the nuclear backscatter meters over IR thermography was judged insignificant by the NBS researchers.[1]

Electrical-resistance moisture meters exploit the principle that changes in electrical resistance indicate water content. With its inevitable impurities, water is a good conductor, and electrical resistance in nonconductive insulating materials varies inversely with their moisture content. This method is more accurate for determining the moisture content in wood than in roofing materials, but it can nonetheless indicate the presence or absence of water. Obviously it is less accurate than the other three discussed moisture-detection techniques. It is not used for general moisture-detection surveys.

Electrical-resistance moisture meters are small, battery-activated, hand-held instruments weighing only 18 oz or so (see Fig.18-1). The probes must puncture the membrane to get a reading, and this requirement disqualifies these meters for the nondestructive classification.

---

[1]Lawrence Knab, Robert Mathey, and David Jenkins, "*Laboratory Evaluation of Nondestructive Methods to Measure Moisture in Builtup Roofing Systems*," *Build. Sci. Ser.* 131, NBS, January 1981, p. xiii.

TWENTY

Waterproofed Deck Systems

Under the economic pressure of soaring land costs and the consequent need for more efficient use of building space, waterproofed plazas and roof terraces (called waterproofed decks for our purpose) have become more popular over the last several decades. Underground storage areas, access tunnels, and parking garages, even computer rooms and office space, are located below plazas, planters, and sidewalks. Rooftop parking garages, swimming pools, and terraces are other examples of waterproofed decks.

Waterproofed decks, besides meeting all the requirements of roofs, must satisfy one further requirement: accommodation of traffic loads or plant growth. This more rigorous structural requirement eliminates compressible insulations—e.g., fiberglass boards—that provide satisfactory service for conventional roof construction. A waterproofed deck is like a protected membrane roof (PMR) system, which similarly intensifies the requirements of the insulation while generally alleviating the requirements of the membrane. (In both a PMR and a waterproofed deck system the membrane is shielded from weathering and ultraviolet radiation.) As a consequence, there is a wide spectrum of suitable waterproofing membrane materials: both built-up bituminous and single-ply synthetics, sheet- and fluid-applied.

A waterproofed deck system contains (listed in top-down order, in PMR arrangement), some or all of the following components:

- Wearing surface
- Aggregate percolation stratum
- Insulation
- Protection boards
- Membrane
- Deck

Of these components, the aggregate percolation stratum and insulation are the only nonessential ones. And in this era of energy-conscious design, insulation can be considered virtually essential over heated or cooled space.

For a conventional waterproofed deck assembly, with insulation below the membrane, there is another possible component, a vapor retarder designed to prevent vapor flow upward into the insulation.

WEARING SURFACE DESIGN

Because of their often heavily trafficked locations, waterproofed deck systems have the following surfacing design requirements:

- Structural strength to bear traffic loads

- Durability under heavy wear and weathering

- Esthetic appearance (on plazas and roof terraces)

- Heat reflectivity (to avoid summer temperature buildup)

The first two criteria are mandatory; the latter two are optional.

Cast-in-place or precast concrete, ceramic, or masonry units generally satisfy these requirements. Dark colors—especially black—are normally avoided because of their high heat absorption. (See Table 20-1 for solar absorptance values of several surfacing materials.)

Joints on a roughly 10-ft maximum grid are normally required to accommodate thermal contraction and expansion in surfacing units. They can experience daily temperature extremes of 80°F, annual extremes of 180°F in the most severe climates. Use of snow-melting equipment (embedded electrical-resistance cables, hot water or steam piping) complicates the problem of thermal design of the surfacing.

Table 20-1   Solar Reflectance of Surfacing Materials

Surfacing material	Solar reflectance*
White or light cream brick	0.50–0.70
Yellow, buff brick or stone	0.30–0.50
Concrete, red brick or tile	0.20–0.35
Green grass	0.25
Crushed rock	0.20
Gravel-surfaced bituminous roofing	0.15
Asphalt paving	0.05–0.10

*Most tabulated values are computed from the *ASHRAE Handbook and Product Directory, 1977 Fundamentals*, table 3, p. 2.9, which tabulates values for solar absorptance.

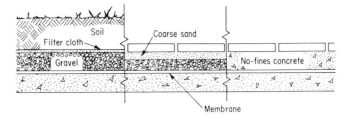

Fig. 20-1 Percolating strata under open-jointed surfacing units or soil can be chosen to suit subsurface drainage requirements. (Maxwell C. Baker, National Research Council of Canada.)

Rapid temperature changes accompanying intermittent operation of the snow-melting system or expansion of corroding embedded pipes can crack surfacing slabs.[1]

A basic difference in waterproofed deck design concept concerns the surfacing joints. There are two types:

- Open-jointed construction

- Monolithic (i.e., closed-joint) construction

AGGREGATE PERCOLATION VS. OPEN SUBSURFACE DRAINAGE

Open-jointed surfacing units—precast concrete, tile or masonry units— usually sit on a bed of loose aggregate that conducts water downward to the waterproofing membrane, where it runs to drains. Fine gravel, clean coarse sand (held on a No. 30 sieve), or no-fines (i.e., porous) concrete can serve as the percolating stratum underlying the open-jointed surfacing units (see Fig. 20-1). Loose aggregate can accommodate freeze-thaw cycling of entrapped water, with freezing expansion contained harmlessly within the aggregate interstices.

In a less frequently specified, but generally superior alternative to loose aggregate percolation, the surfacing units can sit on pedestals, corner posts, or sleepers, providing faster, less obstructed subsurface drainage. Surfacing soffits can also be grooved, corrugated, or otherwise contoured to create open paths for subsurface drainage.

This alternative pedestal-supported, open-jointed system has several advantages:

- Faster, more efficient drainage, with better surface-dirt removal

[1] G. K. Garden, "Roof Terraces," *Can. Build. Dig.*, no. 75, May 1968, p. 75-3.

- Ventilative drying of subsurface areas

- Easier access for maintenance and repair of subsurface components

In return for these benefits, this alternative poses a more complex design/construction problem and probably higher initial cost.

   Monolithic (closed-joint) construction changes the waterproofed deck design philosophy, demoting the membrane from the primary to the secondary line of waterproofing defense. The membrane nonetheless remains an essential component of the system because the surfacing of waterproofed decks is normally exposed to traffic loads, drastic temperature changes, and freeze-thaw cycling, with expansion of water trapped in cracks, holes, or other imperfections. This closed-joint waterproofed deck construction is often called "sandwich slab," denoting the sandwiching of the membrane between the structural slab, on the bottom, and the wearing surface, on top. Insulation may be interposed between the membrane and the wearing surface material, with a setting bed for the surfacing units: precast, ceramic or masonry tiles, or manufactured conglomerates.

   Regardless of whether the surfacing is open- or closed-jointed, the peripheral joint adjoining a building wall should be open, to prevent ponding against the wall. Surfacing should also slope away from

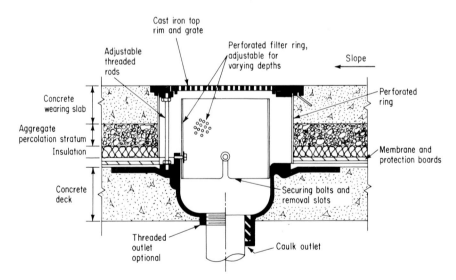

Fig. 20-2   Drain for monolithic, closed-jointed wearing surface provides two-level drainage (at surface and membrane levels). Two concentric, perforated steel rings filter dirt from lower-level drain water. (Josam Manufacturing Co.)

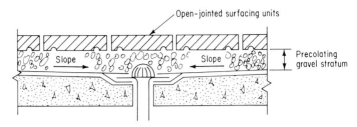

Fig. 20-3 Open-jointed surfacing waterproofed deck system requires drains only at subsurface membrane level. (Maxwell C. Baker, National Research Council of Canada.)

adjoining walls, to direct water away from them, both at and below the surfacing.

DRAINAGE DESIGN

Reliable drainage is no less important for a waterproofed deck than for a roof, although for somewhat different reasons. The primary threat from ponded water on a roof is to the membrane, whereas ponding on a waterproofed deck primarily threatens the wearing surface, especially from freeze-thaw cycling in monolithic surfaces. Both roofs and waterproofed decks suffer the same internal threat from ponded water because the quantity of water penetrating a membrane puncture depends on the time the puncture remains submerged.

For monolithic, or closed-joint, surfacing, the drains must provide access at two levels (see Fig. 20-2). With open-jointed construction, pedestal-supported construction or open-jointed pavers on an aggregate percolation stratum, drains must be located at the subsurface membrane level (see Fig. 20-3).

Minimum recommended slope for a waterproofed deck drainage is ⅛ in./ft, although ³⁄₁₆ or ¼ in. is preferable. Less design slope is usually tolerable for a waterproofed deck than for a roof because a deck's supporting structure seldom has the long spans characteristic of modern roofs with their greater deflections.

LOCATING INSULATION: ABOVE OR BELOW THE MEMBRANE?

With the advent of skyrocketing energy costs, thermal insulation became a more important factor for waterproofed decks as well as roofs. The old practice of simply relying on the aggregate or earth fill to insulate the occupied space is seldom justifiable, unless the space is neither heated nor cooled. Gravel, earth, and coarse sand provide high heat capacity, thus stabilizing heat gains or losses. But their thermal resistances (R values) are low compared with insulating materials of equal thickness.

Generally, the PMR concept—insulation above membrane—is the better design for a waterproofed deck, for several reasons. Protection of the membrane's waterproofing integrity is even more important in a waterproofed deck than in a roof system. Moreover, because the insulation in a waterproofed deck system normally carries higher traffic loads than a roof system, it already requires high compressive strength. By making the insulation moisture-resistant as well, you can conveniently satisfy the requirements for the above-membrane location.

The above-membrane location for insulation in waterproofed deck construction limits the material choice to essentially the same as that in a PMR. As a practical matter, that choice is currently (1981) limited to extruded polystyrene board, the only material possessing the two required properties of adequate compressive strength and moisture resistance. Foamglass is the only other material satisfying both the compressive strength and moisture resistance required for below-membrane use, and its vulnerability to freeze-thaw cycling makes it too risky as an above-membrane insulation, unless the climate is mild enough to preclude freeze-thaw cycling.[1]

Another possible location for the insulation, under the deck, has several disadvantages:

- Subjection of membrane and deck to extreme daily and seasonal temperature cycling

- Possible condensation of upward-migrating water vapor in the insulation

(For further discussion of the below-deck location for insulation, see Chapter 5, "Thermal Insulation.")

PROTECTION BOARDS

Protection boards are essential for protecting the waterproofing membrane during the construction period. Waterproofed deck system membranes require such protection because they are normally vulnerable to puncture during both construction and service life. Special protection boards are generally ⅛ or ¼ in. thick. One manufacturer's two types of protection board comprise (1) a core of stabilized asphalt sandwiched

---

[1]Chester W. Kaplar, "Moisture and Freeze-Thaw Effects on Rigid Thermal Insulations," Tech. Rept. 249, U.S. Army Corps of Engineers, Cold Regions Research and Engineering Laboratory, Hanover, N.H., April 1974, p. 21.

between coated organic felts or (2) a core of stabilized mineral-asphalt mastic reinforced with glass mesh and sandwiched between glass-reinforced asbestos-saturated felt. For the thinner, ⅛-in. thickness, the manufacturer reports 153-psi puncture resistance via ASTM D774 for both types of protection board.

MEMBRANE DESIGN

Conventional built-up waterproofing membranes contain the same felts used in roofing membranes, but different bitumens. Waterproofing bitumens—ASTM D449, Type A and coal-tar bitumen, Type II—have lower viscosity than roofing bitumens (minimum 115°F softening point for ASTM D449, Type A asphalt) and consequently greater waterproofing quality, which is paramount in waterproofed decks.

These less viscous bitumens are acceptable because membranes on waterproofed deck systems run a lesser risk of slippage down slopes and drippage through deck joints than roofing bitumens. Plaza decks are seldom sloped more than ¼ in. Also, the sheltered location of the membrane in a waterproofing system (always below a wearing course and sometimes below the insulation) keeps it at a much lower temperature than the peak temperatures reached by roofing membranes in their conventionally exposed location. The risk of drippage through deck joints is also less prevalent in waterproofing membranes because of the predominant use of poured monolithic concrete instead of steel decks, with their relatively closely spaced joints.

Varying the felt-laying pattern from the conventional shingling pattern used in built-up roofing membranes may be advantageous for waterproofed deck systems. Shingling offers several advantages in multi-ply membrane construction, notably improved slippage resistance (see Chapter 7, "Elements of the Built-up Membrane," for extended discussion). But these advantages of shingling are less important in a waterproofing membrane. Slippage is seldom a problem in waterproofed decks, not only because the slope is limited but because the membrane temperature is also limited.

Because waterproofing quality is relatively more important for a waterproofing membrane than for a roof membrane, the felt pattern can be phased, to improve waterproofing quality. In one such pattern, a four-ply felt-fabric system comprises a bottom-ply base felt with 4-in. edge laps covered with two shingled plies of fabric (19-in. edge lap) and then a final single-ply top felt. Protection boards go on top of the top felt.

Another useful felt pattern comprises two two-ply shingled segments, which also partially exploits the advantage of shingled application (see Fig. 20-4). Note, however, that even with this so-called

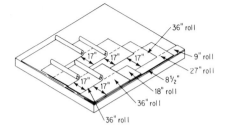

Fig. 20-4 Four-ply waterproofing membrane comprising two shingled plies on top of two other shingled plies, offset 8½ in. (half the 17-in. felt exposure), reduces the number of edge lap joints at membrane's top surface by half. (Maxwell C. Baker, National Research Council of Canada.)

"phased pattern," the entire membrane should, like roof membrane application, be completed on the same day.

The phased felt-laying pattern corrects a serious waterproofing deficiency in the shingling pattern. A fishmouth or curled felt lap opens a direct path diagonally through an entire shingled membrane's cross section. The phased, or ply-on-ply, pattern exposes fewer laps at the membrane's top surface. Moisture penetration through these laps leads only to the next felt layer, not through the entire membrane.

A waterproofed membrane may require a slip sheet to isolate it from thermally induced relative lateral movement between the surfacing above or the structural deck below. Protection boards on top of the membrane can both isolate the membrane from shear stress from traffic surfacing and protect it from construction damage. In a PMR arrangement, a slip sheet below the membrane can isolate it from the structural slab.

Single-Ply Synthetics

Single-ply synthetic membrane materials span the same spectrum as roofing materials and for the same reasons: safety and economy in tall buildings, where hoisting kettles or hot materials is both hazardous and expensive, and conformance with air pollution controls, which may either prohibit or inconveniently limit hot asphalt or coal-tar-pitch operations.

Fluid-applied waterproofing membranes, restricted chiefly to monolithic structural concrete substrates, include

- Polychloroprene (neoprene)
- Chlorosulfonated polyethylene (Hypalon)
- Polyurethane
- Polysulfides (Thiokol)
- Silicone

- Epoxy

- Polyester

- Bituminous blends (rubberized or modified asphalt)

Sheet membranes include

- Polychloroprene (neoprene)

- Butyl (polymerized isobutylene)

- EPDM (ethylene propylene diene monomer)

- Chlorosulfonated polyethylene (Hypalon), with asbestos fiber backing, foam insulation, or other laminate

- Rubberized or "modified" asphalt-polyethylene laminates, with self-adhering, pressure-rolled field or propane-torched, heat-welded field seams

FLASHINGS

Waterproofed deck flashings resemble roof membrane flashings. Like roof-membrane base flashings, waterproofed deck, built-up bituminous base flashings require cant strips to eliminate 90° bends in the stiff flashing felts. If a cant strip is omitted, the stiff flashing felt forms a void at the 90° bend, and pressure from the aggregate fill or topping may break the felts along this vulnerable bend line.

FLOOD-TESTING

Flood-test a waterproofed deck system whenever practicable because a failed waterproofing system is generally more destructive and more expensive to correct than a failed roof. A roof tearoff-replacement usually exceeds $3 psf (1980) cost. But tearoff-replacement of a defective waterproofed deck system may cost twice as much because it often includes jackhammer removal of concrete surfacing. Thus it is advisable to flood-test a waterproofed deck after the membrane and flashing have been applied and before the above-membrane component is applied. However, it must dry thoroughly before the next component is applied. If drying is impracticable, so is flood-testing.

Flood-test with minimum 1-in. water depth over the entire surface. Plug drains, and use permanent or temporary curbs to retain water for 24 h. Sheet steel angles, 26 gage thick, adhered to the waterproofing membrane, can dam the flood-test water.

When drains are not connected, make special provisions for pumping and disposal of water. Cover membrane with protection boards as soon as possible after drying the membrane surface.

ALERTS

Design

Deck

1. Provide minimum 1/8-in. slope (preferably 1/4-in.) in structural deck.

2. Slope deck away from walls.

3. Specify two-level drains for closed-joint waterproofed deck systems.

4. Detail mechanical equipment supports; do not rely on manufacturer to furnish these details.

5. Over precast concrete decks, provide a 3000-psi concrete topping, minimum 3 in. thick, reinforced with welded wire fabric. Precast concrete units must be securely tied together (to prevent relative movement and consequent cracking of the topping).

Built-up Bituminous Membranes

1. Specify priming of concrete decks for built-up bituminous membranes, with primer satisfying ASTM D41 for asphalt membranes, ASTM D43 for coal-tar-pitch membranes, and provide overnight drying.

2. For coal-tar-pitch waterproofing membranes, specify
   (a). coal tar pitch, per ASTM D450, Type II, 25 lb/square interply mopping
   (b). coal-tar-pitch felts, per ASTM D227

3. For asphalt waterproofing membranes, specify
   (a). asphalt, per ASTM D449, Type A, 20 lb/square interply mopping
   (b). asphalt primer, per ASTM D41, 1 gal/square
   (c). felts, per ASTM D226, D250, or D2178, Type III or IV

4. For either coal-tar-pitch or asphalt waterproofing, fabrics per ASTM D173 (bitumen-saturated burlap fabrics), ASTM D1668 (treated glass fabrics).

5. Specify two plies of fabric (glass or cotton) reinforcing at corners (90 and 270°).

Fluid-Applied Membranes

1. Obtain manufacturer's signed approval confirming the suitability of the product for the project use. Also require manufacturer's certification of applicator.

2. Do not specify fluid-applied membrane over lightweight insulating concrete.

3. Specify firm wood-float or comparable light trowel, power machine finish for concrete deck surface.

4. Require concrete contractor, not the roofing contractor, to repair defective concrete surface. Require concrete patching with epoxy mortars.

5. Require water or paper curing of concrete decks, with minimum 20 days' curing time before membrane is applied.

Field

Concrete Decks

1. Check concrete-deck surface with 10-ft straightedge. Permit maximum "gradual" offset of ¼ in./ft (per American Concrete Standard 301-72).

2. Deck surface must be clean, smooth, free of loose particles, grease, oil, and other foreign matter.

Sheet Membranes

1. Maximize sheet size (to minimize vulnerable field-spliced lap seams).

2. Omit adhesive (on adhered systems) over substrate joints or cracks.

3. Lap joints minimum 3 in., or manufacturer's recommendation.

4. Seal exposed ends of sheets to each other and to adjacent surfaces with filler beads of caulking compound.

5. Do not reposition sheet after contact adhesion is made. (Attempts to smooth wrinkles or fishmouths can damage the sheet.)

   For other field alerts, see Chapter 4, "Structural Deck," Chapter 7, "Elements of the Built-up Membrane," and Chapter 11, "Synthetic Single-Ply Membranes."

Glossary of Roofing Terminology

Aggregate Gravel, crushed stone, slag, or mineral granules either (1) embedded in a conventional built-up membrane's bituminous flood coat or (2) applied to a loose-laid roof system as a protective ballast.

Alligatoring Deep shrinkage cracks, progressing down from the surface, in smooth-surfaced membrane coatings and sometimes in bare spots of aggregate-surfaced membranes. It is a consequence of photo-oxidative hardening.

Asphalt Dark brown to black, highly viscous, hydrocarbon produced from the residuum left after the distillation of petroleum, used as the waterproofing agent of a built-up roof.

Backnailing "Blind" (i.e., concealed by overlapping felt) nailing in addition to hot mopping to prevent membrane slippage.

Ballast Aggregate, concrete pavers, or other material designed to prevent wind uplift or flotation of a loose-laid roof system.

Base sheet A felt (often coated) placed as the first (nonshingled) ply in a multiply built-up roof membrane.

Bitumen Generic term for an amorphous, semisolid mixture of complex hydrocarbons derived from petroleum or coal. In the roofing industry there are two basic bitumens: asphalt and coal tar pitch. Before application, they are (1) heated to a liquid state, (2) dissolved in a solvent, or (3) emulsified.

Bitumen trap See *Envelope.*

Blister Spongy, humped portion of a roof membrane, formed by entrapped air-vapor mixture under pressure, with the blister chamber located either between felt plies or at the membrane-substrate interface.

Blocking Continuous wood components anchored to the deck at roof perimeters and openings and doubling as cross-sectional fillers and anchorage bases, used in conjunction with nailers.

Brooming Field procedure of pressing felts into a layer of fluid hot bitumen to ensure continuous adhesion—i.e., elimination of blister-originating voids—of the bitumen film.

457

Btu (British thermal unit)   Heat energy required to raise 1 lb of water 1°F.

Btuh   Btu per hour.

Built-up roof membrane   Continuous, semiflexible roof covering of laminations or plies of saturated or coated felts alternated with layers of bitumen, surfaced with mineral aggregate or asphaltic materials.

Bull   Roofer's term for plastic cement.

BUR   Abbreviation for built-up roof membrane.

Cap flashing   See *Flashing.*

Cap sheet   Mineral-surfaced coated felt used as the top ply of a built-up roof membrane.

Centistoke (cSt)   Unit of viscosity (antonym of fluidity). Water has a viscosity of roughly 1 cSt, light cooking oil 100 cSt.

Channel mopping   See *Strip mopping,* under *Mopping.*

Coal tar pitch   Dark brown to black solid hydrocarbon obtained from the residuum of distilled coke-oven tar, used as the waterproofing agent of dead-level or low-slope built-up roofs.

Coated felt (or base sheet)   A felt that previously has been saturated (impregnated with asphalt) and later coated with harder, more viscous asphalt, which increases its resistance to moisture.

Cold-process roofing   Bituminous membrane comprising layers of coated felts bonded with cold-applied asphalt roof cement and surfaced with a cutback or emulsified asphalt roof coating.

Condensation   Process through which water vapor (a gas) liquefies as air temperature drops or atmospheric pressure rises. (See *Dew point.*)

Counterflashing   See *Flashing.*

Coverage   Surface area that should be continuously coated by a specific unit of a roofing material, after allowance is made for a specified lap.

Crack   Membrane fracture produced by bending, often at a ridge (see *Ridging*).

Creep   (1) Permanent elongation or shrinkage of the membrane resulting from thermal or moisture changes. (2) Permanent deflection of structural framing or structural deck resulting from plastic flow under continued stress or dimensional changes accompanying changing moisture content or temperature.

Cricket   Ridge built up in a level valley or perimeter to direct rainwater to a drain.

Curing   Final step in the irreversible polymerization of a thermosetting plastic, usually requiring some combination of heat, radiation, and pressure.

Curled Felt Membrane defect characterized by a continuous, open longitudinal seam with top felt rolled back from underlying felt.

Cutback Solvent-thinned bitumen used in cold-process roofing adhesives, flashing cements, and roof coatings.

Cutoff Detail designed to prevent lateral water movement into the insulation where the membrane terminates at the end of a day's work or at an isolated roof section, usually removed before work proceeds.

Dead level Absolutely horizontal, of zero slope (see *Slope*).

Deck Structural supporting surface of a roof system.

Delamination Separation of felt plies in a built-up membrane; separation of insulation boards into horizontal strata.

Dew point Temperature at which water vapor starts to condense in cooling air at existing atmospheric pressure and vapor content.

Double pour Doubling of the flood-coat, graveling-in operation, to provide additional waterproofing integrity to the membrane.

Edge stripping Application of felt strips to cover a joint between flashing and built-up membrane.

Elastomer Macromolecular material that rapidly regains its original shape after release of a light deforming stress.

Elastomeric Having elastic properties, capable of expanding or contracting with the surfaces to which the material is applied without rupturing.

Emulsion Intimate mixture of bitumen and water, with uniform dispersion of the bitumen globules achieved through a chemical or clay emulsifying agent.

Envelope Continuous edge formed by folding an edge base felt over the plies above and securing it to the top felt. The envelope thus prevents bitumen seepage through the exposed edge joints of the laminated, built-up roofing membrane.

Equilibrium moisture content Moisture content of a material at a given temperature and relative humidity, expressed as percent moisture by weight.

Equiviscous temperature (EVT) Temperature at which asphalt has the correct viscosity (50–150 cSt) for hot mopping.

Ethylene propylene diene monomer (EPDM) Thermosetting, synthetic rubber used in single-ply elastomeric sheet roof membranes.

Expansion joint Structural separation between two building segments, designed to permit free movement without damage to the roof system.

Exposure Transverse dimension of a felt not overlapped by an adjacent felt in a built-up roof membrane. Correct felt exposure in a shingled, built-up membrane is computed by dividing the felt width minus 2 in. by the number of plies—e.g., for four plies of 36-in.-wide felt, exposure = (36-2)/4 = 8½ in.

Fabric Woven cloth of organic or inorganic filaments, threads, or yarns.

Fallback Reduction in bitumen softening point, sometimes caused by refluxing or overheating in a closed container.

Felt Flexible sheet manufactured by interlocking fibers with a binder or through a combination of mechanical work, moisture, and heat.

Felt layer Spreader-type, wheel-mounted equipment for laying felt and simultaneously dispensing hot asphalt in a single operation.

Finger wrinkling Wrinkling of exposed felts in small, finger-sized ridges parallel to the longitudinal direction of the felt roll, caused by transverse moisture expansion of the felt.

Fishmouth Membrane defect consisting of an opening in the edge lap of a felt in a built-up membrane, a consequence of an edge wrinkle.

Flash point Temperature at which a test flame ignites the vapors above a liquid surface.

Flashing Connecting devices that seal membrane joints at walls, expansion joints, drains, gravel stops, and other places where the membrane is interrupted. *Base flashing* forms the upturned edges of the watertight membrane. *Cap* or *counterflashing* shields the exposed edges and joints of the base flashing.

Flashing cement Trowelable, plastic mixture of bitumen and asbestos (or other inorganic) reinforcing fibers and a solvent (a stiffer, more sag-resistant material than plastic cement).

Flux Bituminous material used as a feed stock for further processing and as a material to soften other bituminous materials.

Glaze coat Thin, protective coating of bitumen applied to the lower plies or top ply of a built-up membrane when application of additional felts or the flood coat and aggregate surfacing is delayed.

Grain Weight unit equal to 1/7000 lb, used in measuring atmospheric moisture content.

Granule See *Mineral granules.*

Gravel Coarse, granular aggregate resulting from natural erosion or crushing of rock, used as protective surfacing or ballast on roof systems.

Gravel stop Flanged device, usually metallic, with vertical projection above the roof level, designed to prevent loose aggregate from rolling or washing off the roof and to provide a finished edge detail for the roof.

Gravelling in Embedding aggregate surfacing into a built-up bituminous membrane flood coat.

Holiday Area where interply bitumen mopping or other fluid-applied coating is discontinuous.

Hood Sheet metal cover over piping or other rooftop equipment.

Hot stuff or hot   Roofer's term for hot bitumen.

Hygroscopic   Attracting and absorbing atmospheric moisture.

Ice dam   Drainage-obstructive ice formation at eave of snow-covered sloped roof.

Inorganic   Comprising matter other than hydrocarbons and derivatives, not of plant or animal origin.

Insulation   See *Thermal insulation.*

Lap   Dimension by which a felt covers an underlying felt in a multi-ply built-up bituminous membrane. *Edge* lap indicates the transverse cover; *end* lap indicates the cover at the end of the roll. These terms also apply to single-ply membranes.

Loose-laid roof system   Design concept in which insulation boards and membrane are not anchored to the deck but ballasted by loose aggregate or concrete pavers.

Membrane   Flexible or semiflexible roof covering, the waterproofing component of the roof system.

Mineral granules   Natural or synthetic aggregate particles, ranging in size from 500 μm (1 μm $=$ 10^{-6} m) to ¼-in. diameter, used to surface cap sheets, asphalt shingles, and some cold-process membranes.

Mineral-surfaced sheet   Asphalt-saturated felt, coated on one or both sides and surfaced on the weather-exposed side with mineral granules.

Monomer   Class of molecules with molecular weight ranging roughly between 30 and 250, capable of combining into huge, polymeric macromolecules, 100 to 10,000 times as large as the basic monomeric molecules, through chainlike repetition of the basic monomeric chemical structure.

Mop-and-flop   Application technique in which roof-system components (insulation boards, felt plies, cap sheets, and so on) are first placed upside down adjacent to their final locations, coated with adhesive, turned over, and adhered to the substrate.

Mopping   Application of hot, fluid bitumen to substrate or to plies of built-up membrane with a manually wielded mop or a mechanical applicator.

   Solid mopping   A continuous coating.

   Spot mopping   Pattern of hot bitumen application in roughly circular areas, generally about 18-in. diameter, on a grid of unmopped, perpendicular bands.

   Sprinkle mopping   Random pattern of bitumen beads hurled onto the substrate from a broom or mop.

   Strip mopping   Mopping pattern featuring parallel mopped bands.

Nailer   Wood member bolted or otherwise anchored to a nonnailable deck or wall to provide nailing anchorage of membrane roof felts or flashings.

Neoprene   Synthetic rubber (chemically polychloroprene) used in fluid- or sheet-applied elastomeric single-ply membranes or flashing.

One-on-one   Nonshingled application pattern of a single ply of felt followed later by application of a second ply (see *Phased application*).

Organic   Comprising hydrocarbons or their derivatives, or matter of plant or animal origin.

Parting agent   Powdered mineral (talc, mica, and so forth) placed on coated felts to prevent adhesion of concentric felt layers in the roll (sometimes called a *releasing agent* or *antistick compound*).

Perlite   Aggregate used in lightweight insulating concrete and preformed insulating board, formed by heating and expanding silicaceous volcanic glass.

Perm   Unit of water-vapor transmission, defined as 1 grain water vapor/(ft^2 · h) per inch of mercury pressure difference. (1 in.Hg = 0.491 psi.)

Permeance   Index of a material's resistance to water-vapor transmission. (See *Perm.*)

Phased application   Applying the felt plies of a built-up roof or waterproofing membrane in two or more operations, separated by a delay normally of at least 1 day.

Picture framing   Rectangular membrane ridging pattern formed over insulation-board joints.

Pitch pocket   Flanged, open-bottomed metal container placed around a column or other roof-penetrating element and filled with bitumen or plastic cement to seal the joint.

Plastic cement   Trowelable, plastic mixture of bitumen and asbestos (or other inorganic) stabilizing fibers and a solvent, used mainly for horizontal surfaces as opposed to *flashing cement*, which is designed for vertical surfaces requiring sag resistance.

Plasticizer   High-boiling-point solvent or softening agent added to a polymer to facilitate processing or increase flexibility or toughness in the manufactured material.

Ply   Layer of felt in a built-up roof membrane; a four-ply membrane has at least four plies of felt at any vertical cross section cut through the membrane.

Polymer   Long, chain macromolecules produced from monomers, for the purpose of increasing tensile strength of sheets used as membranes or flashing.

Polyvinyl chloride (PVC)   Thermoplastic polymer, formulated with a plasticizer, used as a single-ply sheet membrane material or liquid coating.

Primer   Thin, liquid, bituminous solvent applied to seal a surface, absorb dust, and promote adhesion of subsequently applied bitumen.

Protected membrane roof (PMR)   Roof assembly with insulation on top of the membrane instead of vice versa, as in the conventional roof assembly (also known as an inverted or upside-down roof assembly).

Re-covering Covering an existing roof assembly with a new membrane instead of removing the existing roof system before installing the new membrane.

Rake Edge of a roof at its intersection with a gable.

Reglet Horizontal groove in a wall or other vertical surface adjoining a roof surface for anchoring flashing.

Relative humidity (RH) Ratio (expressed as percentage) of the mass per unit volume (or partial pressure) of water vapor in an air-vapor mixture to the saturated mass per unit volume (or partial pressure) of the water vapor at the same temperature.

Rep Unit of vapor permeance resistance; reciprocal of perm.

Reroofing Removing and replacing an existing roof system (as opposed to mere *Re-covering*); also called *tearoff-replacement*.

Resin Basic raw material for manufacturing polymers, a synthetic polymer containing no deliberately added ingredients.

Ridging Membrane defect characterized by upward displacement of the membrane, usually over insulation-board joints (see *Picture framing*).

Roll roofing Coated felts, generally mineral-surfaced, supplied in rolls and designed for use without field-applied surfacing.

Roof system Assembly of interacting components designed to weatherproof, and normally insulate, a building's top surface.

Roofer Roofing subcontractor.

Saddle See *Cricket*.

Saturated felt Felt that has been immersed in hot bitumen.

Scupper Channel through parapet, designed for peripheral drainage of the roof, usually as safety overflow system to limit accumulation of ponded rainwater caused by clogged drains.

Scuttle Curbed opening, with hinged or loose cover, providing access to roof (synonymous with *hatch*).

Self-healing Property of the least viscous roofing bitumens, notably coal tar pitch, that enables them to seal cracks formed at lower temperatures.

Selvage joint Lapped joint detail for two-ply, shingled roll roofing membrane, with mineral surfacing omitted over a transverse dimension of the cap sheets to improve mopping adhesion. For a 36-in.-wide sheet, the selvage (unsurfaced) width is 19 in.

Shark fin Curled felt projecting up through the aggregate surfacing of a built-up membrane.

Shingling Pattern formed by laying parallel felt rolls with lapped joints so that one longitudinal edge *overlaps* the longitudinal edge of one adjacent felt, whereas the other longitudinal edge *underlaps* the other adjacent felt. (See *Ply.*) Shingling is the normal method of applying felts in a built-up roofing membrane.

Single-ply membrane Membrane, either sheet- or fluid-applied, with only a single layer of material, designed to prevent water intrusion into the building.

Skater's cracks Curved cracks observed in smooth-surfaced built-up membranes.

Slag Porous aggregate used as built-up bituminous membrane surfacing, comprising silicates and alumino-silicates of calcium and other bases, developed with iron in a blast furnace.

Slippage Relative lateral movement of adjacent felt plies in a built-up membrane. Occurs mainly in sloped roofing membranes, exposing the lower plies, or even the base sheet, to the weather.

Slope Tangent of the angle between the roof surface and the horizontal, in inches per foot. The Asphalt Roofing Manufacturers' Association ranks slopes as follows:
Level: ½-in. maximum
Low slope: over ½ in. up to 1½ in.
Steep slope: over 1½ in.

Smooth-surfaced roof Built-up roofing membrane surfaced with a layer of hot-mopped asphalt or cold-applied asphalt-clay emulsion or asphalt cutback, or sometimes with an unmopped, inorganic felt.

Softening point Temperature at which bitumen becomes soft enough to flow, as measured by standard laboratory test in which a steel ball falls through a measured distance through a disk made of the tested bitumen.

Softening-point drift Change in softening point during storage or application. (See also *Fallback.*)

Solid mopping See *Mopping.*

Split Membrane tear resulting from tensile stress.

Spot mopping See *Mopping.*

Sprinkle mopping See *Mopping.*

Spudder Heavy steel implement with a dull, bevel-edged blade designed to remove embedded aggregate from a built-up membrane surface (also called *scraper*).

Stripping (1) Technique of sealing the joint between base flashing and membrane plies or between metal and built-up membrane with one or two plies of felt or fabric and hot- or cold-applied bitumen. (2) Taping joints between insulation boards or deck units.

Substrate Surface (structural deck, insulation, or vapor retarder) upon which the roof membrane is placed. Also, the deck, vapor retarder, or membrane surface upon which insulation, or other roof system component, is placed.

Sump Depression in roof deck around drain.

System See *Roof system.*

Tearoff Removing a failed roof system down to the structural deck.

Thermal conductance (C) Heat energy in *Btu* per hour (Btuh) transferred via conductance only through a 1-ft^2 area of homogeneous material per °F temperature difference from surface to surface. The unit is Btuh/[ft$^2 \cdot$°F] [in metric terms, W/(m$^2 \cdot$K)].

Thermal conductivity (k) Heat energy (Btuh) transferred via conductance only through a 1-in.-thick, 1-ft^2 area of homogeneous material per °F temperature difference from surface to surface. Unit for k is Btuh/(in.$\cdot$ft$^2 \cdot$°F).

Thermal resistance (R = 1/C) Material's resistance to conductive heat flow, in °F/(Btuh$\cdot$ft^2)—that is, for a 5°F temperature difference surface to surface, 1 Btuh would flow through a 1 ft^2 specimen with R = 5.

Thermal shock Stress-producing phenomenon resulting from sudden temperature change in a roof membrane when, for example, a rain shower follows brilliant, hot sunshine.

Thermal shock factor (TSF) Mathematical expression for calculating the theoretical temperature drop required to split a rigidly held membrane test sample under tensile contractive stress.

Thermoplastic Changing viscosity under thermal cycling (fluid when heated, solid when cooled).

Thermosetting Hardening permanently when heated, owing to cross-linking of polymeric resins into a rigid matrix.

Through-wall flashing Water-resistant membrane or material assembly extending through a wall's horizontal cross section and designed to direct water flow through the wall toward the exterior.

Vapor barrier See *Vapor retarder.*

Vapor migration Flow of water vapor from a region of high vapor pressure to a region of lower vapor pressure.

Vapor retarder Roof component designed to obstruct water vapor flow through a roof or wall.

Vent Opening designed to convey water vapor, or other gas, from inside a building or building component to the atmosphere.

Vermiculite Aggregate used in lightweight insulating concrete, formed by heating and consequent expansion of mica rock.

Viscoelastic Characterized by changing mechanical behavior, from nearly elastic at low temperature to plastic, like a viscous fluid, at high temperature.

Viscosity Index of a fluid's internal resistance to flow, measured in centi-stokes (cSt) for bitumens. (Water has a viscosity of roughly 1 cSt; light cooking oil 100 cSt.)

Walking in Manually forcing insulation boards against previously installed boards to tighten the joints.

Wrinkling See *Ridging.*

Appendix

Table 1   SI Conversion Factors

To convert from	To	Multiply by
°F	°C	0.556 (°F − 32)
°C	°F	1.8 (°C) + 32
°C	K	°C + 273.15
mm	in.	0.03937
in.	mm	25.400
g/m²	lb/square (100 ft²)	0.0205
lb/square (100 ft²)	g/m²	48.83
pcf (lb/ft³)	kg/m³	16.02
kg/m³	pcf	0.0624
psf (lb/ft²)	kg/m²	4.882
kg/m²	psf	0.205
psi (lbf/in.²)	kPa	6.895
kPa	psi	0.1450
lbf/in.	kN/m	0.1751
kN/m	lbf/in.	5.710
Perm (vapor permeance) [grain/(ft²·h·in.Hg)]	ng/(m²·s·Pa)	57.23
R[°F/(Btuh·ft²)]	R[K/(W·m²)]	0.176
R[K/(W·m²)]	R[°F/(Btuh·ft²)]	5.678
U[Btuh/(ft²·°F)]	U[W/(m²·K)]	5.678
U[W/(m²·K)]	U[Btuh/(ft²·°F)]	0.176

Key to abbreviations:

  °F = degrees Fahrenheit
  °C = degrees Celsius
  K = degrees Kelvin
Btuh = Btu per h
  lbf = pound-force
  N = newton (= 0.2248 lbf)
  kN = kilonewton = 10^3N

  m = meter
  mm = millimeter
  g = gram
  kg = kilogram (= 10^3g)
  ng = nanogram (= 10^{-6}g)
  Pa = pascal (= N/m²)
  kPa = 10^3Pa
  W = watt (= 3.414 Btuh)

Table 2   Important ASTM Standard Specifications for Roofing and Waterproofing

ASTM designation	Standard specification for
	BITUMENS
D312	Asphalt Used in Roofing
D449	Asphalt Used in Dampproofing and Waterproofing
D450	Coal Tar Pitch for Roofing, Dampproofing, and Waterproofing
	FELTS
D224	Smooth-Surfaced Asphalt Roll Roofing (Organic Felt)
D226	Asphalt-Saturated Organic Felt Used in Roofing and Waterproofing
D227	Coal-Tar Saturated Organic Felt Used in Roofing and Waterproofing
D249	Asphalt Roll Roofing (Organic Felt) Surfaced with Mineral Granules
D250	Asphalt-Saturated Asbestos Felt Used in Roofing and Waterproofing
D371	Asphalt Roll Roofing (Organic Felt) Surfaced with Mineral Granules; Wide Selvage
D2178	Glass Felt, Asphalt-Impregnated, Used in Roofing and Waterproofing
D2626	Asphalt-Saturated and Coated Organic Felt Used in Roofing
D3158	Asphalt-Saturated and Coated Organic Felt Used in Roofing
D3378	Asphalt-Saturated and Coated Asbestos Felt Base Sheet Used in Roofing
D3672	Venting, Inorganic Felt Base Sheet in Built-up Roofing
D3909	Asphalt Roll Roofing (Glass Felt) Surfaced with Mineral Granules
	FABRICS
D173	Bitumen-Saturated Cotton Fabrics Used in Roofing and Waterproofing
D1327	Bitumen-Saturated Woven Burlap Fabrics Used in Roofing and Waterproofing
D1668	Glass Fabrics (Woven and Treated) for Roofing and Waterproofing
	PRIMERS AND CEMENTS
D41	Asphalt Primer Used in Roofing, Dampproofing and Waterproofing
D43	Creosote Primer Used in Roofing, Dampproofing and Waterproofing
D2822	Asphalt Roof Cement
D3019	Lap Cement Used with Asphalt Roll Roofing
	SURFACING
D1227	Emulsified Asphalt Used as a Protective Coating for Roofing
D1863	Mineral Aggregate Used on Built-up Roofs
D2823	Asphalt Roof Coatings
D2824	Aluminum-Pigmented Asphalt Roof Coatings
D3805	Application of Aluminum-Pigmented Asphalt Roof Coatings
	SINGLE-PLY MEMBRANE
D3468	Liquid-Applied Neoprene and Chlorosulfonated Polyethylene Used in Roofing and Waterproofing

Table 2 *(cont.)*

ASTM designation	Standard recommended practice or test method
D61	Softening Point of Pitches (Cube-in-Water Method)
D2398	Softening Point of Bitumen in Ethylene Glycol (Ring-and-Ball)
D2829	Sampling and Analysis of Built-up Roofs
D3617	Sampling and Analysis of New Built-up Membranes
D2523	Testing Load-Strain Properties of Roofing Membranes
D3746	Comparative Impact Resistance of Bituminous Roofing Systems
D3105	Elastomeric and Plastomeric Roofing and Waterproofing Materials
E96	Water Vapor Transmission of Materials in Sheet Form
E84	Surface Burning Characteristics of Building Materials
E108	Fire Tests of Roof Coverings
E119	Fire Tests of Building Construction and Materials

Table 3 Important ASTM Standard Specifications for Thermal Insulation

ASTM designation	Standard specification for
C552	Cellular Glass Block and Pipe Thermal Insulation
C578	Preformed, Block-Type Cellular Polystyrene Thermal Insulation
C591	Preformed Cellular Urethane Thermal Insulation
C726	Mineral Fiber Roof Insulation Board
C728	Perlite Thermal Insulation Board

ASTM designation	Standard test method or recommended practice for measuring
C165	Compressive Properties of Thermal Insulation
C177	Steady-State Thermal Transmission Properties by Means of the Guarded Hot Plate
C203	Breaking Load and Calculated Flexural Strength of Preformed Block-Type Thermal Insulation
C518	Steady-State Thermal Transmission Properties by Means of the Heat-Flow Meter
C855	Thermal Resistance Factors for Preformed, Above-Deck Roof Insulation

Name Index

Subject Index